Surface Chemistry
and Catalysis

FUNDAMENTAL AND APPLIED CATALYSIS

Series Editors: M. V. Twigg
Johnson Matthey
Catalytic Systems Division
Royston, Hertfordshire, United Kingdom

M. S. Spencer
Department of Chemistry
Cardiff University
Cardiff, United Kingdom

CATALYST CHARACTERIZATION: Physical Techniques for Solid Materials
Edited by Boris Imelik and Jacques C. Vedrine

CATALYTIC AMMONIA SYNTHESIS: Fundamentals and Practice
Edited by J. R. Jennings

CHEMICAL KINETICS AND CATALYSIS
R. A. van Santen and J. W. Niemantsverdriet

DYNAMIC PROCESSES ON SOLID SURFACES
Edited by Kenzi Tamaru

ELEMENTARY PHYSICOCHEMICAL PROCESSES ON SOLID
SURFACES
V. P. Zhdanov

METAL–OXYGEN CLUSTERS: The Surface and Catalytic Properties of
Heteropoly Oxometalates
John B. Moffat

PRINCIPLES OF CATALYST DEVELOPMENT
James T. Richardson

SELECTIVE OXIDATION BY HETEROGENEOUS CATALYSIS
Gabriele Centi, Fabrizio Cavani, and Ferrucio Trifirò

SURFACE CHEMISTRY AND CATALYSIS
Edited by Albert F. Carley, Philip R. Davies, Graham J. Hutchings,
and Michael S. Spencer

A Continuation Order Plan is available for this series. A continuation order will bring delivery of each new volume immediately upon publication. Volumes are billed only upon actual shipment. For further information please contact the publisher.

Surface Chemistry and Catalysis

Edited by

Albert F. Carley
Philip R. Davies
Graham J. Hutchings
Michael S. Spencer

Cardiff University
Cardiff, Wales, UK

Kluwer Academic / Plenum Publishers
New York, Boston, Dordrecht, London, Moscow

ISBN: 0-306-47393-3

© 2002 Kluwer Academic / Plenum Publishers, New York
233 Spring Street, New York, N.Y. 10013

http://www.wkap.nl/

10 9 8 7 6 5 4 3 2 1

A C.I.P. record for this book is available from the Library of Congress

Printed in the United States of America

ɑc

Professor Wyn Roberts, Cardiff, 2002

Contributors

K. Al-Shamery Fritz-Haber-Institut der Max-Planck-
Gesellschaft Department of Chemical
Physics, Faradayweg 4-6, D-14195 Berlin,
Germany

M. Bäumer Fritz-Haber-Institut der Max-Planck-
Gesellschaft Department of Chemical
Physics, Faradayweg 4-6, D-14195 Berlin,
Germany

I. M. Ciobîcă Schuit Institute of Catalysis, Eindhoven
University of Technology P.O. Box 513,
5600 MB Eindhoven, The Netherlands

W. Drachsel Fritz-Haber-Institut der Max-Planck-
Gesellschaft Department of Chemical
Physics, Faradayweg 4-6, D-14195 Berlin,
Germany

N. Ernst Fritz-Haber-Institut der Max-Planck-
Gesellschaft Department of Chemical
Physics, Faradayweg 4-6, D-14195 Berlin,
Germany

F. Frechard Schuit Institute of Catalysis, Eindhoven
University of Technology P.O. Box 513,
5600 MB Eindhoven, The Netherlands

Contributors

H.-J. Freund
Fritz-Haber-Institut der Max-Planck-Gesellschaft Department of Chemical Physics, Faradayweg 4-6, D-14195 Berlin, Germany

D. Wayne Goodman
Department of Chemistry, Texas A&M University, P.O. Box 30012, College Station, TX 77842-3012, USA

Jerzy Haber
Institute of Catalysis and Surface Chemistry, Polish Academy of Sciences, ul.Niezapominajek 8,30-239 Krakow,Poland

H. Hamann
Fritz-Haber-Institut der Max-Planck-Gesellschaft Department of Chemical Physics, Faradayweg 4-6, D-14195 Berlin, Germany

C. G. M. Hermse
Schuit Institute of Catalysis, Eindhoven University of Technology P.O. Box 513, 5600 MB Eindhoven, The Netherlands

Graham J. Hutchings
Department of Chemistry, Cardiff University, P.O. Box 912, Cardiff, CF10 3TB, UK

A. P. J. Jansen
Schuit Institute of Catalysis, Eindhoven University of Technology P.O. Box 513, 5600 MB Eindhoven, The Netherlands

H. Kuhlenbeck
Fritz-Haber-Institut der Max-Planck-Gesellschaft Department of Chemical Physics, Faradayweg 4-6, D-14195 Berlin, Germany

G. U. Kulkarni
Chemistry and Physics of Materials Unit, Jawaharlal Nehru Centre For Advanced Scientific Research, Jakkur P.O., Bangalore-560 064, India

Xiaofeng Lai
Department of Chemistry, Texas A&M University P.O. Box 30012, College Station, TX 77842-3012, USA

J. Libuda
Fritz-Haber-Institut der Max-Planck-Gesellschaft Department of Chemical Physics, Faradayweg 4-6, D-14195 Berlin, Germany

Contributors

Sir Ronald Mason KCB FRS	Aylesbury, Bucks, HP22 4NH, UK
Keith McCrea	University of California, Berkeley: Lawrence Berkeley National Laboratory, CA 94720-1460, USA
Douglas C. Meier	Department of Chemistry, Texas A&M University P.O. Box 30012, College Station, TX 77842-3012, USA
Jessica Parker	University of California, Berkeley: Lawrence Berkeley National Laboratory, CA 94720-1460, USA
C. N. R. Rao FRS	Chemistry and Physics of Materials Unit, Jawaharlal Nehru Centre For Advanced Scientific Research, Jakkur P.O., Bangalore-560 064, India
T. Risse	Fritz-Haber-Institut der Max-Planck-Gesellschaft Department of Chemical Physics, Faradayweg 4-6, D-14195 Berlin, Germany
G. Rupprechter	Fritz-Haber-Institut der Max-Planck-Gesellschaft Department of Chemical Physics, Faradayweg 4-6, D-14195 Berlin, Germany
Wolfgang M. H. Sachtler	V. N. Ipatieff Laboratory, Center for Catalysis and Surface Science, Department of Chemistry, Northwestern University, Evanston, IL 60208 USA
Norman Sheppard FRS	School of Chemical Sciences, University of East Anglia,Norwich NR4 7TJ UK
Gabor Somorjai	University of California, Berkeley: Lawrence Berkeley National Laboratory, CA 94720-1460, USA
Sir John Meurig Thomas ScD, FRS	Davy Faraday Research Laboratory, The Royal Institution of Great Britain, 21 Albemarle Street, London, W15 4BS, UK
R. A. van Santen	Schuit Institute of Catalysis, Eindhoven University of Technology P.O. Box 513, 5600 MB Eindhoven, The Netherlands

C. P. Vinod   Chemistry and Physics of Materials Unit,
                                     Jawaharlal Nehru Centre For Advanced
                                     Scientific Research, Jakkur P.O., Bangalore-
                                     560 064, India

Peter B. Wells   Department of Chemistry, Cardiff University,
                                     P.O. Box 912, Cardiff, CF10 3TB, UK

PREFACE TO THE SERIES

Catalysis is important academically and industrially. It plays an essential role in the manufacture of a wide range of products, from gasoline and plastics to fertilizers, herbicides and drugs, which would otherwise be unobtainable or prohibitively expensive. There are few chemical - or oil - based material items in modern society that do not depend in some way on a catalytic stage in their manufacture. Apart from manufacturing processes, catalysis has found other important and ever-increasing uses; for example, successful applications of catalysis in the control of pollution and its use in environmental control are certain to increase in the future.

The commercial importance of catalysis and the diverse intellectual challenges of catalytic phenomena have stimulated study by a broad spectrum of scientists, including chemists, physicists, chemical engineers, and material scientists. Increasing research activity over the years has brought deeper levels of understanding, and these have been associated with a continually growing amount of published material. As recently as seventy five years ago, Rideal and Taylor could still treat the subject comprehensively in a single volume, but by the 1950s Emmett required six volumes, and no conventional multivolume text could now cover the whole of catalysis in any depth. In view of this situation, we felt there was a need for a collection of monographs, each one of which would deal at an advanced level with a selected topic, so as to build a catalysis reference library. This is the aim of the present series, *Fundamental and Applied Catalysis*.

Some books in the series deal with particular techniques used in the study of catalysts and catalysis: these cover the scientific basis of the technique, details of its practical applications, and examples of its usefulness. An industrial process or a class of catalysts forms the basis of other books, with information on the fundamental science of the topic, the use of the process or catalysts, and engineering aspects. Single topics in catalysis are also treated in the series with books giving the theory of the underlying science, and relating it to catalytic practice. We believe that this approach provides a collection that is of value to both academic and industrial workers. The series editors welcome comments on the series and suggestions of topics for future volumes.

Martyn Twigg
Michael Spencer

PREFACE

In 2001 Wyn Roberts celebrated both his 70th birthday and 50 years of working in surface science, to use the term "surface science" in its broadest meaning. His notable scientific achievements are referred to in the chapters by Sir Ronald Mason and Sir John Meurig Thomas, also in the list of publications (Appendix 1), and we will not describe them again here. The four editors wished to mark the anniversary with a contribution of lasting value, something more than the usual festschrift issue of a relevant journal.

The aim of the book is to cover some of the areas of surface chemistry and catalysis in which Wyn has always had interests. The chapters, in their different ways, all deal with significant features in these areas. The authors were chosen from some of the many eminent scientists who have worked with Wyn in various ways. Their serious contributions, and indeed their readiness to contribute, demonstrate the high esteem with which Wyn is held in the community. We asked the authors, in view of the millennium, to look forward in their science as well as backwards, but, understandably cautious in view of the unpredictable nature of science, there is little gazing into the crystal ball.

We had much discussion over the title of the book but eventually we agreed that "Surface Chemistry and Catalysis" was the best succinct description of Wyn's interests. After the introductory chapters the book is divided into three sections: Surface Science (Sheppard, Somorjai and van Santen); Model Catalysts (Freund, Goodman and Rao) and Catalysis (Sachtler, Hutchings, Haber and Wells).

As editors we are very pleased with the authors' contributions. We feel that the book is not only a worthy contribution to celebrate Wyn's anniversary but we hope it will also be a major contribution to the scientific literature, a work describing much of surface chemistry and catalysis at the start of the 21st century.

Albert Carley
Phil Davies
Graham Hutchings
Mike Spencer

Cardiff, August 2001

Contents

CHAPTER 4.   HIGH-PRESSURE CO DISSOCIATION AND CO
             OXIDATION STUDIES ON PLATINUM SINGLE
             CRYSTAL SURFACES USING SUM FREQUENCY
             GENERATION SURFACE VIBRATIONAL
             SPECTROSCOPY
             Keith McCrea, Jessica Parker and Gabor Somorjai

CHAPTER 5.   MODELING HETEROGENEOUS CATALYTIC
             REACTIONS
             *I. M. Ciobîcă, F. Frechard, C. G. M. Hermse, A. P. J.
             Jansen and R. A. van Santen*

CHAPTER 6.   MODEL SYSTEMS FOR HETEROGENEOUS
             CATALYSIS: QUO VADIS SURFACE SCIENCE?
             *H.-J. Freund, N. Ernst, M. Bäumer, G. Rupprechter, J.
             Libuda, H. Kuhlenbeck, T. Risse, W. Drachsel, K. Al-
             Shamery, H. Hamann*

Chapter 1

TEYRNGED I MEIRION WYN ROBERTS
A tribute to Meirion Wyn Roberts

Ronald Mason

Wyn Roberts' contributions to surface science have been of real significance to the development of structure–reactivity relations and, in particular, to catalysts. I am at some disadvantage in introducing this commemorative volume since it is sometime since I played an active 'hands on' role in surface science researches. I can hope, however, that I bring an objectivity to this review, relatively uninfluenced by current emphases or fashion!

The major theme, of remarkable continuity over 40 years or more, has been Roberts' studies of the chemisorption of oxygen and of the mechanisms of oxidation at metal surfaces. K.W. Sykes asked Roberts to investigate, as part of his doctoral thesis studies, the role of sulphur in catalysing nickel carbonyl formation[1] – one more contribution to the seemingly eternal debate on electronic or intermediate mechanisms. The need for chemically and physically well defined surfaces, to achieve reproducible results, was not lost on Roberts. After a spell as a postdoctoral fellow with F.C. Tompkins, at Imperial College, Roberts returned to oxidation mechanisms with a study of the oxidation of iron films[2]. A mixed regime of associative and disassociative processes was demonstrated and must have had a significant influence on Roberts' later work which was to define the various states of oxygen at metal surfaces.

At Queen's University of Belfast a discernible change of pace, coinciding with Martin Quinn joining the Roberts group, can be recognised from the studies, based on surface potential measurements, work function and a

Surface Chemistry and Catalysis, Edited by Carley *et al.*
Kluwer Academic/Plenum Publishers, New York, 2002

photoelectric investigation, of the nickel – oxygen system. A quote from the publication[3] is appropriate: "the escape depth of an excited electron after oxygen interaction is estimated to be not greater than 20Å". I shall return to surface sensitive techniques which have dominated Roberts' researches for more that 30 years. Roberts and Wells gave us insight on the aluminium – oxygen system[4,5] and a study[6] of nitrogen on tungsten wrapped up, as it were, a series illuminating, highly formative papers based on 'conventional' physiochemical measurements.

I use the description, formative, for they provided essential data, along with those from the observations of other workers, to exploit, with insight and maximum effect, the emergence of surface sensitive spectroscopies. In that regard, I can do no better than to remind the reader of the seminal paper on 'Some observations on the surface sensitivity of photoelectron spectroscopy' which Brundle and Roberts published in 1972[8]. Data were reported for the sensitivity of photoelectron spectroscopy for H_2O, CO_2 and CO absorbed on gold and, later[8], for the interaction of mercury with gold. These experiments at the University of Bradford put the measurement of surface coverages and chemical shifts (surface species) on a quantitive basis and were the forerunners of comparisons of the surface sensitivities of low - and high – energy photoelectron spectroscopy.

From these early days, surface sensitive spectroscopies and low energy electron diffraction have been used to probe reactions and atomic/molecular arrangements of ligands at a variety of metal surfaces. A review (Photoelectron Spectroscopy and Surface Chemistry[9]) summarises a near–decade of exciting developments and I would add a few personal highlights. Kishi and Roberts[10] described the associative (low temperature) and diassociative (higher temperatures) sorption of carbon monoxide at iron surfaces, observations which had a direct impact on our views of Fischer – Tropsch catalysis. Related, in the sense of demonstrating the dependence of surface reactions on physical parameters such as temperature and structure, were the studies by the author of the reactivity of carbon monoxide on Pt(111) and 'stepped' surfaces. These studies were predicated by Gabor Somorjai's investigations of low – and high – Miller index surfaces of platinum. In 1979, Johnson, Matloob and Roberts[12] described the adsorption of nitric oxide on Cu (100) and Cu (111) surfaces. Intriguing results, particularly those obtained at low temperatures, could be related to structural studies of metal nitrosyls which had shown linear and non-linear bonding of the nitrosyls to a metal centre. Johnson *et al.* showed that the non-linear species at a copper surface dissociated slowly at 80K and that, subsequently, the mobile N-adatoms reacted with linear metal-nitrosyls surface species to form nitrous oxide – the latter being retained at the surface. Much later, Carley et al. have applied[13] scanning tunnelling microscopy to the

NO-Cu(110) surface and demonstrated a two phase system of chemisobed nitrogen and oxygen adatoms; both adatoms form well ordered structures.

It is not clear to this author that there has been an adequate recognition, within the surface community, of the implications of these results – demonstrating, as they do, versatility and variety of metal-ligand bonding; unequivocal demonstration of reaction mechanisms dominated by surface-mobile species; data relevant to intra – and interlayer potential energy functions. But they were certainly not lost in Cardiff, particularly from the standpoint of Roberts *et al.* taking their studies of oxygen-metal interactions to the point where these and other issues have provided observations and arguments which have a profound significance for our views on surface catalysis. Roberts in his splendid article[14] "Heterogeneous catalysis since Berzelius: some personal reflections" quoted J.M.Thomas "Everything is interesting: not everything is important". We might add the problem, the real difficulty at times, is one of recognising importance, of recognising how some observations can carry interpretations which can form a turning point in the development of a topic such as catalysis. Carley *et al*[13]. and, earlier, Roberts[15] have given us summaries of the researches which have highlighted the role of short-lived transients and precursors in the mechanisms of surface reactions. The early identification of defect states, O^- and Ni^{3+} in nickel oxide, together with the lack of reactivity of a 'perfect' nickel oxide[16] overlayer led to the coherent set of investigations of the activation of absorbates at metal surfaces, to the studies of chemical reactivity involving multicomponent reaction systems. The combination of scanning tunnelling microscopy and X-ray photoelectron spectroscopy in one spectrometer has produced a wealth of information, not least on the mobility of metal and adatoms and of surface reconstruction following gross metal movement. What was often inferred from classical kinetic analyses is now a matter of direct observations. The value of the studies of ammonia oxidation at a Cu (110) surface cannot be overemphasised – they require new thinking on surface catalyzed reactions with metastable kinetically controlled oxygen states playing a major role in the selective oxidation of hydrocarbons. My reading of the earlier work on the oxidation of ethene and propene[18,19], suggests that it was this that led Roberts to explore the experimentally more tractable ammonia oxidation systems with the very recent studies showing that five distinctly different oxygen states can be recognised by STM at a Cu (100) surface[13,15,17]. Roberts[14] has set an exacting target: "Understanding, control and exploitation of *self-organising* (my ital) processes on solid surfaces may well become one of the key technologies of the next century." (Italicization is used to draw attention to parallels with biomolecular reactions).

I hope that this brief review will serve as a reminder of the long, well-constructed road that Wyn Roberts has travelled over some 50 years. His 'bonding' with oxygen has been remarkable; also remarkable is the regard and respect shown by his many friends and colleagues, captured so elegantly and authoritatively in this commemorative volume. We all know that Wyn still has some bon mots which he is keeping for the future – bydd y dyfeisgarwch yn parhau.

REFERENCES

1. M.W. Roberts and K.W. Sykes Proc. Roy. Soc. A242 534 (1957)
2. M.W. Roberts Trans. Farad. Soc. 57 99 (1961)
3. C.M. Quinn and M.W. Roberts Trans. Farad. Soc. 61 1775 (1865)
4. M.W. Roberts and B.R. Wells Surf. Sci. 8 453 (1967)
5. idem. ibid 15 325 (1969)
6. C.S. Mckee and M.W. Roberts Trans. Farad. Soc. 63 1418 (1967)
7. C.R. Brundle and M.W. Roberts Proc. Roy. Soc. A 383 (1972)
8. ibid Chem. Phys. Letters 18 380 (1973)
9. M.W. Roberts Adv. in Catalysis 29 (1980)
10. K. Kishi and M.W. Roberts. J. Chem. Soc. Farad. Trans. 71 1715 (1975)
11. Y. Iwasawa, R. Mason, M. Textor and G. Somorjai Chem. Phys. Lett. 44 468 (1976)
12. D.W. Johnson M.H. Matloob and M.W. Roberts J. Chem. Soc. Farad. Trans. 2143 (1979)
13. A.F. Carley, P.R. Davies, R.V. James K.R. Harikumar, G.U Kulkarni and M.W. Roberts Top. in Catal. 11/12 299 (2000)
14. M.W. Roberts, Catalysis Letters 67, 1 (2000)
15. M.W. Roberts. Chem. Soc. Review 6 437 (1996); Surf. Sci., 'The first 30 years', Editor: C.B. Duke, 299/300, 769 (1994)
16. A.F. Carley, P.R. Chalker & M.W. Roberts, Proc. Roy. Soc. – London, SerA. 399, 167 (1985)
17. B. Afsin, P.R. Davies, A.Pashusky, M.W. Roberts & D. Vincent, Surf. Sci. 284, 109 (1993)
18. C.T. Au and M.W. Roberts J. de Chimie Physique 78 921 (1981)
19. C.T. Au, Li Xing-Chang, Tang Ji-an and M.W. Roberts J. Catalysis 106, 538 (1987)

Chapter 2

TECHNIQUE AND PROGRESS IN SURFACE AND SOLID-STATE SCIENCE

John Meurig Thomas

Key words: Field ion microscopy; time-of-flight mass spectrometry; photoelectron spectroscopy; ESCA; XPS; UPS; high-resolution transmission electron microscopy; high-resolution scanning transmission electron microscopy (HRSTEM); electron tomography Z-contrast microscopy.

Abstract: Field-ion microscopy (and cognate techniques such as single atom probe detection), as well as photoelectron spectroscopy (both of the U.V. –stimulated and X-ray-stimulated kind) are chosen as typical examples of important new techniques which transformed surface chemistry and the scientific foundations of catalysis. To each of these large areas of research, M. W. Roberts has made seminal contributions. In particular, through his work on carbon monoxide adsorbed at metal surfaces, he provided the first direct proof that Fischer-Tropsch catalysts (for the synthesis of alkanes, alkenes and alkanols from mixtures of CO and H_2) function by causing dissociation of the CO into elemental surface species. Other techniques, especially electron-microscopic ones, used by the author to achieve deeper understanding of surface and solid-state chemistry, *are also* highlighted.

1.   INTRODUCTION

In most disciplines it is generally true that faster and more enlightening progress in proceeding along the tortuous path from genesis to maturity of a particular sector of a subject is achieved *via* the agency of new techniques and tools than through the formulation of new concepts and ideas. Few scientific revolutions are concept-driven: the growth and burgeoning of most experimental disciplines in physical, biological, medical or engineering

Surface Chemistry and Catalysis, Edited by Carley *et al.*
Kluwer Academic/Plenum Publishers, New York, 2002

5

science usually depend on the arrival of new instruments, new techniques and the discoveries that emerge from their application. Where would structural chemistry, molecular biology, organometallics and materials science be without X-ray crystallography? Where would modern astronomy be without the charge-couple device? Where would natural product and other facets of organic chemistry or biochemistry be without mass spectrometry, nuclear magnetic resonance and two-dimensional gel permeation chromatography or two-dimensional electrophoresis?

Statements and rhetorical questions such as those enumerated above have always been valid in the history of science. We need think only of the contributions made *via* microscopy following the pioneering endeavours in the seventeenth century of Leeuwenhoek in The Netherlands and Hooke in the U.K., or of the striking advances in the realms of analytical chemistry, astronomy and astrophysics following, first, the discovery in 1814 of Fraunhofer lines, and, later in 1861, their interpretation and application to the determination of the composition of the sun and other stars by Kirchhoff and Bunsen.

I begin this article with these philosophical musings because I know that, throughout his scientific career, Wyn Roberts has taken particular delight in exploring new ways — new techniques and new tools — of doing things in surface science and cognate subjects. I also recall with pleasure that ever since we each began our separate academic careers (after having emerged from the same stable — see Personal Reminiscences, Section 5 below), whenever we met in the intervening years we would always describe to one another the merits and excitement of the particular new techniques and tools that we were each developing in our own laboratories.

2.   OVERLAPPING ADVENTURE IN MICROSCOPY: FROM THE IMAGING OF ATOMS TO TOMOGRAPHIC ANALYSIS

Shortly after Wyn Roberts began his independent academic research career, first in Imperial College, London with F.C. Tompkins, later with C. Kemball in Queen's University, Belfast, the world of surface science was galvanized by the remarkable micrographs published by the German-American physicist Erwin Müller. Müller had convincingly demonstrated[1] that field-ion microscopy (FIM) could produce remarkable pictures, in atomic detail, of the surfaces of minute tips and domes of refractory metals and alloys using ionisable noble gases as the imaging media. Views such as that shown in Fig. 1 left no doubt about the reality of the existence of atoms.

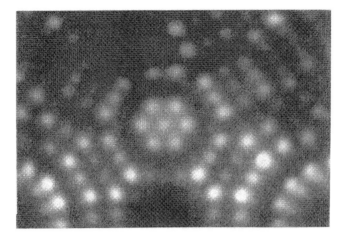

Figure 1. Field ion micrograph of a 111-oriented tungsten specimen. Each white dot is the magnified image of an individual atom (After Miller[4])

More than that, the FIM technique was soon harnessed, especially by Ehrlich[2], to determine points of fundamental scientific importance concerning the adsorption energies and diffusivities of one metal on another. Moreover, FIM enabled surface specificities and anisotropics be charted in quantitative detail. Soon Müller himself, later others, showed how the insights achievable through FIM could be further extended by using the atom probe mass spectrometer (Figure 2)[3].

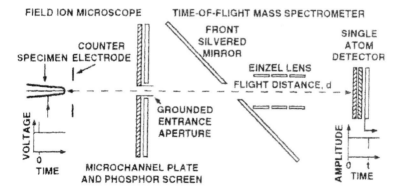

Figure 2. Schematic diagram of a classical atom probe which features a linear time-of-flight mass spectrometer with a single atom detector (After Miller[4])

At first the atom-probe FIM technique was largely of interest to surface physicists and chemists, but soon it was realised by metallurgists and materials scientists that, potentially, APFIM was one of the most powerful techniques to determine the precise atomic distribution of 'foreign' or 'impurity' species in the vicinity of dislocation cores and along grain (or sub-grain) boundaries. The recent work of Smith and others[3-6] illustrate beautifully the power and present day relevance of this technique. By stripping away, layer by layer, of atoms from the surface downwards, APFIM has greatly enlarged our knowledge of the strength of materials (particularly intermetallic solids). Specifically this procedure, which may be legitimately described as (destructive) *atom probe tomography*, is capable of identifying the so-called Cottrell atmosphere surrounding the core of a dislocation. (It is to be recalled that it was A.H. Cottrell who first suggested that the preferential accumulation of impurity atoms at the core of a dislocation played a crucial role in governing the nature of the stress-strain curve of a solid, and especially the transition from elastic to plastic deformation). Later in this article (Section 4.1) we shall say more about Tomography, but of a non-invasive (non-destructive) kind.

Wyn Roberts pursued field-ion and field-emission microscopy while he was at Belfast, but it was another, much more widely applicable technique that commanded most of his energies and imagination during that phase of his career: photoemission studies. These were arguably the most pioneering studies that he pursued in Belfast: they certainly attracted world-wide attention because they provided unique insights into the nature of the chemisorbed bond, into the phenomenon of place-exchange and incorporation — burial of initially chemisorbed layers within the underlying bulk solid — and into electron transfer at surfaces. They were to constitute the platform on which he later built (at Bradford) his pioneering studies of ultra violet and X-ray induced photoelectron spectroscopy, UPS and XPS.

3.   THE EMERGENCE OF PHOTOELECTRON SPECTROSCOPY AND ITS IMPACT IN SURFACE SCIENCE

Ever since B.M.W. Trapnell's book[7] on chemisorption was published in the mid-1950s, I took a keen interest in the possibility of applying Einstein's photoelectric equation, which had, as Trapnell had outlined, been earlier used intermittently by several physicists to probe photoemission from surfaces prior to and following the uptake of adsorbed species. Wyn Roberts had probably been aroused to a greater degree by these observations than I had been, for he set about, in Belfast, to construct a state-of-the-art facility that, in his expert hands, yielded results of high relevance in the context of gas-solid (especially gas-oxide) interactions. By the mid 1960s Roberts' group was leading the way in charting molecular and atomic events at oxide and other surfaces; and this work attracted much attention, notably from eminent scientists such as N.F. Mott and J.W. Linnett. It is noteworthy to recall that at the Faraday Discussion in Liverpool in 1966, Roberts gave one of the key papers, which was heard by two of his heroes, H.S. Taylor and E.K. Rideal.

My own interest in photoelectron spectroscopy was kindled afresh by a paper published by Linnett in *Nature* in 1968[9]. This was one of the earliest reports concerning the determination of the nature of the chemisorption of diatomic molecules at a metal surface in which a monochromatic (HeI) u.v. source was used to liberate electrons from valence orbitals at the solid surface. Not only was I struck by the potential power of Linnett's technique, I was exhilarated to discover the existence of Kai Siegbahn's "ESCA" studies, to which reference was made in Linnett's paper. Thanks to an efficient Inter-Library Loan Service operated at the University College of North Wales, Bangor, where I was then employed, I was able to obtain a copy of this monumental publication (prepared as an extended report by the Uppsala group to its main sponsor, the U.S. Air Force). On reading — over a period of several weeks — the content of that report, my education in solid state science as well as in spectroscopy and theoretical chemistry was greatly enlarged. Not since the time that I had read, some five or six years earlier, Mott and Gurney's seminal monograph on ionic crystals[11] had I been exposed to such a wealth of experimental technique married to digestible theory pertaining to electronic energy levels within and the band structures of solids.

Those who are acquainted with Siegbahn *et al's* landmark tome[10] will recall that, *inter alia*, they demonstrated, using Langmuir monolayers of iodostearic acid films that 'ESCA' was potentially capable of detecting monolayers of a foreign species on a solid support. Both Wyn Roberts and I, and doubtless others (of whom we were unaware at the time — and since!), immediately felt that ESCA, or PES (photoelectron spectroscopy) as it was later to be designated, would be of enormous value to the surface chemist. Each of us had been tutored in an era (early to late 1950s) when it appeared that no experimental technique was available that was sufficiently sensitive to detect directly an adsorbed monolayer on a solid possessing a surface area less than <u>ca</u> $1m^2g^{-1}$. Here was PES (X-ray induced PES in particular, XPS) that potentially had the ability to *detect* a monolayer, or a sub-monolayer on a flat surface of a few cm^2. It also held the promise of being able, *via* chemical shifts in core-electron binding energies, to identify the nature of the chemical bond linking the adsorbate to the adsorbent.

My own group at Aberystwyth, in collaboration with Mickey Barber and his at the AEI Research and Development Centre (now of blessed remembrance) in Trafford Park, Manchester, set about studying single crystals of graphite[12] as well as various kinds of high-grade carbon fibres[13] using an early AEI electron spectrometer (ES100). We established beyond doubt that no oxygen could be detected at freshly cleaved, basal surfaces ((000l) planes) of graphite: the O1s signal (at a predicted binding energy in the vicinity of 530eV) was no higher than the background noise. When, however, the incoming monochromatic soft X-rays (AlK$_\alpha$) impinged upon the prismatic surfaces ((10$\bar{1}$0) and (11$\bar{2}$0)) there was a clear O1s signal, which we could plausibly assume was from a complete monolayer of bound oxygen at these prismatic faces[12]. This was the first experiment to establish the monolayer sensitivity of XPS as a technique for studying solid surfaces.

In a later experiment[14] also involving cleaved surfaces of single-crystal graphite, my colleagues and I determined the sticking coefficients, at room temperature, of molecular oxygen at the basal (000l) and the prismatic surfaces. We discovered that the sticking coefficient at the (000l) surface is fifteen orders of magnitude less than at the prismatic surfaces! We note in passing the recent work of Dahl *et al*[15], where they measure the sticking coefficient of nitric oxide on Ru(000l) surfaces to be nine orders of magnitude (at least) lower than at monatomic steps on this surface. My own group measured[16-18] by an electron microscopic technique (using gold decoration to detect monatomic steps on (000l) graphite surfaces) that the rate at which oxygen molecules dissociatively adsorb and gasify carbon atoms from a graphite surface is greater by a factor of at least 10 at monatomic step edge than at the terraces of basal planes.

Whereas my own interests in XPS and surface studies focused initially on graphite, carbon fibres and diamond, before branching out to many other problems in surface and solid-state chemistry (see below), Wyn Roberts' work dealt primarily with the surface chemistry of metals. He also was the principal architect in designing one of the first really versatile "ESCA 3" commercial instruments, which enabled either sequential or parallel recording of XPS, UPS and Auger electron spectra from a single surface[19-21]. Early on he demonstrated that reliable data pertaining to adsorption, at sub-monolayer amounts, on metals could be retrieved (using the state-of-the-art equipment assembled in his laboratories in the University of Bradford).

Some of the key advances registered by Roberts and his group in the next decade or so included:

(a) work function studies which enabled a distinction to be made between oxygen chemisorption and the onset of surface reconstruction leading to oxidation through monitoring the electron energy distribution and electron yield in photoemission studies[19];

(b) establishing from low energy electron diffraction, optical diffraction and Auger electron spectroscopy studies[20] that oxygen chemisorption on W(112) and Cu(210) surfaces involved precursor states, that oxygen dissociation occurred only at a limited number of surface sites and that after bond cleavage, the oxygen adatoms undergo extensive surface diffusion before becoming finally chemisorbed; (we note in passing that this conclusion harmonises with his more recent work[22] on oxygen transients that he and others, notably Ertl *et al*, have picked up using scanning tunnelling microscopy);

(c) providing the first definitive evidence[23] that carbon-oxygen bond cleavage during the chemisorption of carbon monoxide is a facile process, occurring on iron at low temperatures. This work by the Roberts group, deploying UPS and XPS to their utmost advantage, was of profound significance in heterogeneous catalysis generally and in Fischer-Tropsch syntheses of alkanes, alkenes and alcohols in particular. Roberts' work during the period 1975 to 1980 was pre-eminent in establishing reliable spectral methods (*via* photoelectron spectroscopy) for the quantitative analyses (and assignments) of surface concentrations of individual surface species; and

(d) proof that the activation of CO and CO_2 at certain metal surfaces (*e.g.* Cu(110)) occurred freely at low temperatures, and that disproportion of dimers seemed the most plausible route to carbonate formation at the surfaces of copper, aluminium and magnesium[24].

Among the advances made in surface and solid-state chemistry in my own group were the following:

(a) the elucidation of the nature of the bound oxygen on the faces of diamond single crystals: it transpires[25] that approximately half of it is attached in ketonic form (as >C=O), and the remainder is bound mainly in ether and hydroxyl linkages;

(b) determination of the relative number of exposed carbon atoms at a diamond surface that are at flat regions (terraces) at steps or at kinks. (This is achieved by using F atoms as a probe and recording the intensities of C1s core XPS peaks corresponding to CF, CF_2 and -CF_3 surface groups);

(c) the identification[13] of sub-monolayer amounts of nitrogen and oxygen, and also of even smaller traces of chlorine, bound to the cylindrical surfaces of technological grade carbon fibres;

(d) the determination[26] of the XPS cross-sections for various core levels and the corresponding electron escape depths of all the elements from lithium to uranium: this enabled XPS to be used as a quantitative, non-

destructive analytical tool for any solid consisting of any of ninety or
so elements. The method was shown[27] to be a reliable means of
chemical analysis of aluminosilicates and other complex solids;

(e) the unambiguous determination of the valence state of certain
transition-metal ions from the occurrence or otherwise of 'shake-up'
peaks in the core-level XPS peaks of the element in question[28]. In
compounds of copper, for example, cupric ions exhibit 'shake-up'
peaks (in 2p levels) but cuprous ions do not, because a d^9 electronic,
unlike a d^{10} configuration, possesses the necessary electronic vacancy
in the d-electron manifold to permit promotion of an inner p electron
thereby generating 'shake-up' satellites;

(f) determination[29] of the electronic band structures, and in particular the
nature of the d-band, of transition metal chalcogenides such as MoS_2
from the angular variation of photoemission using HeII radiation as
the stimulating source; and

(g) establishing[30] a sensitive method, based on photoelectron diffraction
for locating the sites of impurities (substitutional or otherwise) in
complex minerals, and determining[30] the surface structure and
composition of layered silicate minerals.

4.   THE ELECTRON MICROSCOPE AND ADVANCES IN SURFACE AND SOLID-STATE CHEMISTRY

Nowadays, the electron microscope is an indispensable instrument in the
portfolio of techniques which the modern chemist, especially the materials-
oriented and solid-state chemist, must possess to elucidate the structure and
properties of new substances. Forty years ago, when I began to turn to
microscopy (mainly optical at that time) as a means of understanding the
surface reactivity of solids, there were very few — only two in fact —
Departments of Chemistry in the Universities of the U.K. that had electron
microscopes as part of their experimental armoury. In the intervening years,
by judicious use of electron optical techniques, numerous vital items of
information of a structural, mechanistic, compositional and often of an
electronic kind have been retrieved. Not only are insights gained (through
electron microscopy) into the existence of whole new families of structures,
hitherto unsuspected, but one also uncovers the structural characteristics of
imperfections in solids. Moreover, it is often the case that these
imperfections either reflect or suggest altogether new structure, again
hitherto unforeseen.

Electron microscopy is, therefore, a powerful agent for aiding chemical synthesis of new materials, a fact which is particularly important in the field of heterogeneous catalysis, because new types of catalytic materials may be defined and described conceptually and then, if the preparative tactics are successful, identified and characterised. There is also the ever-improving role of the electron microscope as an analytical tool — very few techniques known to the chemist and materials scientist can rival it in its sensitivity and detection limits. We show in Section 4.1 that scanning electron microscopes now permit the imaging and the identification of nanoclusters consisting of just a few atoms. There are many other sectors of chemistry besides surface science and catalysis where electron microscopy proves invaluable in elucidating inter-relationships between structures and structure-property correlations. (I have outlined these in a recent review[32] entitled "Chemistry with the Electron Microscope: Some Highlights in a Lifetime's Journey").

My own interest in electron microscopy arose from a desire to learn more about the role of dislocations and stacking faults in governing the chemical properties of crystalline solids[33-35]. When, in the early 1960s, I began my endeavours, I focussed initially on the reactivity of solids, in particular on the oxidation (gasification) of layered minerals such as graphite and molybdenite. Why were there highly localised centres of enhanced reactivity at certain regions in the basal surfaces? Apart from showing[16,36,37], first by optical then by electron microscopy that emergent, non-basal dislocations of either a screw or edge character were directly responsible for enhanced reactivities at the basal (0001) faces of graphite and MoS_2, I also devised the so-called etch-decoration technique, which proved particularly illuminating in (a) detecting isolated vacancies (which were also centres of enhanced reactivity) in the basal plane and (b) enabling the kinetics of oxidation of individual monolayers parallel to the basal places to be determined. Figures 3 to 5 show the essence and applicability of the etch-decoration technique, which entails deposition of gold atoms from the gas phase on to the layered mineral and allowing the deposited atoms to migrate on the surface and to congregate (nucleate) into minute microcrystallites (which are visible in a low-power electron microscope) at topographical irregularities such as steps and terraces of monatomic height.

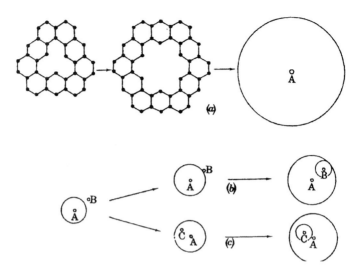

Figure 3. (a) Schematic illustration of the development by anisotropic oxidation of vacancies in the surface layer of graphite. The oxidizing conditions may be arranged so as to yield circular, rather than hexagonal, monolayer (3.3Å) depressions; (b) An indigenous vacancy on the second layer (B) would, after further development, show up as another depression tangential to the first; (c) whereas a vacancy created (as at C) by the oxidizing gas would not be tangential (After Thomas *et al*[16])

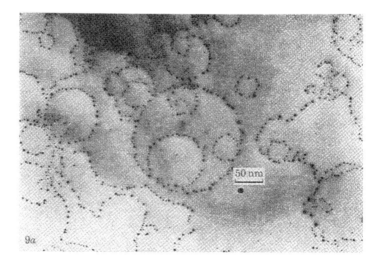

Figure 4. Typical example showing how the vacancy distribution in the surface and successive layers may be determined by the etch-decoration technique[16]

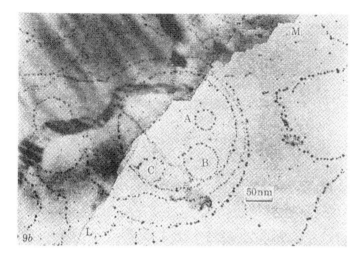

Figure 5. A created vacancy (A) is clearly visible on the fourth layer from the surface of an oxidized crystal of graphite. The indigenous vacancy at B has been developed to a larger radius than at C (both on the fourth layers) because it was uncovered first by the receding third layer. (The largely undecorated ledge LM is on the underside of the crystal)[16]

In the very nature of things, a growing confidence in the versatility and elegance of electron microscopy led one to the exploration of other properties and structures of solids and their surfaces. And with the technological advances that made if possible for commercial companies to offer first 100kV, then 200 (or higher) kV high-resolution transmission electron microscopes (HRTEM) it soon became possible to image, in real space, the ultrastructural features of many silicates, several other minerals and an increasing number of both simple and complicated solids.

Whereas X-ray based techniques of structural analysis had yielded a good deal of ambiguous information concerning the spatial character of polytypic intergrowths in, for example, chain silicates (of the pyroxenoid family, general formula $ASiO_3$ with $A \equiv Ca^{2+}$, Mg^{2+}, Fe^{2+}, Mn^{2+} such as wollastonite, rhodonite and pyroxmangite), HREM, in contrast, yielded precise and unambiguous information[38,39]. Non-random disorder in rather exotic sheet silicates (*e.g.* chloritoid[40]) as well as direct evidence for "staging" in graphite intercalates[41] could also be directly probed by HRTEM.

With the arrival in the late 1970s of a new generation of commercial and custom-built electron microscopes, the situation for the experimental surface and solid-state chemist further improved. Difficulties associated with beam-damage of sensitive specimens (such as zeolites) became less serious with accelerating voltages of 200kV and beyond; moreover, there were improved vacua in these microscopes, thereby further diminishing beam-damage. All this meant that structures of, and imperfections in, a wide range of well-known, as well as many new zeolites that were hitherto beyond reach, could be systematically explored[42,43]. A key point to note is that microporous solids such as zeolites are best envisaged as materials possessing three-dimensional surfaces. This means that, fortunately, all the techniques of diffraction, imaging and spectroscopy that are usually employed in solid-state structural elucidation automatically yield, when the solids under investigation are microporous or zeolitic, information about the surfaces of this remarkable category of solid.

Recently, further advances have become possible owing to the use of charge-couple devices (CCDs)[44]. Since these are more than a hundred times as sensitive to electrons than photographic plates or films, the problem of electron-beam sensitivity of specimens has also been further greatly reduced. Moreover, far lower electron beam intensities than hitherto may now be used to record electron diffraction data — so much so that one may satisfy the so-called kinematic rather than the dynamic conditions of recording. In effect, this means that multiple scattering of electrons may be avoided; and so it now becomes feasible to invoke electron crystallographic methods (akin to direct methods in X-ray crystallography) in undertaking structure determination (of hitherto unknown materials) by electron microscopy. This approach has been pioneered by Terasaki and co-workers at Tohuku University, and he has solved the structures of many mesoporous solids — which may also be envisaged as possessing three-dimensional surfaces — in this fashion[44-46].

4.1   High-Resolution Scanning Transmission Electron Microscopy (HRSTEM)

Chemists in general, and surface chemists in particular, are perhaps less aware of the extraordinary range of important structural, compositional, spatial and tomographic information about solids, especially those of catalytic significance, that may be obtained from application of HRSTEM.

Fig. 6 summarises the salient features of the disparate kinds of signals that may be gathered in HRSTEM when mono-chromatic high-energy electrons are scanned over the surface of a thin specimen of the solid under investigation using a finely-focused probe (*ca* 8Å dia).

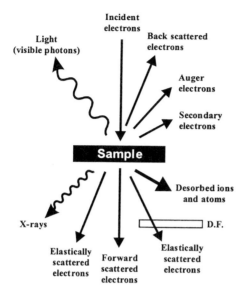

Figure 6. Schematic diagram illustrating the wealth of information derivable from high-resolution scanning transmission electron microscopy (HRSTEM) (see text)

Although HRSTEM images exhibit slightly diminished resolution compared with analogous HRTEM ones — because of the finite source size of the probe and reduced detection efficiency imposed by the geometry of the collection — they yield many more insights than their HRTEM counterparts. First, there is the ultra-high microanalytical resolution from the X-ray (or Auger-electron or electron-energy loss) signals that are produced; second, all such signals may be collected simultaneously, together with back-scattered electrons; and third, by recording high-angle annular dark field (HAADF) (in other words, Rutherford scattered) electrons one may record Z-contrast images. In addition, as my colleague Paul Midgley in Cambridge has recently demonstrated[47,48], electron tomography may be accomplished by taking a series of two-dimensional (2D) images with the specimen oriented at a series of angles with respect to the incoming beam.

Z-contrast (or atomic number) imaging exploits the fact that the intensity of electrons scattered through high angles from this specimens follow the Z^2 dependence of Rutherford's law. Thus, one platinum atom scatters as strongly as about a hundred oxygen atoms or thirty-two silicon atoms. This technique is therefore 'tailor-made' for detecting clusters of catalytically active metals such as Pd, Ru, Pt, Re, or clusters thereof on light supports such as mesoporous silica, zeolites or carbon.

Valuable as the information yielded by HRSTEM instruments is (in providing HAADF (Z-contrast) or X-ray emission (XRE) mapping), especially in investigating nanoclusters as well as other nanofeatures or interest to the surface chemist, it is limited to providing 2D projections of 3D arrangements. The need exists to analyse specimens, non-destructively — in contrast to the APFIM described in Section 2 above —fully in three dimensions, especially if we seek to understand the factors responsible for the loss of activity, selectivity and stability of nanoparticle catalysts.

A suitable approach is electron tomography[49], where 3D structure is reconstructed from a tilt series of 2D projections. In practice, this is achieved by tilting the specimen under investigation on one axis using a microscope goniometer and taking a micrograph at a range of tilts, typically every 1 to 2°, correcting for specimen movement, focus change and astigmatism throughout[47]. The series of micrographs can then, typically, be processed offline using routines based on the so-called Radon Fourier-slice therein[50], to reconstruct the full 3D structure.

In conventional electron Tomography (see for example the recent work of de Jong *et al*[51] on "steamed" zeolites) such a tilt series is acquired using bright field (BF) images which are dominated by Bragg-scattered electrons. However, the large beam currents that of necessity must be used in conventional HRTEM means that such an approach cannot be used on specimens that are beam sensitive. It so happens that much of my research effort in designing powerful, highly selective catalysts (using the principles of solid-state and surface chemistry[52,53]) entails the use of mesoporous silica, and, in particular, bimetallic nanoparticles such as Pd_6Ru_6, Ru_6Sn and $Cu_4Ru_{12}C_2$, which are very good selective hydrogenation catalysts. It is therefore essential to develop a tomographic technique that does not incur severe beam damage during its implementation. The framework structure of mesoporous silica degenerates rather rapidly under conventional HRTEM illumination, and so is unsuitable for the prolonged acquisition times required for a tilt series. With HRSTEM, however, far smaller total beam exposure is involved. Another advantage arises from the scattering geometry, in that HAADF imaging (*via* Rutherford scattering) excludes almost all the electrons scattered coherently through low angles. This means that the resulting images are free from the misleading complications of electron phase contrast and, as such, they are true projections of the structure.

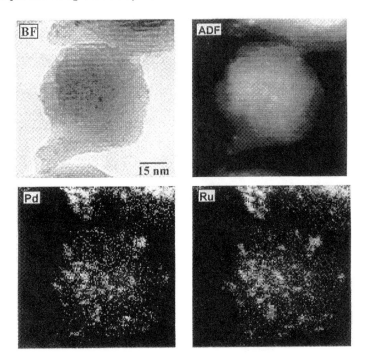

Figure 7. Typical HRSTEM images, taken under Z-contrast conditions of nanoparticle bimetallic catalysts. Each which dot represents a cluster (*ca* 10Å diameter) of Pd_6Ru_6 aligned along the mesopores (*ca* 30Å diameter) of the silica support.

As 'Rutherford' images are formed primarily from incoherently scattered electrons, the combination of HRSTEM (HAADF) imaging, coupled with high accelerating voltages (*e.g.* 300kV in the Philips field-emission-gun scanning transmission microscope CM300), which means that there are minimal inelastic interactions and thus minute beam damage, is an ideal basis for the tomography of supported nanocatalysts.

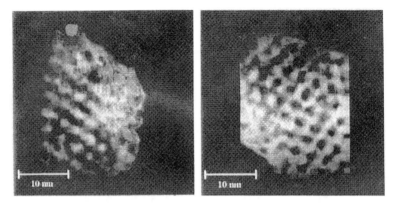

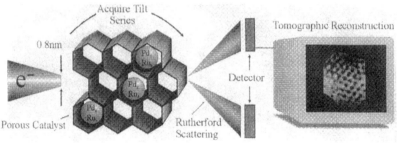

Figure 8. A graphical summary of the essence of the Z-contrast tomography described in the text (After Midgley *et al*[47]).

My colleagues Paul Midgley, Mathew Weyland, and I[47] have illustrated the power of Z-contrast tomography by examining Pd_6Ru_6 bimetallic catalysts[54] supported on mesoporous silica. A HAADF tilt series was acquired on a CM300 field-emission gun scanning transmission electron microscope (FEGSTEM) equipped with an appropriate HAADF detector[47]. The series was acquired from +60° to –48° with an image every 2° giving a total of 55 images. The resulting reconstructions (which were effected using routines programmed in the IDL image processing language running on an IRIX operating system) were displayed using multi-level voxel (*i.e.* a three-dimensional pixel) — full details are given in ref. 47, and a 3D animation demonstrating the results may be viewed at http://www-hrem.msn.cam.ac.uk/imw259/Work/Tomo.html. Fig. 8 illustrates (statically) the essence of what may be achieved with this novel technique.

5.   PERSONAL REMINISCENCES

Wyn Roberts and I first met when we were schoolboys in South Wales, he being a pupil in the Amman Valley Grammar School, Ammanford, I in the Gwendraeth Valley Secondary School, some ten miles further west in the anthracite coalfields of Carmarthenshire. We each represented our respective schools in the rugby teams, and were opponents on the playing fields on more than one occasion in the late 1940s. We also met during the summer months on the athletic track, he being a sprinter, I a race-walker. But it was at the University College of Swansea that we first got to know one another properly. He had entered the College to read chemistry in 1949, I following in 1951. He was deeply immersed in his revisions for the Honours exam in early 1952 when I first had a long conversation with him about our studies. It was at the sea front, overlooking Swansea Bay, one early summer's evening that I happened to be strolling along the promenade when he came towards me from the opposite direction. I vividly recall talking about a book which I had just purchased — and which he had bought earlier I soon realised — namely C.A. Coulson's *"Valence"*. He told me how much this monograph had helped him to understand more fully the short course on quantum mechanics then being given in the Department by a man with whom we were both, later, to do our postgraduate research: K.W. (Keble) Sykes, one of C.N. Hinshelwood's former high-flyers from Oxford.

In the early 1950s Swansea began to be mentioned on the world stage, not because of the quality of the academic work being pursued at the University College (good as it was), nor because of the quality of its rugby and soccer teams (good as they were also), nor even because of the skills of the Glamorganshire cricket XI, who frequently played their matches against the Australians, Indians, West Indians, South Africans and New Zealanders at the St. Helens Ground, Swansea. No, it was because of the genius and flamboyance of the Anglo-Welsh writer and poet, Dylan Thomas, who was a Swansea man born and bred. When he died prematurely in a New York City hospital in 1953, Swansea became even more famous. It was in New York City that Dylan Thomas' *Under Milk Wood* was first performed. It is a play for voices, and had been written deliberately for a radio audience. Whenever I listen to or see that play, and especially when I hear the description:

> *'At the sea end of the town, Mr. and Mrs. Floyd, the cocklers, are*
> *asleep, quiet as death, side by wrinkled side, toothless, salt and brown,*
> *like two old kippers in a box'*

I think of Swansea, and especially does my mind roam to that conversation I had with Wyn about Coulson's *"Valence"* 'at the sea end of the town'.

Wyn played an important part in the direction that I took at the outset of my postgraduate career. Like so many first generation chemists, I was so intoxicated by the power, beauty and romance of the subject after I obtained my initial degree, that I did not feel that one sub-discipline of chemistry attracted me more than any other as a topic for PhD research: they all seemed interesting. The Head of Department at Swansea, Professor Charles Shoppee, a highly successful steroid chemist, persuaded me that his general line of research would be profitable. And, indeed, I spent two days or so in his research laboratory (preparing epicholesterol iodide). At that time Wyn Roberts and I entered into deep conversation about his research, which he was carrying out with Keble Sykes. The more I listened to what Wyn was endeavouring to accomplish, the more I felt the attraction to embark on physico-chemical aspects of the study of solid surfaces. Wyn played a key role in conveying this information to our superiors, and in the spirit of harmony that prevailed in Swansea at the time, Sykes and Shoppee agreed to transfer me to another laboratory, where I soon became immersed in preparing "clean" surfaces of carbon (for chemisorption studies) by evaporation in high vacuum. We shared the same laboratory (with three other PhD students) for one year, before Wyn left for London and Aldermaston.

Regularly — on average once every month or so — we have been in touch as fellow scientists and friends ever since. When he got married in 1956, I was his best man; and when I married in 1959 he was my best man.

Wyn's pioneering studies in the surface chemistry of solids has, from the beginning, but especially since he began running his own Department in Bradford, elicited the admiration of many. As mentioned earlier, my immediate predecessor at Cambridge, the 1920 Professor of Physical Chemistry, Jack Linnett, had the highest admiration for Wyn's work; so also did Nevill Mott, one of the greatest physical scientists of the twentieth century. The attached copy of the letter that Mott sent from the Cavendish Laboratory to Wyn at Bradford on 14 September 1966 was, I know from Wyn's reaction at the time, an inspiration. It led to much exchange of information between them, and Mott — an exceptionally busy man with a multiplicity of irons in the fire both nationally and internationally — made a point of visiting Wyn's laboratory in the late 1960s, to each's mutual advantage. Many seminal scientific ideas flowed from their friendship. And it also was the main reason why Wyn and his able former PhD student Clive McKee wrote their timely and compendious monograph (as part of a series on the physics and chemistry of materials under the general editorship of H. Fröhlich, P.B. Hirsch and N.F. Mott) for the Clarendon Press, Oxford.

UNIVERSITY OF CAMBRIDGE
DEPARTMENT OF PHYSICS

From   CAVENDISH LABORATORY
PROFESSOR SIR NEVILL MOTT, F.R.S.   FREE SCHOOL LANE
                                                  CAMBRIDDGE
                                                  Telephone: Cambridge 54481

14th September, 1966

Dear Professor Roberts

Thank you for sending me your reprints. There are two points in the theory of Oxidation about which I felt I might do some theoretical work during the coming year, following up my work with Cabrera of fifteen years ago. One is the question of nucleation and island formation. I have the impression that much less has been done about this than in the case of evaporated metal films, a subject on which we have an experimental programme here.

The other problem is the old hypothesis of Pilling and Bedworth. Why does an oxide film frequently crack up after a brief initial stage of growth if its specific volume is less than that of the metal? I have a few ideas about this which might be interesting to work out.

Perhaps there is some recent work in these fields which I do not know about; if so I wonder if you could let me know?

It would be a pleasure to meet you sometime this coming winter and to discuss these problems. Is anything likely to bring you to Cambridge? If so, do please let me know.

Yours sincerely,

N.F. Mott

Professor M. W. Roberts,
Department of Physical Chemistry,
Bradford Institute of Technology,
Bradford 7.

Wyn's literary skills are considerable, as anyone who has consulted his reviews will know. Two recent examples are especially noteworthy. The first is the 70-page special issue of *Catalysis Letters*[55] that appeared in 2000 entitled "Heterogeneous Catalysis since Berzelius: Some Personal Reflections". This was widely distributed at the Granada International Congress on Catalysis in July 2000, and it has been universally praised. In the opinion of many it is a masterpiece, and it beautifully traces the evolution of heterogeneous catalysis from the birth of the concept to its present-day florescence in the industrial world. The second is his kind, penetrating and well-judged biographical memoir of the late Charles Kemball[56], a great surface chemist whom Wyn recalled with affection possessed the attributes of both a brigadier and a teddy bear!

Others, besides me, have long cherished Wyn's wisdom and insights into metal-gas interfaces. I well recall meeting Ron Mason shortly after he and I had started some joint work on Auger electron spectroscopy in 1971. He had just been talking to Wyn, after having read many of the papers that had originated at Bradford. I have forgotten precisely what Ron's words were, but I recall vividly his saying that "if there is anything that you want to know about the surface properties of metals, Wyn Roberts is the person to whom you should speak".

I have benefited from speaking to him for over half a century. He is as vitally interested now in molecular events at surfaces as he ever was: his intellectual curiosity, productivity and human decency are as evident now as they were when he was a student. Long may his contributions flourish.

6.   REFERENCES

1.   E.W. Müller, 4th Int. Conf. Electron Microscopy, Springer-Verlag, Berlin, 1958
2.   G. Ehrlich and F.G. Hudda, *J. Chem. Phys.*, 36, 3233 (1962)
3.   M.K. Miller and G.D.W. Smith, "Atom Probe Microanalysis: Principles and Applications to Materials Problems", Materials Research Soc., 1989, Pittsburgh, PA. p.37
4.   M.K. Miller, "Atom Probe Tomography: Analysis at the Atomic Level", Kluwer Academic-Plenum Publications, New York, 2000
5.   A. Cerezo, T.J. Godfrety and G.D.W. Smith, *Rev. Sci. Inst.*, 59, 862 (1988)
6.   D. Blavette and A. Menand, *Recherche*, 253, 464 (1993)
7.   B.M.W. Trapnell, "Chemisorption", Butterworths, London, 1955
8.   E.K. Rideal and H.S. Taylor, "Catalysis in Theory and Practice", Macmillan, London, 1926
9.   W.T. Bordass and J.W. Linnett, *Nature*, 222, 660 (1969)
10.   K. Siegbahn, C. Nordling *et al*, "ESCA: Atomic Molecular and Solid-State Structure Studied by Means of Electron Spectroscopy", Almquist and Wiksells, Uppsala, 1967

11. N.F. Mott and R.W. Gurney, "Electronic Processes in Ionic Crystals", Oxford Univ. Press, 1940
12. J.M. Thomas, E.L. Evans, M. Barker and P. Swift, *Trans. Faraday Soc.*, 67, 1875 (1971)
13. M. Barber, P. Swift, E.L. Evans and J.M. Thomas, *Nature*, 227, 1131 (1970)
14. M. Barber, E.L. Evans and J.M. Thomas, *Chem. Phys. Lett.*, 18, 423 (1973)
15. S. Dahl *et al, Phys. Rev. Lett.*, 83, 1814 (1999)
16. J.M. Thomas, E.L. Evans and J.O. Williams, *Proc. Roy. Soc. A*, 331, 417 (1972)
17. J.M. Thomas, R.M.J. Griffiths and E.L. Evans, *Science*, 171, 174 (1971)
18. J.M. Thomas, *Carbon*, 8, 413 (1970)
19. C.M. Quinn and M.W. Roberts, *Trans. Farad. Soc.*, 61, 1775 (1965)
20. C.S. McKee, D.L. Perry and M.W. Roberts, *Surf. Sci.*, 39, 176 (1973)
21. C.R. Brundle and M.W. Roberts, *Proc. Roy. Soc.*, A331, 383 (1972)
22. M.W. Roberts, *Surf. Sci.*, 299-300, 769 (1994)
23. K. Kishi and M.W. Roberts, *J. Chem. Soc. Farad. Trans. 1*, 71, 1721 (1975)
24. A.F. Carley, M.W. Roberts and A.J. Strutt, *J. Phys. Chem.*, 98, 9175 (1994)
25. S. Evans and J.M. Thomas, *Proc. Roy. Soc.*, A353, 103 (1977)
26. S. Evans, R.G. Pritchard and J.M. Thomas, *J. Elec. Specy.*, L4, 34 (1978)
27. J.M. Adams, S. Evans, P.I. Reid, J.M. Thomas and M.J. Walters, *Analytical Chem.*, 49, 2001 (1977)
28. S. Evans, R.G. Pritchard and J.M. Thomas, *J. Phys.*, C10, 2483 (1977)
29. R.H. Williams, J.M. Thomas, M. Barber and N. Alford, *Chem. Phys. Lett.*, 17, 142 (1972)
30. J.M. Thomas, S. Evans and J.M. Adams, *Phil. Trans. Roy. Soc. A*, 292, 563 (1979)
31. S. Evans, E. Raferty and J.M. Thomas, *Surf. Sci.*, 89, 64 (1979)
32. J.M. Thomas, *Microscopy and Microanalysis*, 2002 (in press)
33. E.E.G. Hughes and J.M. Thomas, *Nature*, 193, 838 (1962)
34. J.M. Thomas, C. Roscoe, K.M. Jones and G.D. Renshaw, *Philos. Mag.*, 10, 325 (1964)
35. J.M. Thomas, Chemistry and Physics of Carbon, 1, 122 (1966)
36. J.M. Thomas and E.L. Evans, *Nature*, 214, 167 (1967)
37. O.P. Bahl, E.L. Evans and J.M. Thomas, *Surf. Sci.*, 8, 473 (1967)
38. J.M. Thomas, D.A. Jefferson, D.J. Smith and E.S. Crawford, *Chemica Scripta*, 14, 167 (1978/79)
39. J.M. Thomas, *Nature*, 281, 523 (1979)
40. D.A. Jefferson and J.M. Thomas, *Proc. Roy. Soc.*, 361, 399 (1978)
41. E.L. Evans and J.M. Thomas, *J. Solid State Chem.*, 14, 99 (1975)
42. J.M. Thomas and G.R. Millward, *J. Chem. Soc. Chem. Commun.*, 1380 (1982)
43. J.M. Thomas, M. Audier, SA. Ramdas, L.A. Bursill and G.R. Millward, *Farad. Disc.*, 72, 345 (1981)
44. O. Terasaki, T. Ohsuna, N. Ohnishi, K. Hiraga, *Currt. Opin. in Solid State & Materials Science*, 2, 94 (1997)
45. Y. Sakamoto, M. Kaneda, O. Terasaki *et al, Nature*, 408, 449 (2000)
46. J.M. Thomas, O. Terasaki, P.L. Gai, W. Zhou and J.M. Gonzalez-Calbet, *Accnts. Chem. Res.*, 34, 2001 583
47. P.A. Midgley, J.M. Thomas, M. Weyland and B.F.G. Johnson, *Chem. Commun.*, 2001 907
48. M. Weyland, P.A. Midgley and J.M. Thomas, submitted
49. J. Frank (ed.), "Electron Tomography", Plenum Press, New York, 1992
50. S.R. Deans, "The Radon Transform and Some of its Applications", J. Wiley, Chichester, 1983

51. A.J. Koster, U. Ziese, A.J. Verkleij, A.H. Janssen and K.P. de Jong, *J. Phys. Chem. B*, 104, 9368 (2000)

52. J.M. Thomas, Angew. Chemie Intl. Edn. Eng., 27, 1673 (1988)

53. J.M. Thomas, *ibid*, 38, 3588 (1999)

54. R. Raja, G. Sankar, S. Hermans, D.S. Shephard, T. Maschmeyer, S. Bromley, J.M. Thomas and B.F.G. Johnson, *Chem. Commun.*, 1571 (1999)

55. M.W. Roberts, *Catalysis Lett.*, 67, 1 (2000)

56. M.W. Roberts, Biographical Memoirs of Fellows of the Royal Society, 46, 285 (2000)

Chapter 3

50 YEARS IN VIBRATIONAL SPECTROSCOPY AT THE GAS/SOLID INTERFACE
Some Personal and Group Endeavours

Norman Sheppard

Key words: Infrared spectroscopy, silica, oxide surfaces, hydrogen-bonding, adsorbed
                   hydrocarbons, substituted alkenes, hydrogenation, Reflection-Absorption
                   Infrared Spectroscopy, RAIRS, Metal-Surface Selection Rule, MSSR,
                   Vibrational Electron Energy-Loss Spectroscopy, VEELS, ethylidyne.

Abstract: An account is given of the origin of the study of the infrared spectra of
                  adsorbed molecules in Cambridge in the 1950s and of the pioneering
                  contributions of this technique to the determination of the structures of
                  chemisorbed species. More specifically it describes the endeavours of the
                  Cambridge/Norwich research group by this means to determine the structures
                  of such species from the adsorption of ethylene and substituted ethylenes on
                  silica-supported metal catalysts. Success in this field proved to be elusive until
                  the development of vibrational spectroscopic techniques that enabled the
                  measurement of spectra of single monolayers of hydrocarbons on flat single-
                  crystal surfaces. The account describes the stage-by-stage development of
                  these researches which depended on the successive introduction of new ultra-
                  high vacuum techniques within surface science.

1.   INTRODUCTION

I am pleased to contribute to this volume in honour of the fine contributions that Wyn Roberts has made to the field of surface science, starting in the pioneering days of the surface photoelectron spectroscopies. I offer an account of my personal experiences and encounters, particularly acquired during my research group's development of vibrational spectroscopy for the determination of the structures of the adsorbed molecules chemisorbed on oxide-supported metal catalysts. To provide a specific historical perspective I have described the step-by-step progress made in identifying the adsorbates formed from ethylene chemisorbed on platinum catalyst particles supported on high-area silica, Pt/SiO_2. This particular example has the merit that its long term progress covered the period of, and therefore involved, many of the advances in surface science techniques that occurred over more than four decades. A number of these advances have enabled vibrational spectroscopy to continue as a major present-day contributor to the determination of the structures of adsorbed species.

2.   EARLY DAYS IN SURFACE VIBRATIONAL SPECTROSCOPY

I first thought of the possibility of studying adsorption on surfaces by infrared spectroscopy on reading a paper by George Pimentel and colleagues. They had taken a powdered sample of silica gel and obtained spectra before and after interaction with D_2O so as to replace the surface OH groups by OD [1]. Silica gel has the advantage of having an exceptionally high surface area, and hence a high proportion of surface groups within the total sample. However, the necessity of reducing heavy scattering of radiation by the samples in air required the incorporation of each of them in paraffin wax. Hence, although the changes in the spectra due to the change of surface features (from OH to OD) was successfully measured in this very favourable case, the example highlighted the experimental difficulties that would have to be overcome if work on adsorbed species were to develop further.

My second thoughts were stimulated by David Tabor, a member of the Tribophysics Research Group in the Cavendish Laboratory, after my return to the Department of Colloid Science in Cambridge from a postdoctoral period in the USA. Tabor asked me if it would be possible to obtain the infrared spectrum of a monolayer, such as of a long chain carboxylic acid, on a surface. Again this was a favourable case to consider as both the multiple CH_2 groups and C=O groups of such a molecule give exceptionally strong infrared absorption bands. Using a rule of thumb that exceptionally strong absorbers give acceptable spectra at thicknesses of about 10 micrometers, and realising that a monolayer thickness of such a substance would be of the order of 10 Å (10^{-9} compared with 10^{-5} m), I calculated that the stronger features of such a spectrum would be of intensity of a small fraction of one percent at normal resolution. I suggested that an improvement of about two orders of magnitude in detector sensitivity, or other equivalent gain in experimental techniques, would be necessary for such measurements to become feasible on a routine basis. This conclusion reinforced a pessimistic outlook for surface infrared spectroscopy.

As a result of these prognostications when David J.C. Yates, a postdoctoral colleague in the Department, asked if I could help him to understand an adsorption process by infrared spectroscopy, I was not encouraging. He had been studying the adsorption of ammonia on a tubular piece of porous silica glass by measuring its change in length, by interferometric methods, as a function of the amount adsorbed. He had found that a first contraction was followed by the more normal expansion with increasing adsorption and wondered whether infrared spectroscopy could throw light on the different processes concerned. On further discussion it turned out that the porous glass not only had the advantage of being of very high area but that it also scattered very little radiation in the visible region, in marked contrast to the silica gel studied by Pimentel. By implication it scattered even less in the infrared region because the pores in the material were of dimensions substantially less than the wavelength of the radiation. The scattering of radiation by a high-area material increases rapidly when the surface dimensions exceed the wavelength of the radiation.

We therefore tried a preliminary experiment by simply placing a cylinder of the material, as supplied, in the infrared beam. The result was a complete blackout up to near 4000 cm^{-1}. I knew that, with the thickness of the glass tube of *ca.* 1 mm, silica itself should not absorb above about 2000 cm^{-1}, taking into account its fundamental and overtone lattice modes. Above this limit, the blackout seemed probably to be associated with the presence of water molecules adsorbed on the surface or filling the pores. As a result of this promising start we designed, and David Yates built, a metal infrared cell in which the porous glass sample could be cleaned from occluded water by heating and evacuation, and then dosed with an adsorbate down to approximately liquid nitrogen temperatures [2]. The spectrum of the cleaned adsorbent showed that, additional to the silica blackout absorption, there was another similar region between 4000 and 3500 cm^{-1}, which must originate in OH groups attached to the surface by chemical bonds. Thus the spectra showed that much of the surface was covered with OH groups which would themselves provide, possibly preferred, sites for adsorption processes. Later infrared work showed that chemically bound OH groups are a general feature of 'oxide' surfaces.

Between the silica and OH blackouts a window was available that was convenient for studying the CH or NH stretching-mode absorptions of adsorbates. The adsorption of ammonia, David's initial interest, led to a very strong band centred near 3200 cm^{-1} which, from its great breadth, we realised had to arise principally from very strong hydrogen-bonding of surface OH groups to the ammonia molecule, rather than from the weaker

NH absorptions of the adsorbate [3]. Methane was adequately adsorbed at near liquid-nitrogen temperature and this gave a stronger absorption at 3006 and a weaker, sharper, one at 2985 cm^{-1} from the adsorbed molecule [2]. The first of these corresponded clearly to the infrared-active v_3 triply-degenerate CH stretching mode of methane and the second to the Raman active v_{CH} 'breathing' mode. The latter is infrared forbidden for isolated molecules in the gas phase, but had become allowed through the one-sided distortion of the tetrahedral molecular structure caused by the adsorption forces. The well-marked difference in the breadths of the two bands led to the interpretation of the considerable breadth of the 3006 cm^{-1} band as being consistent with blurred-out rotation wings associated with interrupted rotation about a single axis perpendicular to the surface; the band was too narrow to be consistent with free rotation about all three axes. The induced dipole moment of the v_1 mode, which is modulated in magnitude with the vibration frequency, is of necessity fixed in direction perpendicular to the surface. and therefore is not modulated during molecular rotation. Later experiments over a considerable temperature range with adsorbed methyl iodide confirmed the one-dimensional rotation model in that case also [4]. Yates and I also showed that surface distortions could cause the vibration of adsorbed hydrogen molecules to become infrared active and, from quantitative intensity measurements, we were able to estimate the strength of the surface field-gradient [2].

Having embarked on our own successful experiments we made a more detailed survey of the literature and found that Terenin and colleagues in Leningrad had from the late 1940s published work which was also based on the use of porous silica glass as adsorbent. They had, however, principally studied the effects of adsorbates on the first overtone of the surface OH stretching mode [5,6]. At that time, the Leningrad School had not published anything in the infrared fundamental region. However, before our results appeared in print, A.N. Sidorov from that laboratory also started to publish such work [7]. He chose to study the adsorption of larger molecules such as phenol at room temperature, so that his experiments were complementary to ours. At this time, all the published spectroscopic results involved physical adsorption, usually through hydrogen-bonding to the surface OH groups.

3. INFRARED SPECTRA OF CHEMISORBED MOLECULES: EARLY TEXACO AND CAMBRIDGE CONTRIBUTIONS

In the 1950s I was very fortunate to be invited to give papers at a new series of biannual Gordon Conferences in chemistry, held in New Hampshire USA, on the theme of infrared spectroscopy. At the first of these I gave a paper on our infrared work on conformational (rotational) isomerism but to the second, probably held in 1954, I described the work by David Yates and myself, discussed above, on physical adsorption on porous glass. This went down very well. After my lecture I was approached by Stanley Francis of the Texaco Research Laboratory who told me that a colleague of his, Robert Eischens, had recently managed to obtain infrared spectra of carbon monoxide adsorbed on metal-on-silica samples. I was very interested in this because this was in the field of chemisorption rather than physisorption, closely related to catalysis and doubtless the reason for the interest of Texaco in such work. For the first time this implied the possibility, at least in principle, of determining the structures of surface species which are active in catalytic reactions. Hitherto such species could only be inferred indirectly from kinetic studies of reaction products. If the spectroscopic method was to become feasible I could foresee many applications, such as the possibility of checking on the actual structures of surface hydrocarbon species suggested by my friend Charles Kemball to be the intermediates in H/D isotopic substitution experiments of ethylene or ethane over metal surfaces [8].

Eischens' achievement was an impressive one because the surface area of the metal particles was bound to be much less than that of the silica support or of porous glass, *i.e.* a few tens rather than several hundreds of $m^2 g^{-1}$. Such metal catalysts appear opaque to visible radiation but in the infrared region they transmitted radiation down to the blackout limit of the silica support, *ca.* 1300 cm^{-1} at the thickness of sample that he used, thus allowing interaction with the surfaces of the metal particles en route. Francis told me that Eischens had boosted the sensitivity of his spectrometer by cutting out all but one of the teeth of the attenuation comb of his Perkin-Elmer spectrometer so that full scale on the chart corresponded to 10% absorption.

When two years later, on my way to the next infrared Gordon Conference (probably in 1956), I visited Eischens, he told me that the metal-on-silica samples that he had used were prepared according to procedures that he had learnt from his PhD. supervisor, P.W. Selwood. The latter used such preparations of nickel catalysts to measure the changes in magnetisation of metals such as Ni as a result of chemisorption [9]. In the meantime Eischens, with S.A. Francis and W. Pliskin [10], in 1954 had published their spectra of CO on several silica-supported metals. Pliskin and Eischens had followed this up in 1956 with much weaker spectra from ethylene and acetylene adsorbed on nickel [11]. By the time of our meeting Yates and I, with a research student Leslie Little, had also started to measure infrared spectra of ethylene and acetylene on Ni, Pd and Pt, using our porous glass as the support. This was later published in 1960 [12]. Our thicker silica tubes cut out transmission below 2000 cm^{-1} but our use of a diffraction-grating spectrometer gave us enhanced resolution. Hence by then Eischens and I had matters of mutual interest to discuss.

[During my visit to Eischens I saw an advertisement by Varian Associates for a nuclear magnetic resonance (NMR) spectrometer illustrating a spectrum of ethanol; on my return to Cambridge I told Professor A.R. Todd (later Lord Todd), who had recently appointed me as Assistant Director of Research in Spectroscopy in the University Chemical Laboratory, that this looked likely to be the next fruitful form of spectroscopy for molecular structure analysis. So it proved to be!]

4.   ALKENES CHEMISORBED ON SILICA-SUPPORTED METAL CATALYSTS: THE CAMBRIDGE/EAST ANGLIA CONTRIBUTIONS

In his thesis in 1958 [13] Little described an infrared cell for reducing metal salts on the porous glass to metal particles in a furnace section, from which the cylindrical metal/silica sample could be moved along a tubular guide so that spectra at room temperature could be obtained in a section with infrared-transmitting windows. Separate spectra could be obtained of the gas phase and of the gas-phase plus adsorbate sample.

Infrared spectra of adsorbed hydrocarbons enable general distinctions to be made between saturated and unsaturated surface species but, as we shall see below, nearly three decades were to pass before the spectra of the initially adsorbed C_2 surface species were to find definitive structural interpretations. The spectra obtained after hydrogenation were usually stronger and more susceptible to analysis. Little's thesis contained spectra from ethylene on Pd/SiO$_2$. Before hydrogenation the spectrum was weak with saturated and unsaturated v_{CH} absorptions; after hydrogenation a spectrum was obtained which was identified as from an *n*-butyl surface species. He also obtained spectra from acetylene on SiO$_2$-supported Pd and Cu. In each case the initial weak spectrum was from a predominantly unsaturated species but after hydrogenation long-chain alkyl species were clearly present which must have resulted from polymerisation. Little in 1966 also published a long-appreciated book that first collected together the experimental techniques used, and the typical results obtained, in the infrared spectroscopic study of adsorbed species [14].

In 1958, in a review with Pliskin in Advances in Catalysis [15], Eischens published room-temperature spectra from adsorbed species from longer-chain monoalkenes on Ni/SiO_2 before and after hydrogenation. For the hexenes, he showed that all the spectra obtained were independent of the position of the C=C along the chain, *i.e.* that adsorption on this catalyst led to double-bond migration. In all cases the spectrum after hydrogenation were clearly those of the corresponding *n*-alkyl group attached to the surface by the end carbon atom. Until this time Eischens and Pliskin led the field in the study of hydrocarbons adsorbed on metal catalysts but, when their research publications in this area ceased in 1958, they had confined their interests to Ni/SiO_2 as the adsorbent catalyst. Based on their work on CO, we had entered the field experimentally by the time of their first publication on hydrocarbons in 1956 and subsequently widened the field to include work on silica-supported Ni, Pd, Pt, Cu [12], and later on Ir and Rh catalysts. From 1967 Erkelens in the Netherlands [16] and Shopov and Palazov in Bulgaria [17] also joined in this type of work using several silica-supported metals

Michael Clark continued the work that Leslie Little had commenced [18]. He developed an improved method of preparing the supported metal catalysts by direct reduction of appropriate metal salts, deposited from solution within the porous glass (nickel nitrate, palladium chloride, hexachloroplatinic acid for Ni, Pd and Pt respectively) for reduction by hydrogen at *ca.* 300°C. Little had first converted salts to the metal oxides but reduction of these had proved to be a more difficult process. Clark made a wide exploration of spectra before and after the addition of hydrogen to the adsorbed species for ethylene, propene, butenes, butadiene, and acetylene and methylacetylene, on Pt/SiO_2; for ethylene and acetylene on Pd/SiO_2; and for ethylene and butene-1 on Ni/SiO_2. No spectra of adsorbed molecules were observed from saturated hydrocarbons at room temperature, but some poorly-defined spectra from ethane, propane, the butanes and cyclopropane on Pt/SiO_2 were obtained after adsorption at 140°C or higher.

Clark showed that a Pt sample adsorbed between 2 and 3 times less moles of ethylene than hydrogen. Assuming that hydrogen dissociates on two sites this suggested that on average an adsorbed ethylene occupied 4 to 6 surface sites, implying dissociative adsorption. This was consistent with a large increase in intensity of the spectra on hydrogenation of the adsorbed species. These conclusions were in general agreement with Beeck's earlier results on evaporated metal films [19]. Except for the case of acetylene (see below) the initially adsorbed species gave the stronger absorptions from saturated hydrocarbon groups in the bond-stretching v_{CH} region, although it had to be borne in mind that ethylenic v_{CH} absorptions are, bond for bond, *ca.* 4 times weaker than from alkyl groups [12, 20]. The spectra in general were easier to interpret after hydrogenation and (on Pt, Pd or Ni) was similar to those of *n*-butyl groups, showing that dimerisation had occurred before or during hydrogenation.

4.1   Problem: Different hydrogenated spectra from substituted alkenes on different metals.

Butene-1 or *cis*-butene-2 gave the same spectroscopic results over Pt, confirming Eischen's conclusions (for Ni) about double-bond migration. After hydrogenation an *n*-butyl group was obtained on Ni but, surprising, over Pt or Pd the hydrogenated spectra were methyl- rather than methylene-rich from the adsorbed butenes and from propene. They bore close resemblances to those of 2-butyl or isopropyl groups respectively [18,21]. Why was there this difference between the different metals? Acetylene over Pt and Pd gave initial spectra which implied a predominance of unsaturated adsorbed species and, as previously found by Little on Pd [12,13], hydrogenation led to long-chain alkyl groups clearly formed by polymerisation of the initial C_2 species.

Because of the small pores in our porous silica glass only limited amounts of metal could be deposited internally without fracture. Powdered silica supports as used by Eischens enabled much higher concentrations of metal to be incorporated. These gave relatively much reduced background absorption from the silica, thereby enabling absorptions from the surface hydrocarbon species to be observed additionally in the $vC=C$ and $\delta CH_3/CH_2$ angle deformation regions, 1700-1300 cm^{-1}. Eischens had prepared his metal catalysts on finely divided silica powders (Cab-O-Sil) which rested on a horizontal CaF_2 plate. When Richard Twemlow in his M.Sc. thesis [22] tried to repeat his method with loose powder the catalyst sample flew in all directions on attempting to pump out hydrogen from the cell - a fine example of a fluid bed! I later learnt that Eischens' method of deposition from solution caused the powder to adhered to the plate. We therefore decided to attempt a more convenient version of Eischens' method whereby pressed discs of the silica-plus-metal-salt could be mounted vertically in the infrared cell and reduced *in situ* to metal catalysts. The success of this method required that the disc remained porous to hydrogen or the hydrocarbon adsorbents. This proved to be the case.

By this means, John W. Ward, in his Ph.D. thesis of 1962 [23], obtained improved and extended versions of many of the spectra recorded by Michael Clark on Ni and Pt catalysts. He extended the work to include room temperature spectra from adsorbed isobutene, cyclopropane, cyclopentane, cyclohexane, benzene and toluene on these two catalysts. Once again, using the usual group-characteristic absorptions of organic molecules, the spectra obtained after hydrogenation proved easier to interpret than those obtained after initial adsorption. In the case of cyclopropane essentially the same spectra were obtained as from propene, showing that ring opening had occurred on the surface. For the other cyclic hydrocarbons hydrogenation showed that cyclic structures were retained after hydrogenation of the initially adsorbed species. Benzene and cyclohexane showed a mutual relationship on the metal surfaces in that the former gave cyclohexane and cyclohexyl adsorbed species after hydrogenation, whereas the latter gave adsorbed benzene after dehydrogenation by evacuation. Repeated hydrogenation/dehydrogenation cycles led to a gradual release of the final hydrogenated product to the gas phase.

Alkyl groups normally give stronger v_{CH} absorptions in the 3000-2900 compared with the 2900-2800 cm^{-1} regions. However initially adsorbed ethylene or the *n*-butenes gave stronger absorptions below 2900 cm^{-1} whereas the propene spectra conformed to the normal pattern. These differences were difficult to understand. Furthermore Ward's spectra did show that more pronounced absorptions were observed in the gas phase after hydrogenation of a given initial species adsorbed on Pt compared with Ni [24]. This later proved to be a significant observation.

In order to obtain more information in a difficult-to-understand situation, I decided that we should explore the temperature-dependence of the surface species as reflected in their spectra. In his PhD. thesis, presented in 1965 [25], Barry A. Morrow described the construction of an infrared cell in which the sample under investigation could be heated to 500°C (enabling, for example, the *in situ* reduction of the salts to metal particles in hydrogen at 350°C) or cooled to near liquid nitrogen temperatures of -196°C (77 K). Because of the heating effect of the absorbed infrared beam and the poor conductivity of the silica discs, the latter temperatures could only be reached during spectral measurements if an inert gas was present in the cell [26a].

Morrow described the temperature-dependent spectra from ethylene and butene-1 adsorbed on Ni and Pt catalysts over a range of temperatures between 150° and −145 °C (423 and 128 K) [26]. Temperature as a variable parameter very soon proved its worth in accounting for many of the previously-puzzling differences between the spectra of the surface species obtained after hydrogenation on these catalysts (methylene- and methyl-rich v_{CH} spectra on Ni and Pt respectively). Whereas the methylene-rich (*n*-butyl like) spectrum obtained after hydrogenation of butene-1 on Ni at room temperature changed little on lowering the temperature, the methyl-rich spectrum on Pt was greatly strengthened under the same circumstances [26b]. This phenomenon was reversible and it soon became clear that the methyl-rich spectrum was caused by the temperature-dependent physical adsorption of *n*-butane with its two end-methyl groups. The difference between the Ni and Pt cases was, after all, simply that at room temperature the C_4 hydrogenated product remained bonded to the surface as an n-butyl group on Ni, but on Pt it was fully hydrogenated to gas phase or physically adsorbed butane! More generally the spectra for the initially adsorbed and hydrogenated species of both ethylene and butene-1 were similar for Pt at room temperature and for Ni at -78°C (198 K). The surface-chemical reactions proved to be closely similar on the two metals but with different temperature regimes.

4.2   Problem: A too simple C_2H_4/SiO_2 low temperature spectrum.

There remained however a difficulty in interpreting satisfactorily the spectra of the initially adsorbed species in molecular terms. The spectrum from ethylene on Ni at room temperature led principally to C_4 *n*-butyl species after hydrogenation. We therefore concentrated on the spectrum obtained on Pt because hydrogenation at room temperature gave mostly ethane, implying the predominant presence of C_2 surface species. The spectrum on Pt at 20°C consisted of a weak band near 3015 cm^{-1} from an unsaturated species and four bands from saturated species between 3000 and 2800 cm^{-1}, one of which was weak and clearly from an overtone of an angle deformation mode of a CH_2 or CH_3 group [26a]. We interpreted the two strongest bands near 2920 and 2880 cm^{-1} as vCH_2 asym. and vCH_2 sym. of an MCH_2CH_2M (di-σ adsorbed) surface species (M=metal), although with some misgivings about the inversion of the usually expected intensity relationship of these two bands. An overlapping band near 2920 cm^{-1}, revealed by higher temperature spectra, we assigned to the dissociatively adsorbed species $M_2CH-CHM_2$, and the single absorption in the v_{CH} region we assigned to MCH=CHM. I had been earlier asked by G.C. Bond at a conference if there was evidence in the spectrum for the π-bonded $(CH_2=CH_2)M$ species but had replied that there was no support for such a species because of the absence of a stronger absorption near 3080 cm^{-1}, normally expected for the $v=CH_2$ asym. mode of such a species. The remaining very weak absorption at 2960 cm^{-1}, if from a CH_3 group, implied a very minor fraction of M_2CH-CH_3 or MCH_2-CH_3 species.

Greater difficulty however attended the interpretation of the spectrum of ethylene on Pt/SiO$_2$ taken at -145°C (128 K) which had a single dominant and broad absorption in the region of alkane groups near 2910 cm^{-1} [26a]. This implied a species with at most two CH bonds, *e.g.* a structural assignment to the dissociatively adsorbed species $M_2CH-CHM_2$, which seemed very strange at low temperatures, conditions which would be expected instead to favour the associatively adsorbed MCH_2-CH_2M. We were therefore left with a major unsolved problem which was not to be understood for a considerable number of years.

In the period between Barry Morrow's experimental work and its publication, I and my research group moved from Cambridge to the, then new, University of East Anglia (UEA). This is an appropriate place to break off the scientific narrative to consider a number of the new experimental techniques of surface science, developed from the 1960s, which were to play significant roles in our resumed investigations.

5. CONTRIBUTIONS FROM SURFACE SCIENCE

5.1 Low Energy Electron Diffraction (LEED)

In the early 1970s I was invited, with Robert Eischens and others, to a Gordon Conference to assess the impact on surface chemistry of the ultra-high vacuum (UHV) techniques that had been developed as part of the NASA space programme. A principal speaker was L.H. Germer, who had shared with C. Davisson (and G.P. Thomson) a Nobel prize for the discovery of electron diffraction. He described his surprise that the adsorption of molecules or atoms on metal surfaces could change the diffraction of normal-incidence low-energy electrons to give patterns which seemed to him to imply that the metal atoms moved their positions. This was the birth of Low Energy Electron Diffraction (now universally known as LEED) and was to provide a very successful second method for the identification of species on surfaces through their adsorption sites. Perhaps Germer imagined that only the metal atoms were strong enough scatterers to register their positions. We now realise that the adsorbed molecules sufficiently modify the electron density patterns in the metal surface, as well as contributing their own scattering contributions, to show up their relative positions on the surface. Ultimately - with more difficulty - the atomic positions of the adsorbates themselves could be determined. Later LEED studies did identify surface reconstructions of metal surfaces themselves, either as a result of temperature changes or of adsorption processes. My first introduction to UHV techniques was to see such work being carried out in Wyn Roberts' group when I visited Queens University Belfast in the early 1960s.

For me the truly exciting aspect of that Gordon Conference was a realisation that adsorption could be studied on surfaces of the same metal with different atomic arrangements as cut from a single crystal of a metal (Figure 1). The concept of adsorption on a surface could then be explored with greater sophistication, with the species observed being dependent on the particular surface chosen. Would it become feasible in the future to find sufficient sensitivity to obtain vibrational infrared spectra from adsorbed species under such well defined conditions? This possibility was an improvement on a poorly successful attempt that we had made to obtain spectra of adsorbed molecules on evaporated metal films [27]. It also made redundant an idea that I had conceived about the possibility of obtaining infrared spectra from adsorption on mats of single-crystal metal whiskers.

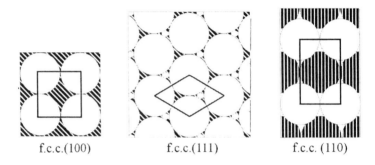

f.c.c.(100)   f.c.c.(111)   f.c.c. (110)

Figure 1. The different arrangements of metal atoms on the (100), (111) and (110) faces of a face-centred close-packed metal lattice.

5.2   Reflection-Absorption Infrared Spectroscopy (RAIRS) and the Metal-Surface Selection Rule (MSSR)

In 1959 Francis and Ellison [28] had carried out infrared experiments of the type that I had earlier envisaged, namely of multilayers of surfactants on a flat metal surface, studied by the reflection of the infrared beam off the underlying metal surface. Although (as I had calculated) a monolayer of the surfactant gave barely detectable absorptions, a number of multilayers gave well-defined spectra but with unexpected relative band intensities within the spectra of the long hydrocarbon chain. Francis and Ellison correctly interpreted this effect, which later became known as the result of the operation of the Metal Surface Selection Rule (MSSR), as caused by a well known phase change of 180° accompanying reflection off a metal surface for electric vectors of the incident radiation that are parallel to the surface. This phase change does not apply for incident electric vectors perpendicular to the surface, which can be optimised by using an incident beam at near-grazing incidence with radiation polarised in the plane perpendicular to the surface. The practical effect on the spectrum of the adsorbate is that only those modes with vibrational dipole-moment components perpendicular to the surface can be observed. For symmetrical adsorbates this causes substantial proportions of vibrations to be forbidden in the spectra. The experimental technique is termed reflection-absorption infrared spectroscopy, with the abbreviations RAIRS or IRAS. The technique was put on a sound quantitative basis by R.G. Greenler [29]; J. Pritchard first used this method to obtain spectra on metal single crystals under UHV [30]; and M.A. Chesters added the high sensitivity available through the use of Fourier Transform techniques (see below). The latter subsequently enabled RAIRS to play a full role in the vibrational spectroscopy of weak infrared absorbers such as adsorbed hydrocarbons [31].

5.3   Temperature-Programmed Desorption (TPD)

Also in the early 1960s I was invited to a Conference on Physical Electronics in Milwaukee, Wisconsin, where the new technique of Temperature Programmed Desorption (TPD) [termed alternatively, but less appropriately, Thermal Desorption Spectroscopy, TDS] was described. The idea was that, as the temperature of the sample was raised in a continuously programmed manner, different adsorbed species would be driven off in the inverse order of their strengths of adsorption, *e.g.* the physically adsorbed species before any chemisorbed ones, and that their natures and concentrations could be measured by mass spectrometry. An analysis of the kinetics could distinguish between first order complete-molecule desorption and second order recombination reactions, and could measure the activation energies for the various processes. My first reaction to this proposal was that the physicists who had developed this technique had much underestimated the extent to which temperature is a potent force for change within chemistry *i.e.* could itself changing the nature of the species under investigation. In due course this became understood. Although, in comparison with LEED and RAIRS which determine the species actually on the surface, this is an indirect method it has nevertheless become a very valuable adjunct to the others and provides valuable information about the energetics of the surface reactions and the empirical formulae of adsorbates.

5.4   Vibrational Electron Energy-Loss Spectroscopy (VEELS)

The above-mentioned advances were followed about a decade later by the development of Electron Energy Loss Spectroscopy. This, as applied to vibrational spectroscopy (VEELS), was the first to provide a spectroscopic method of sufficient sensitivity to be capable of wide application to single adsorbed monolayers on flat metal surfaces. In the literature this technique is alternatively abbreviated as HREELS where HR stands for high resolution. It is high resolution in electron spectroscopy terms at the wavenumber-equivalent resolution of 30 cm^{-1} at that time, later improved to 10 cm^{-1}, but is of overall low resolution compared with infrared spectroscopy. Hence the alternative abbreviation VEELS seems to be preferable. At that time work in RAIRS was only capable of work on very highly-absorbing adsorbates such as CO but the much wider success of VEELS provided the incentive for Chesters to further improve RAIRS sensitivity as described above [31].

VEELS was first described as an experimental technique in a paper by Propst and Piper published in 1969 [32]. My attention was drawn to this paper by my colleague David King and I confirmed that the energy losses from the monoenergetic incident beam of electrons indeed corresponded to the absorption of quanta of vibrational energy by the adsorbate molecules. I wrote to Professor Propst to express my strong interest in such work. Surprisingly I received a less-than-enthusiastic reply and indeed no further such publications were to come from his laboratory. My next contact with such work occurred when I spent a number of months in 1975 at the Centre de Recherches sur la Catalyse, in Villeurbanne near Lyon, during a period of study leave. There I had conversations with Jean Claude Bertolini who had built an EELS spectrometer and was beginning to use it for surface-chemical work. On my return to Norwich I looked into the literature and discovered that the scientific seed planted by Propst and Piper had led a number of other laboratories to further develop the technique, notably that of Harald Ibach in Germany who had applied it to a variety of problems in surface chemistry. His early choice for the study of adsorbates on metal single-crystal surfaces had been CO. This gives strong energy losses and the likely nature of the adsorbed species is restricted by the sites available on the chosen surface plane [33]. With Tam Nguyen, I shortly afterwards made very effective use of this VEELS literature, together with RAIRS studies by Alex Bradshaw and Hoffmann [34], in reviewing the overall work on CO on metal catalysts [35]. In this review we resolved the controversy concerning whether the several types of CO spectra on metal catalysts were due to linear or 2- or 3-fold bridged species as advocated by Eischens [10,15], or to all-linear species on sites of different degrees of coordination as suggested by George

Blyholder [36]. As I had long supposed through comparison with the spectra of CO-containing metal cluster compounds, the former explanation proved to be the correct one, with the latter the cause of minor variations within the range characteristic of the linear ('on-top') species.

Ibach had also published work on hydrogen adsorbed on tungsten (111) and had assigned two absorptions at 1251 and 1049 cm^{-1} with increasing coverage to 'on top' and 2-fold bridged species [37]. I wrote to Ibach to say that the former assignment was inconsistent with the literature on metal cluster compounds which showed that the 'on top' species should absorb at a much higher value near 2000 cm^{-1}. This led to an invitation to visit Ibach's laboratory and to my later acceptance by the VEELS community as an expert on the literature on the vibrational features of metal ligands. During the visit Ibach also showed me a recent spectrum that he had obtained of ethylene adsorbed on Pt(111) which will be discussed further below.

5.5 Fourier-Transform Infrared Spectroscopy (FTIR)

This method of obtaining infrared spectra uses interferometry rather than the dispersion of radiation by a prism or diffraction grating. All the different wavelengths contribute to the interferogram which can be subsequently analysed into the wavelength (or wavenumber) component parts by Fourier transformation mathematical procedures. The method was first applied in the 1960s to the energy-deficient far infrared region when the interferogram, then recorded on paper tape, was analysed by large mainframe computers. In the early 1970s the much more attractive possibility arose of incorporating a minicomputer within the spectrometer of sufficient capacity and speed that even the much more extensive mid-infrared region could be explored in something approximating to 'real time'. At the resolution of a few cm^{-1} the multiplex advantage of FTIR provides a gain in sensitivity of considerably more than an order of magnitude. Furthermore, the incorporation of the computer in the spectrometer provides many conveniences. These include alternative ways of presenting the spectra and the important capacity to subtract background spectra, *e.g.* of catalysts before adsorption, accurately in absorbance terms (or the equivalent ratioing in transmission terms). Seeing the advantages for adsorption work, I was fortunate to make a successful application to the U.K. Science Research Council for a grant for a (then expensive) FT spectrometer. This led to many friendly contacts with other research groups interested in the spectra of adsorbed species and who visited me in order to measure trial spectra.

5.6 The Photoelectron and Auger Spectroscopies

A number of other techniques of importance for the study of surfaces, but with which I had less direct contact, were developed in the 1960s and 1970s. These included the photoelectron spectroscopies using ultraviolet excitation for the study of valence-electron orbitals (UPS) or X-ray excitation for the investigation of core orbitals (XPS) of adsorbates. Wyn Roberts made pioneering contributions to both of these areas and subsequently very effective use of XPS in studying many adsorption systems. Auger spectroscopy became important for studying the elements present in an adsorption system and for assessing surface cleanliness

6. THE QUEST FOR SPECTRAL UNDERSTANDING RENEWED

After my research group had settled into their new quarters in UEA we returned to the question of the identification of species from the adsorption of ethylene and substituted ethylenes on silica-supported metal catalysts.

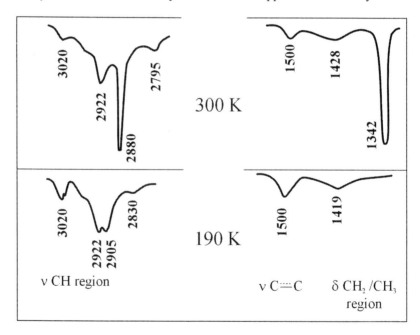

Figure 2. The infrared spectra of ethylene adsorbed on Pt/SiO₂ at room temperature (*ca.* 300 K), upper spectrum, and 190 K, lower spectrum

Aditham Ravi studied spectra on Ir/SiO$_2$ [38,39], Howard A. Pearce on Rh/SiO$_2$ [40,41,42], Aldona Lesiunas [42,43,44] and Janet D. Prentice on Pd/SiO$_2$ and Pt/SiO$_2$ systems respectively [42,44,45], David James on Pd/SiO$_2$ [46,47], and Terry Grimm III on Pt/SiO$_2$ [48] The iridium investigation gave similar results to those on Pt but the investigations by Pearce and Prentice had the advantage of the use of the newly-acquired FTIR spectrometer and gave correspondingly new information.

In addition to higher sensitivity, this also provided much improved digital subtraction of the metal/SiO$_2$ background spectra in the 1500 to 1300 cm^{-1} region (preceding the SiO$_2$ blackout at *ca.* 1300 cm^{-1}) which includes characteristic absorptions from νC=C and CH$_3$/CH$_2$ angle deformation modes. On all three metals this revealed a very broad absorption near 1430 and a sharper and particularly strong one near 1340 cm^{-1} (Figure 2). Particularly on Pt/SiO$_2$ a clear absorption also occurred near 1500 cm^{-1} which is the position expected for a π-bonded H$_2$C=CH$_2$, on the analogy of the spectrum of the well known Zeise's salt, [H$_2$C=CH$_2$]PtCl$_3^-$K$^+$. The latter was a surprising result as we had previously rejected the presence of such a species on the grounds of a lack of the absorption near 3080 cm^{-1} that is expected from a C=CH$_2$ group.

6.1   Problem: what unexpected species is related to the 1340 cm^{-1} absorption?

Our earlier assignment of the di-σ MCH$_2$-CH$_2$M species to the two absorptions at *ca.* 2920 and 2880 cm^{-1} on Pt/SiO$_2$ (with slightly different values on Rh and Pd) also ran into difficulty in the lower wavenumber region. Although the *ca.* 1430 cm^{-1} band could be from the δCH$_2$ mode of such a species, the *ca.* 1340 cm^{-1} band seemed not to be interpretable on such a basis. What was the origin of the latter band? Was there a third species present? Soma, who had agreed with our assignment of the band near 1500 cm^{-1} to the π-bonded species [49], also published spectra on Pt/Al$_2$O$_3$ showing the 1340 cm^{-1} band [50] while we were considering its interpretation, but her assignment of it to the π-complex seemed unsatisfactory. In order to check on the possibility of the presence of a third species, we decided to see if one or more of them could be displaced by postadsorption of CO after ethylene adsorption, first using Rh and then Pt catalysts. Similar results were obtained on the two metals. On Pt the CO postadsorption led to a virtual elimination of the 3020 and 1500 bands; at least a very great weakening of the 2880 cm^{-1} absorptions [40-42]; a shift of the 1340 cm^{-1} band to a slightly higher value with no appreciable change in intensity; the 2920 cm^{-1} and *ca.* 1430 cm^{-1} bands were little affected. These observations contradicted our assignment of the *ca.* 2920 and 2880 cm^{-1}

absorptions in the v_{CH} region to the same surface species. We assumed that, as the dominant v_{CH} absorption in the alkyl region, the 2880 cm^{-1} band was from the expected di-σ bonded (CH$_2$-CH$_2$) species.

It might have been expected that the dominant bands in the v_{CH} and angle deformation regions at *ca.* 2880 and 1340 cm^{-1} respectively would be from the same surface species, and the relative intensities of the corresponding bands were very similar in the spectra on different metals. Nevertheless this also seemed to be strongly contradicted by their very different behaviours on CO postadsorption and in our following work we assumed the latter evidence to be the more significant.

The evidence that the 2920 and 2880 cm^{-1} absorptions seemed to be from different species implied the presence of three different species absorbing in this region, *each giving only a single absorption.* How could this be consistent with the presence of non-dissociative (CH$_2$CH$_2$) species of either the π or di-σ types? Each of these would normally be expected to have at least three out of the four v_{CH} modes infrared active. In his Ph.D. thesis [40] Pearce suggested that a reason for the presence of a single v_{CH} absorption for the expected di-σ species might be the operation of the metal-surface selection rule (the MSSR). This had previously only been applied to interpret spectra of species adsorbed on flat metal surfaces and is dependent on the presence of mobile conduction electrons. Could it be applicable also to spectra on metal-particle catalysts? The interpretation of the spectra of ligands on metal cluster compounds seemed to have been successful using the usual correlation rules. However our metal particles were much larger, typically (as estimated from electron micrographs) for Pt/SiO$_2$ of diameters of from 50 to 150 Å, they frequently exhibit facets, and were of such a size that the electronic properties of metals might be applicable. I asked Robert Eischens for his views on this possibility. As he responded favourably we published this idea as a hypothesis worthy of further evaluation [41]. Greenler later vindicated our hypothesis by concluding that the MSSR should become of importance for metal particles of diameters greater than about 20 Å [51]. We also took the opportunity to represent the MSSR in terms of the equivalent virtual images set up in the metal by the presence of the oscillating dipoles associated with the vibrations of the adsorbed molecules - a chemist's rather than a physicist's approach to the phenomenon, Figure 3. The 'image' approach became popular, not only amongst vibrational spectroscopists.

In the case of a di-σ species of symmetry C$_{2v}$ the requirement that the vibrational dipole moment has to be perpendicular to the surface means that only one of the four v_{CH} modes, the in-phase v_{CH_2} *sym.* mode, will be infrared active, Figure 4. Likewise, in the angle-deformation region down to the silica cut-off at 1300 cm^{-1} only the in-phase vCH_2 *sym.* mode would be allowed. Later, in a more general consideration of the vibrations of adsorbed species, Erkelens and I [52] considered the MSSR in relation to its application to several types of vibrational spectroscopies and showed that the allowed modes are, more generally, the completely symmetrical modes of the surface complex.

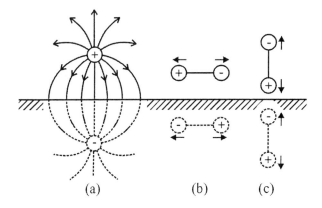

Figure 3. The metal-surface selection rule as related to the virtual images set up within a metal by a charged atom (a), or parallel or perpendicular vibrations of a polar diatomic molecule, (b) and (c)

The spectroscopic implications of the MSSR as applied to the non-dissociated π-(CH$_2$CH$_2$) species are even more drastic as all the infrared active modes of the free ethylene molecule down to 1300 cm^{-1} become forbidden in the surface complex. An out-of-plane mode near 960 cm^{-1} remains active, and is intensified, but this falls in the region of the silica blackout. Only the normally-forbidden completely symmetrical vCH$_2$ *sym.* and the vC=C modes are allowed under the MSSR and these become dipolar active solely due to polarisability distortions of the adsorbed molecule, induced by interaction with the metal atom in the surface. This emphasises that intensities of the absorption bands from this species are likely to give great underestimates of its concentration. This consideration negates the basis of many of our earlier calculations of intensity changes between the initial and hydrogenated spectra which had led to the conclusion that the initially adsorbed species must include a high proportion of dissociative adsorption. The observed absorptions at *ca.* 3020 and 1500 cm^{-1} are very much as expected for the allowed modes; the previously sought 3080 cm^{-1} band for a π complex is one of those forbidden under the MSSR. Now we were at last making progress with spectroscopic conclusions that gave a much better fit to chemical expectations! There remained however the problem of the species to which the 1340 cm^{-1} band could be attributed.

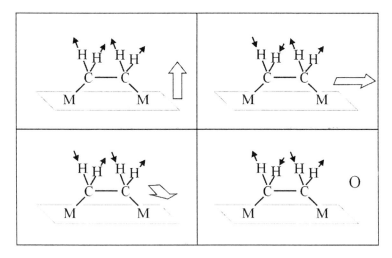

Figure 4. The directions of the vibrational dipole moments of the different CH bond-stretching normal modes of a C_{2v} di-σ(CH$_2$-CH$_2$) surface species.

6.2   Contributions from single-crystal spectra

It was at about this time that, as mentioned earlier, I paid a visit to Prof. Ibach in Jülich, Germany. VEELS work can give single-species spectra on particular surface planes and absorptions can be studied over the complete vibration frequency range. Ibach confirmed to me that for reflection at the specular angle the spectra were expected to be dominated by the vibrational dipolar mechanism and hence have the same selection rules as in RAIRS. At other angles of reflection a second 'impact' scattering mechanism can give features from any of the vibrational modes, although these would normally be observed weakly in any particular direction. The dipolar features become rapidly attenuated away from the specular direction and can be identified on this criterion.

On my congratulating Ibach on the excellent VEELS work that he had done on adsorbed CO, he told me that he had recently obtained an on-specular spectrum of ethylene adsorbed at room temperature on a Pt(111) surface. I was of course excited, but then disappointed, when he told me that he thought that the spectrum implied the presence of an ethylidene surface species, CH$_3$-CHM$_2$. I told him that we had found no real evidence for methyl groups in our work on Pt/SiO$_2$; also that I knew of no suggestions that methyl-containing species were likely to occur from the adsorption of ethylene on metal surfaces, except possibly through self-hydrogenation to ethyl groups. I did notice however that this spectrum had its strongest feature close to the position of our 1340 cm^{-1} absorption. Unfortunately, the VEELS spectrum in the v_{CH} region was of relatively weak and poor quality. This is because increasing energy-losses are associated with also increasing deviations of the electrons concerned from the specular direction. As a result it is difficult to distinguish in this region between dipolar-excited and impact-excited energy-losses.

Soon afterwards Somorjai's group in Berkeley, California, [53] performed a LEED analysis under closely similar experimental conditions and concluded that the CC bond of the surface complex is perpendicular to the surface. This strongly implied that Ibach's spectrum of a saturated species could be from an, at that time very unexpected, ethylidyne surface species, $(CH_3-C)M_3$. I found that an infrared spectrum had been obtained of such a ligand in the cluster complex $(CH_3-C)Co_3(CO)_9$ [54]. This could be compared with the VEEL spectrum of the surface species after the simplifications in the latter spectrum caused by the operation of the MSSR had been taken into account. In fact there was good agreement in wavenumber and intensity terms if the original assignments of the vCC and methyl rocking modes could be inverted, the former but not the latter giving a vibrational dipole change perpendicular to the metal surface. Our reinvestigation of the spectrum of the cluster-compound, including the CD_3-containing isomer, confirmed the corrected assignment [55]. I was able to send the complete assignments for both isotopic species to Ibach as the MSSR simplifications do not apply to the cluster compound spectra. He wrote back to say that he had recently obtained off-specular spectra from the surface complexes and that our data fitted the additional features in the impact-excited spectra extremely well. From then onwards a general agreement was reached that the ethylidyne species, or its higher homologues propylidyne, butylidine etc. (as obtained from the adsorption of propene, butenes etc.) were to be expected at room temperature on (111) surfaces of Pt and many other metals. This also clearly applied to (111) facets of metal-particle catalysts; indeed the ethylidyne spectrum is now used to characterise the presence of such facets [56].

As the VEEL spectrum on Pt(111) had been unhelpful, there remained the problem of which of the absorptions in the v_{CH} region of the spectra on Pt/SiO$_2$ was to be associated with the ethylidyne species. Spectra from adsorbed species on oxide-supported metal particles are region-limited by the low-wavenumber cutoffs of the oxide support, *i.e.* to >1300 cm^{-1} for silica and >1100 cm^{-1} for alumina. The VEELS technique has the great advantage of obtaining complete vibrational spectra of species adsorbed on flat metal surfaces down to *ca.* 200 cm^{-1}. Encouraged by this advantage, the infrared spectroscopists realised that they could do similar work, and at higher resolution, if only the then limited sensitivity of RAIRS could be improved. Michael Chesters achieved this by very successfully coupling a Fourier Transform infrared spectrometer to a RAIRS system in UHV [31]. When this was used to investigate the adsorption of ethylene on Pt(111) at room temperature it became quite clear that the 1340 cm^{-1} band of the only ethylidyne species present had its companion in the ν_{CH} region at 2880 cm^{-1} with an intensity ratio between these two absorptions essentially the same as found in the spectrum on Pt/SiO$_2$ before the post-adsorption of CO. We confirmed the assignment of the 2880 cm^{-1} absorption to the CH$_3$-containing ethylidyne using mixed-isotope spectra obtained by isotope scrambling [57].

6.3 Spectra and structures: Agreement at last!

Our assumption that the different behaviours of the two absorptions at 2880 and 1340 cm^{-1} to CO-adsorption implied that they came from different surface species was after all found to be false. The misleading CO experiment had proved to be what is sometimes jokingly called 'one experiment too many' and it had held up our correct understanding of the spectra by nearly a decade [58]. This new finding implied that, if the non-dissociatively adsorbed di-σ(CH$_2$-CH$_2$) is present, it had by difference to be associated with the *ca.* 2920 and 1430 cm^{-1} pair of absorptions. We published these structural reassignments in two conference papers [59,60]. At the second of these meetings we learnt that Thomas P. Beebe and John T. Yates Jr. had independently reached the same conclusion about the assignment of the 2880 and 1340 cm^{-1} absorptions to the ethylidyne species [61].

The final assignment of the 2880 cm^{-1} band to the ethylidyne species, and the dominance of this methyl absorption, had the bonus of providing an explanation for the change in intensity patterns in the ν_{CH} region between the spectra from adsorbed propene and butene-1 that was mentioned earlier. For the corresponding alkylidynes, the MSSR favours the νCH$_3$ *asym.* mode, clearly absorbing at 2960 cm^{-1} for propylidyne, and the νCH$_3$ *sym.* mode absorbing at 2870 cm^{-1} for butylidyne.

Ibach and Lehwald had earlier proposed that a VEELS spectrum of ethylene on Pt(111), taken at the low temperature of 97 K, with energy-losses at 2930 and 1430 cm^{-1}, might be from the di-σ species [62]. We too, in the period before we could explore the lower wavenumber region, had found the absorption at *ca.* 2910 cm^{-1} at 98 K. At that time we had been reluctant to assign this band to the chemically-likely di-σ species because we had not realised the spectral simplification that resulted from the operation of the MSSR [26a]; the spectrum had then appeared to be too simple. Ibach and Lehwald had also shown [62] that this spectrum transformed on warming to room temperature into that later associated with the ethylidyne species. Carlos De La Cruz, in work for his Ph.D. thesis [63], therefore decided to repeat the low temperature spectra on Pt/SiO$_2$ with our now much-improved Fourier-Transform spectroscopic facilities. This showed that the previously-observed low-temperature *ca.* 2910 cm^{-1} absorption could be resolved into two overlapping components at 2922 and 2906 cm^{-1}, the second of which was transformed into the 2880 cm^{-1} band of ethylidyne on warming to room temperature [64,65]. This transformation closely mirrored the VEELS findings on Pt(111) and left the remaining band at 2922 cm^{-1} to be confidently assigned also to a di-σ type of species (di-σ*) adsorbed on a different type of facet of the Pt particles, probably (100). Gratifyingly a research group led by Michael Trenary shortly afterwards reached closely similar conclusions from the room temperature spectra of ethylene adsorbed on several alumina-supported metals, the latter support making it possible to observe some additional absorptions between 1300 and 1100 cm^{-1} [66,67]. As they used smaller metal particles some additional absorptions were observed through relaxation of the MSSR selection rule, particularly for the ethylidyne species. The final conclusion therefore was that at room temperature on our Pt/SiO$_2$ catalyst there were approximately equal numbers of ethylidyne and di-σ* species, and a number of π-adsorbed species in considerably greater relative amount than is indicated by the proportionate intensity of the *ca.* 3020 and 1500 cm^{-1} bands in the overall spectra.

We had early shown that all the adsorbed species from ethylene on Pt/SiO$_2$ at room temperature could be hydrogenated off the surface as ethane [26a] and Avery, in a kinetic study in this laboratory of a π-deficient spectrum, had shown that the di-σ* species is removed much more rapidly than what we now know to be ethylidyne [68]. Trenary et al. showed that both the π- and di-σ* species were rapidly removed in an atmosphere of hydrogen [66], *i.e.* that both are highly catalytically active. Although not strictly a 'spectator' species [61], ethylidyne - as might be expected for a dissociatively adsorbed species - is a slow performer in this reaction.

In general the room temperature spectra from ethylene on other oxide-supported metals show remarkably similar features to those that we have explored on Pt/SiO$_2$, if sometimes with different temperature regimes, with some significant differences. For example, typical spectra on Pd show proportionately stronger absorptions from the π-species in agreement with single-crystal work [43,44], and an additional species occurs in the spectrum on Ni at room temperature which leads to *n*-butyl groups on hydrogenation [15,26a]. Also analogous types of species were spectroscopically identified under similar experimental conditions for propene and the butenes on single crystal Pt(111) as shown by Avery and Sheppard [69,70] and on oxide-supported platinum by Gulerana Shahid [71-74]. Avery has done similar work on Pd/SiO_2 [75]. As always in scientific work, success in a particular area leads to further questions for exploration. In the present instance these include:- What is the mechanism of formation of ethylidyne from the di-σ species? Which C_2-species on Ni/SiO_2 at room temperature gives rise to the formation of C_4 species (or on Pt/SiO_2 at higher temperatures)? Why at low temperatures is di-σ bonding found on the (111) faces of Ni and Pt, but π-bonding on the (111) face of Pd? Why is the intensity of the 2880 cm^{-1} band of ethylidyne so remarkably sensitive to adjacently adsorbed CO? etc., etc.

My general interest in spectra/structure relationships for adsorbed hydrocarbons led me to classify and review such spectra in the literature, mostly obtained by VEELS, on single-crystal surfaces in 1988 [76]. With only a few intermediate cases, the single-crystal spectra at low temperatures from adsorbed ethylene showed two distinct patterns with minor variations, which we labelled Type I and Type II, and structurally attributed, with their sp^3 and sp^2 hybrdisations, to di-σ and π species respectively. Ethylene on Pt(111) at low temperatures has a Type I spectrum at one extreme of the series. At higher temperatures the very distinctive spectroscopic pattern from the ethylidyne species was immediately apparent in many cases. With Carlos De La Cruz, I later did an analogous review for the spectra of the oxide-supported metals in 1996 [58] and 1998 [77], the latter together with an up-dating of the single-crystal results.

More recently improvements in LEED, the newer technique of photoelectron diffraction (PED), and the determination of adsorption sites by scanning tunnelling microscopy (STM), have provided alternative means for attributing structures to adsorbed species. The PED technique shares with vibrational spectroscopy the characteristic of being local-site based. So far remarkably good overall agreement has been found between the diffraction-based results and the earlier spectroscopically deduced structures for the C_2 surface species derived from the adsorption of ethylene or acetylene on metal single crystals [78]. (An early disagreement between PED and spectroscopic results for the group characteristic region assigned to the 3-fold bridged species from CO on (111) surfaces of metals has been shown to result from a spectroscopic underestimate of dipolar coupling in the spectra, a phenomenon that is very small within the spectra of hydrocarbons [78].)

7.   CONCLUSIONS

During the period covered by my partial-career in surface studies infrared spectroscopy became the first physical method capable of deducing the structures of many chemisorbed surface species, often those are of catalytic interest. Previously such structures could only be indirectly and, as it turns out, often incompletely inferred from the kinetic analysis of desorbed molecules. The ethylidyne species is a case in point; the existence of such a species had not been envisaged before vibrational spectroscopic evidence became available in conjunction with LEED. As we discovered, on the metal surfaces with which this paper is primarily concerned it is essential to take into account the simplifying effects of the Metal-Surface Selection Rule on the spectra before the usual infrared correlation rules for identifying organic groupings can be used to provide convincing structural conclusions. In many cases these conclusions have been later confirmed in general terms by the more direct, but also more experimentally demanding and therefore less frequently applied, diffraction methods. Within their own experimental limitations the latter are capable in principle of giving more detailed information about the structures of the adsorbed species and about their adsorption sites.

It is generally seen as the ultimate aim of these researches to determine the structures of active species on the finely divided catalysts. However the spectra obtained from such samples typically contain overlapping components from different surface species adsorbed on, for example, different types of facets of the individual metal particles. The development of the UHV-based techniques from the 1960s, often collectively referred to as surface science, enabled adsorbed species to be studied one at a time on particular single-crystal faces, and these results proved to be of great assistance in analysing the more complex spectra on the catalysts. As we have seen, the finally satisfactory conclusions reached above for ethylene adsorbed on Pt/SiO_2 depended on the applications of three new experimental techniques and one new theoretical understanding, all of which were developed since the commencement of the research programme. These are the Fourier Transform infrared technique, which enabled better quality spectra to be obtained from the finely-divided calalysts; VEELS and then the FT-version of RAIRS, which enabled one-species simplified spectra to be obtained on metal single-crystals; and the insight that the MSSR caused simplifications in the spectra on even the finely-divided samples. In turn the interpretations of the spectra of species on the single-crystal surfaces has depended, with good success, on similarities with spectra from analogous ligands on metal-cluster compounds whose structures have been determined directly by X-ray crystallography [55,79,80].

The obtaining of convincing structural conclusions from the spectra turned out to be a long-term and laborious process, but the outcome has been the development of a suite of spectroscopic techniques that are now applied with great efficiency to numerous other systems of adsorbates and adsorbents. During an investigation such as this one is continuously reminded that Nature is very much in charge of the conclusions reached. It is necessary to work hard in terms of experimentation and imagination to find a pattern that fits well with already established results, but there is a correct answer waiting to be discovered The philosophers and sociologists of science sometimes question whether scientific results are not the imaginary constructs of scientists and therefore depend upon the influences of particular individuals. A multistep study, such as the one chronicled in this paper, shows clearly that the scientists are responsible for discovering the truths, but it is Nature which determines what truths there are to be discovered.

I have tried to bring out in the above text the contributions made by the different individuals in the Cambridge/East Anglia surface-spectroscopy research group. I have had the privilege to be the leader of a very pleasant group and I thank all its research student and postdoctoral members for their many skills and original ideas which enabled progress to be made towards successful conclusions. I am also grateful to my colleague Professor Michael Chesters for making possible the single-crystal work of the group and for his much valued general support. The companionship in our joint endeavours has given me much enjoyment and stimulation. It is the scientist's privilege to be able to unravel Nature's secrets. More generally, it is the academic's privilege - during the process of passing on knowledge and skills - to work together with, and benefit from, some of the best brains of the next generation.

8.   REFERENCES

1.   G.C. Pimentel, C.W. Garland and G. Jura, *J. Amer. Chem. Soc.*, 75, 803 (1953).
2.   N.Sheppard and D.J.C. Yates, *Proc.Roy. Soc.*, A 238, 69 (1956).
3.   D.J.C. Yates, N. Sheppard and C.L. Angell, *J. Chem. Phys.*, 23, 1980 (1955).
4.   J.P. Mathieu, N. Sheppard, and D.J.C. Yates, *Z. Elektrochem.*, 64, 734 (1960).
5.   N.G. Yaroslavsky and A.N. Terenin, *Doklady Akad. Nauk S.S.S.R.*, 66, 885 (1949).
6.   L.N. Kurbatov and G.C. Neuimin, *Doklady Akad. Nauk S.S.S.R.*, 68, 341 (1949).
7.   A.N. Sidorov, *Doklady Akad. Nauk S.S.S.R.*, 95, 1235 (1954).
8.   C. Kemball, *Adv. Catal.*, 11, 223 (1959).
9.   P.W. Selwood, "*Adsorption and Collective Paramagnetism*", Academic Press, New York, 1962.
10.   R.P. Eischens, W.A. Pliskin, and S.A. Francis, *J. Chem. Phys.*, 22, 1786 (1954).
11.   W.A. Pliskin and R.P. Eischens, *J. Chem. Phys.*, 24, 482 (1956).
12.   L.H. Little, N. Sheppard and D.J.C. Yates, *Proc. Roy. Soc.*, A 259, 242 (1960).
13.   L.H.Little, Ph. D. thesis, University of Cambridge (1958).
14.   L.H. Little, *Infrared Spectra of Adsorbed Species,* Edward Arnold, London (1966).
15.   R.P. Eischens and W.A. Pliskin, *Adv. Catal.*, 10, 1 (1958).
16.   J. Erkelens and Th. Liefkens, *J. Catal.*, 8, 36 (1967).
17.   D. Shopov and A. Palazov, *Kinet. Katal.*, 8, 862 (1967).
18.   M. Clark, Ph. D. thesis, Cambridge University (1960).

19. O. Beeck, A.E. Smith and A. Wheeler, *Proc. Roy. Soc.*, A 177, 62 (1940).
20. S.A. Francis, *J. Chem. Phys.*, 18, 861 (1950).
21. N. Sheppard, *Pure and Appl. Chem.*, 4, 71 (1962).
22. R. M. Twemlow, M.Sc. thesis, University of East Anglia, (1967).
23. J.W. Ward, Ph.D. thesis, Cambridge University (1962).
24. N. Sheppard, N.R. Avery, M. Clark, B.A. Morrow, R. St. C Smart, T. Takenaka and J.W. Ward in "Molecular Spectroscopy" (P. Hepple, editor), Institute of Petroleum, London, (1968) p. 97.
25. B.A. Morrow, Ph.D. thesis, University of Cambridge, 1965.
26. B.A. Morrow and N. Sheppard, *Proc. Roy. Soc., London*, A311, (a)391, (b)415 (1969).
27. S. van der Walt, Ph.D. thesis, University of Cambridge, 1962.
28. S.A. Francis and A.H. Ellison, *J. Opt. Soc. Amer.*, 49, 131 (1959).
29. R.G. Greenler, *J. Chem. Phys.*, 44, 310 (1966).
30. M.A. Chesters, J. Pritchard and M.L. Sims, *J. Chem. Soc., Chem. Commun.*, 1454 (1970).
31. M.A. Chesters, *J. Electron Spectrosc. Relat. Phenom.*, 58, 123 (1986).
32. F.M. Propst and T.C. Piper, *J. Vac. Sci. Technol.*, 4, 53 (1967).
33. H. Froitzheim, H. Ibach and S. Lehwald, *Phys. Rev. B*, 14, 1362 (1976).
34. F. Hoffmann and A.M. Bradshaw, *Surface Science*, 72, 513 (1978).
35. N. Sheppard and T.T. Nguyen, in "*Adv. Infrared and Raman Spectros.*," (R.J.H. Clark and R.E. Hester, eds.)Vol. 5, Heyden, London (1978), pp. 67-148.
36. G. Blyholder. *J. Phys. Chem.*, 68, 2772 (1964).
37. H. Froitzheim, H. Ibach and S. Lehwald, *Phys. Rev. Lett.*,36, 1549 (1976).
38. Ravi, Ph. D, thesis, University of East Anglia, 1968.
39. Ravi and N. Sheppard, *J. Phys. Chem.*, 76, 2699 (1972).
40. H.A. Pearce, Ph.D. thesis, University of East Anglia, 1974.
41. H.A. Pearce and N. Sheppard, *Surface Science*, 59, 205 (1976).
42. N. Sheppard, D.H. Chenery, A. Lesiunas, J.D. Prentice, H.A. Pearce and M. Primet, "Molecular Spectroscopy of Condensed Phases", Proceedings of the European Congress on Molecular Spectroscopy, Elsevier, Amsterdam, 1975, p. 345.
43. Lesiunas, M. Sc. thesis, University of East Anglia, 1974.
44. J.D. Prentice, A. Lesiunas and N. Sheppard, *J. Chem. Soc., Chem. Commun.*, 76 (1976).
45. J. D. Prentice, Ph. D. thesis, University of East Anglia, (1977).
46. D.I. James, Ph.D. thesis, University of East Anglia, (1983).
47. D.I. James and N. Sheppard, *J. Molec. Struct.*, 80, 175 (1982).
48. T. Grimm III, M.Sc thesis, University of East Anglia, (1981).
49. Y. Soma, *J. Chem. Soc., Chem. Commun.*, 1004 (1976).
50. Y. Soma, *J. Catal.*, 59, 239 (1979).
51. R.G. Greenler, D.R. Snider, D. Witt and R.S. Sorbello, *Surface Science*, 118, 415 (1982).
52. N. Sheppard and J. Erkelens, *App. Spectrosc.*, 38, 471 (1984).
53. L.L. Kesmodel, L.H. Dubois and G.A. Somorjai, *J. Chem. Phys.*, 70, 2180 (1979).
54. W.T. Dent, L.A. Duncanson, R.G. Guy, H.W.B. Reed and B.L. Shaw, *Proc. Chem. Soc. London*, 169 (1961).
55. P. Skinner, M.W. Howard, I.A. Oxton, S.F.A. Kettle, D.B. Powell, and N. Sheppard, *J. Chem. Soc., Faraday Trans. II*, 77, 1203 (1981).
56. M.A. Chesters, C. De La Cruz, P. Gardner, P. Pudney, G. Shahid and N. Sheppard, *J. Chem. Soc., Faraday Trans.*, 86, 2757 (1990).
57. M.A. Chesters, C. De La Cruz, P. Gardener, E.M. McCash, J.D. Prentice and N. Sheppard, *J. Electron Spectrosc. Relat. Phenom.*, 54/55, 739 (1990).

58. N. Sheppard and C. De La Cruz, *Adv. Catal.*, 41, 1-112 (1996).
59. N. Sheppard, D.I. James, A. Lesiunas and J.D. Prentice, *Commun. Dept. Chem. Bulgarian Acad. Sciences*, 17, 95 (1984).
60. B.J. Bandy, M.A. Chesters, D.I. James, G.S. McDougall, M.E. Pemble and N. Sheppard, *Phil. Trans. Roy. Soc. London A*, 318, 141 (1986).
61. T.P. Beebe, Jr., and J.T. Yates, Jr., *J. Phys. Chem.*, 91, 254 (1987).
62. H. Steininger, H. Ibach and S. Lehwald, *Surface Science*, 117, 685 (1982).
63. De La Cruz, Ph.D. thesis, University of East Anglia, 1987.
64. De La Cruz and N. Sheppard, *J. Chem. Soc., Chem. Commun.*, 1854 (1987).
65. De La Cruz and N. Sheppard, *J. Chem. Soc., Faraday Trans.*, 93, 3569 (1997).
66. S.B. Mohsin, M. Trenary and H. Robota, *J. Phys. Chem.*, 92, 5229 (1988).
67. S.B. Mohsin, M. Trenary and H. Robota, *J. Phys. Chem.*, 95, 6657 (1991).
68. N. Sheppard, N.R. Avery, B.A. Morrow and R.P. Young, in "Chemisorption and Catalysis", (P. Hepple, ed.) Institute of Petroleum, London, (1970) p. 135.
69. N.R. Avery and N. Sheppard, *Proc. Roy. Soc. London A*, 405, 1 and 27 (1986).
70. N.R. Avery and N. Sheppard, *Surface Science*, 169, L367 (1986).
71. G. Shahid, M. Phil. thesis, University of East Anglia, 1987.
72. G. Shahid and N. Sheppard, *Spectrochim. Acta*, 46A, 999 (1990).
73. G. Shahid and N. Sheppard, *J. Chem Soc., Faraday Trans.*, 90, 507 (1994).
74. G. Shahid and N. Sheppard, *J. Chem Soc., Faraday Trans.*, 90, 512 (1994).
75. N.R. Avery, *J. Catal.*, 19, 15 (1970).
76. N. Sheppard, *Ann. Rev. Phys. Chem.*, 39, 598-644 (1988).
77. N. Sheppard and C. De La Cruz, *Adv. Catal.*, 42, 181-313 (1998).
78. N. Sheppard and C. De La Cruz, *Catal. Today*, *In press* 2001).
79. C.E. Anson, B.F.G. Johnson, J. Lewis, D.B. Powell, N.Sheppard,A.K.Bhattacharyya, B.R. Bender, R.M. Bullock, R.T. Hembre and J.R. Norton, *J. Chem. Soc., Chem. Commun.*, 703 (1989).
80. C.E. Anson, N. Sheppard, D.B. Powell, B.R. Bender and J.R. Norton, *J. Chem. Soc., Faraday Trans.*, 90, 1449 (1994).

Chapter 4

HIGH-PRESSURE CO DISSOCIATION AND CO OXIDATION STUDIES ON PLATINUM SINGLE CRYSTAL SURFACES USING SUM FREQUENCY GENERATION SURFACE VIBRATIONAL SPECTROSCOPY
A Structure Sensitivity Study

Keith McCrea, Jessica Parker and Gabor Somorjai

Key words: Sum Frequency Generation, CO, Dissociation, Oxidation, Ignition, Platinum, (111), (557), (100), Boudouard Reaction.

Abstract: Using sum frequency generation surface vibrational spectroscopy, platinum single crystal surfaces were investigated at high-pressures and high-temperatures under pure CO or CO and O_2 environments. In 40 Torr of CO, the molecule dissociates on the (111), (557) and (100) surfaces of platinum single crystals at 673 K, 548 K, and 500 K, respectively, indicating CO dissociation is structure sensitive. The Pt(111) surface must be heated to a temperature where the surface is roughened creating step and kink sites, which are known to dissociate CO. The stepped Pt(557) surface does not need to be heated as high as Pt(111) to dissociate CO since there are step sites already available on the surface. The outer most surface atoms of Pt(100) are mobile compared to the low energy (111) surface and so the surface can roughen at a much lower temperature than observed on Pt(111). Under 40 Torr of CO and 100 Torr of O_2, CO oxidation ignition temperature of 620 K, 640 K and 500 K were observed for Pt(111), Pt(557) and Pt(100), respectively, indicating ignition is also structure sensitive. The ignition temperatures for Pt(111) and Pt(557) are similar because the higher concentration of surface atoms on the (111) terraces, common to both surfaces, are more dominant during oxidation than the step sites. Since both CO dissociation and CO oxidation ignition are structure sensitive and follow the same trend of decreasing temperatures for the two processes, it is likely that CO dissociation is important for the onset of ignition.

1.   INTRODUCTION

Professor Wyn Roberts has carried out detailed photoelectron spectroscopy studies of adsorbed oxygen, in both atomic and molecular states, on transition metal surfaces, [1-3]. Such information is of great importance to understand catalytic oxidation processes including partial oxidation as well as combustion. In this paper we discuss our recent findings of the mechanisms of CO oxidation. This is a total oxidation or combustion process that has been studied by experiments at low as well as at high pressures,[4-11] and has been modelled by computer simulations. Platinum surfaces, which are excellent catalysts for this reaction (in addition to that of palladium) were the focus of most of these investigations.

Since the advent of surface science, the primary techniques used to elucidate structure and coverage information about adsorbates on single crystal surfaces have employed electrons under ultra high vacuum (UHV) conditions. Traditional surface specific techniques that have utilized electrons have included low energy electron diffraction (LEED), Auger electron spectroscopy, and high-resolution electron energy loss spectroscopy (HREELS), [14,15]. Because of an electron's short mean free path in the presence of gas, the major limitation of these electron-based techniques is their inability to probe surfaces under high-pressure conditions that are interesting from a catalysis standpoint. Using these techniques, catalysis on single crystals was studied by first characterizing a single crystal surface before a reaction. Information about adsorbate coverage and structure could be determined before a reaction was initiated. The catalyst was then exposed to high-pressure of reactants and reaction kinetics could be calculated by monitoring the gas composition with gas chromatography. After the reaction was completed, the high-pressure of gas was evacuated and the surface was once again characterized [16,17]. This type of study gave information about adsorbates before and after reaction, but any information about adsorbates during reaction could not be determined. In addition to these types of UHV experiments, catalysis was also studied at pressures in which these electron-based techniques could be operated ($< 10^{-6}$ Torr), [11]. In order to understand catalysis *in situ* at the molecular level, new techniques had to be developed which could operate under high-pressure similar to industrial catalytic conditions.

Today, there are several techniques that are being used to probe surfaces under high-pressure conditions. Scanning tunnelling microscopy (STM) is one such technique that is currently being employed for high-pressure catalysis [18]. In addition to STM, techniques that utilize photons can be used to probe catalysis *in situ*. Examples of photon-based techniques include UV Raman spectroscopy, infrared spectroscopy and sum frequency generation surface spectroscopy (SFG)[21]. These techniques can be used if the photons are not significantly absorbed by the high-pressure of gas used during reactions. Sum frequency generation has an advantage over IR and Raman spectroscopies as it is surface specific; neither the bulk or gas phases generate SFG signal. Another advantage of SFG is its ability to detect concentrations of adsorbates below 10% of a monolayer. Because of these properties, SFG is a powerful technique to study catalysis *in situ* as has been shown in previous studies [19,20].

Using SFG to study catalysis, we carried out CO oxidation over platinum single crystal surfaces, both flat ((111) and (100)) and stepped ((557)), to explore the structure sensitivity of the catalytic combustion process. High CO (40 Torr) and O_2 (100 Torr) pressures were used at temperatures in the 300 – 700 K range. These reaction conditions are similar to those utilized for CO combustion in the automobile catalytic converter or other catalytic combustion reactors. The reaction has two regimes, separated by the ignition point that depends both on the CO partial pressure [6,8,10] and the platinum surface structure. Sum frequency generation was employed to monitor the reaction intermediates on the platinum surface during the reaction. Gas chromatography was used to monitor the gas composition as a function of time and temperature, thereby determining the reaction rates

Studies in which the single crystals were introduced to only 40 Torr of CO were also performed. It was found that CO will dissociate and deposit carbon at a particular temperature as monitored by Auger electron spectroscopy. This temperature is different for each single crystal used, and so CO dissociation on Pt is structure sensitive with the (100) and stepped (557) crystal dissociating CO at temperatures lower than (111).

When CO oxidation experiments are performed, we find by SFG that the surface is covered with molecular CO occupying top platinum sites below ignition reaction conditions. At a particular temperature, the onset of ignition occurs, and the temperature can rise by several hundred degrees. This temperature coincides closely with the CO dissociation temperature, and so the ignition temperature is also structure sensitive. The ignition temperature also rises with CO pressure, an added complexity in the reaction mechanism. SFG spectra above the ignition temperature indicate that CO is absent from the surface. This can be interpreted assuming that the reaction is mass transport limited because of the rapid surface reaction rates and the platinum surfaces are oxygen covered, which reacts with impinging CO on impact [6].

2. EXPERIMENTAL

The experiments performed on Pt(111), Pt(100) and Pt(557) single crystal surfaces were performed using the equipment described below.

2.1   UHV/High Pressure System

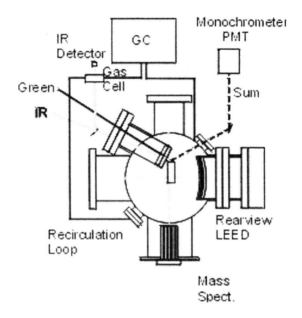

Figure 1. UHV and High-Pressure Chamber

Single crystal samples were mounted in a UHV/Reaction chamber pumped by a turbomolecular pump and an ion pump to a base pressure of 2 $\times$ 10^{-10} Torr. A schematic of the instrument is shown in Figure 1. The chamber was equipped with an Omicron rear view retarding field analyzer (RFA) for low energy electron diffraction (LEED) and Auger electron spectroscopy. A mass spectrometer was attached to the system to monitor the background gases under UHV conditions. By use of resistive heating, the sample could be heated up to a temperature of 1200 K, and cooled under vacuum with liquid nitrogen to 140 K. The crystals were cleaned in UHV by repeated cycles of Ar^+ bombardment followed by exposure of 5 $\times$ 10^{-7} Torr of O_2 at 1125 K for two minutes. The sample was then annealed at 1135 K in UHV for one minute. After the crystal was determined to be clean by Auger spectroscopy, pressures of CO (Scott Specialty) and O_2 were introduced through a gas manifold system. The CO had an initial purity of 99.99% was further purified by passing it through a zeolite trap cooled with liquid nitrogen. During high-pressure catalytic reactions, the sample was isolated from the vacuum pumps by a gate valve. For SFG experiments, the chamber had a window made of CaF_2 to allow the IR light to enter.

2.2   Laser System for SFG

Sum frequency generation is a surface specific vibrational spectroscopy with sub-monolayer sensitivity [21-25]. During a SFG experiment, a visible laser beam (532 nm) is overlapped with a tunable IR beam (1950 – 4000 cm^{-1}). As the IR beam is scanned over the frequency range of interest, a vibrational spectrum of molecules adsorbed on the surface is obtained. SFG is an excellent technique for studying molecules adsorbed on single crystal surfaces under high-pressure catalytic reaction conditions. It is surface specific in that only a medium that lacks inversion symmetry may generate SFG signal under the electric dipole approximation. Because of this selection rule, the gas phase and the bulk of the single crystal do not generate SFG signal, the signal is generated specifically at the surface.

Under the electric dipole approximation, the intensity of the sum frequency signal is proportional to the square of the second-order nonlinear susceptibility $\chi^{(2)}$. The susceptibility is described by

$$\chi^{(2)} = A_{NR} + \frac{A_R}{(\omega - \omega_0 - i\gamma)}$$

where A_{NR} is the non-resonant contribution, γ is the line width, ω_0 is the resonant vibrational frequency, and ω is the IR frequency. The resonant strength, A_R, is proportional to the number of molecules adsorbed on the surface and the infrared and Raman transition moments.

To generate the infrared and visible laser beams used for SFG, a Continuum Nd:YAG laser operated at 20 Hz with a 20 ps pulse at the fundamental frequency of 1064 nm with 35 mJ of energy per pulse was used to pump a commercially built OPG/OPA system provided by Laservision. The fundamental output of the Nd:YAG laser was split with the first portion being passed through a KTP crystal to generate a 532 nm beam. The 532 nm beam was then split, with the first portion being used for the visible beam during the experiment and the second portion was sent to drive a OPG/OPA stage utilizing two counter rotating KTP crystals to generate a near IR beam between 720 and 870 nm. The IR beam was then difference frequency mixed with the second portion of the fundamental beam through two counter rotating KTA nonlinear crystals. This then generated a tunable IR beam between 1950 and 4000 cm^{-1}. The energy at 2100 cm^{-1} was 200 µJ with a fwhm of 10 cm^{-1} and an accuracy of ±5 cm^{-1}.

Both the IR and visible beams are p-polarized and are both spatially and temporally overlapped on a single crystal mounted in the UHV chamber. The visible beam makes an angle of 60° with respect to surface normal while the IR beam is at 70° with respect to surface normal. The generated SFG beam is then sent through a monochromator and the signal intensity is detected by a photomultiplier tube and integrated by a gated integrator. To normalize for the gas phase attenuation of the IR beam, a portion of the IR beam was sent through a gas cell containing the same pressure of CO as in the high-pressure cell and an IR transmission spectrum was acquired. Both the gas phase path lengths (4 cm) in the chamber and in the cell were the same as shown in Figure 1. The raw SFG data was then divided by the IR transmission spectra, which normalized the SFG data.

3.   RESULTS

3.1   CO Decomposition on Pt(111), Pt(100) and Pt(557) Crystal Surfaces

In this section, the results for experiments that explore the properties of Pt single crystals under high-pressure of CO and high-temperatures are discussed. The pressure of CO used in these experiments (40 Torr) is the same used during CO oxidation reactions. These high-pressure CO experiments are done to help understand the interaction of CO with platinum as a function of temperature and the mechanism of ignition under oxidation conditions. After a crystal was cleaned using the procedure described above, it was exposed to 40 Torr of CO. SFG spectra were then acquired at 300 K. Each spectrum shown in this paper is the average of three consecutive scans unless noted otherwise. The samples were then heated sequentially to higher temperatures and at each temperature SFG scans were obtained and averaged. At a particular temperature for each crystal, the SFG spectra evolved with time. The sample was then cooled to room temperature, and more SFG scans were acquired. For each surface an irreversible process occurred in which the peak position shifted to lower frequency. The chamber was then quickly evacuated, and Auger electron spectroscopy showed that the surface was covered with carbon.

The specifics of these experiments for each crystal will now be discussed. It is important to note that previously in our group, Pt(111) has been extensively studied under 400 Torr of CO and so more comprehensive results for Pt(111) have been reported previously [26]. The conclusions reached in the previous paper support the data reported in this current study, and so Pt(111) was not studied as fully as Pt(557) and Pt(100) under 40 Torr of CO.

3.1.1 Study of Pt(111) crystal surface.

Under UHV conditions, CO adsorbs in a c(4 x 2) surface structure under saturation coverage of 0.5 ML [27-29]. The SFG frequency for the top site peak of CO on a Pt(111) surface at 300 K is 2095 cm^{-1}.

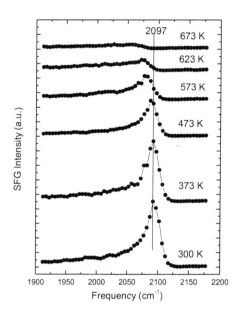

Figure 2. SFG spectra of 40 Torr of CO over Pt(111) as a function of temperature. CO dissociation begins at 673 K.

When Pt(111) is exposed to 40 Torr of CO, the sample could be heated up to 623 K and cooled back down to 300 K without showing any change in the spectra at 300 K. It is important to note that the error in thermocouple measurement is ±10 K. Once the sample was heated to 673 K, the SFG spectra evolved with time, and an irreversible process was observed for the heating cycle. The SFG spectra during the heating cycle are shown in Figure 2. Figure 3 shows the SFG spectra at 300 K before and after heating to 673 K. The peak position of the CO peak is red-shifted, 17 cm^{-1}, from 2097 cm^{-1} before heating to 2080 cm^{-1} after heating. The fitted amplitudes are essentially the same for both spectra, but the peak at 2080 cm^{-1} is broader. This indicates that there are still roughly the same number of CO molecules adsorbed, but due to the shift in the spectrum, it appears that the adsorption environment for the CO molecules has changed.

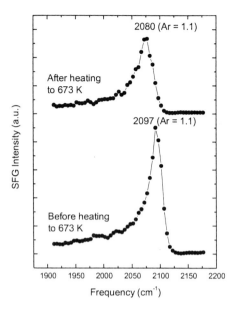

Figure 3. SFG spectra at 300 K of 40 Torr CO on Pt(111) before and after heating to 673 K. An irreversible process has occurred during heating. Ar is the value for the fitted amplitude.

The chamber was then evacuated, and an Auger electron spectrum was acquired. The Auger spectrum was dominated by carbon. This information coupled with the SFG data, indicates that CO dissociates on Pt(111) under 40 Torr of CO pressure and at 673 K.

3.1.2 Study of Pt(557) crystal surface.

CO dissociation was also explored on Pt(557). This is a stepped Pt surface composed of 6 atom wide terraces of (111) orientation with 1 atom high steps of (100) orientation [5,30,31]. The Pt(557) surface essentially introduces defects onto a Pt(111) surface, and allows us to study the effect of step sites on the chemistry. From UHV studies, it is known that CO can adsorb on both step edges and terrace sites and these different adsorption sites can be distinguished by the frequency of the CO top site stretch [32]. The adsorption energy of CO on step edges is higher than CO on terrace sites [5,32,33]. One-dimensional domains of CO adsorbed on step edges show a top CO stretching frequency of 2078 cm^{-1}. Two-dimensional domains of adsorbed CO on both step edge and terrace sites exhibit a CO stretching frequency around 2095 cm^{-1} [32]. Because the CO stretching frequency is lower on step sites, the CO bond is weakened, and it is expected that the dissociation barrier for CO adsorbed on step sites is also lower leading to CO dissociation at a lower temperature on Pt(557).

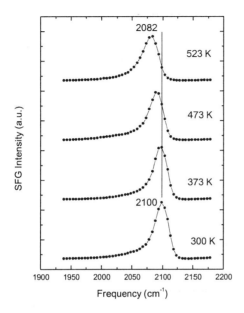

Figure 4. SFG spectra of 40 Torr of CO over Pt(557) as a function of temperature.

When the Pt(557) was exposed to 40 Torr of CO, similar results to those on Pt(111) were observed. A single CO peak was observed at 2100 cm^{-1} (Figure 4). As the sample was heated, a similar shift (to that for Pt(111)) in the CO top site frequency was observed up to 523 K where the CO peak was observed at 2082 cm^{-1} (Figure 4). Once the sample was heated to 548 K, the peak shifted to 2077 cm^{-1} and the intensity of this peak decreased as a function of time (Figure 5). The crystal was then cooled back down to 300 K. The spectra before and after heating in 40 Torr of CO at 300 K is shown in Figure 6. Again, the frequency of the CO peak after the heating cycle was red-shifted to 2087 cm^{-1} while the fitted amplitude remained essentially the same as before the heating cycle. The chamber was then evacuated to 5×10^{-8} Torr and an Auger spectrum was acquired (Figure 7). The Auger spectrum was dominated by carbon indicating CO decomposed at 548 K on Pt(557) under 40 Torr of CO pressure.

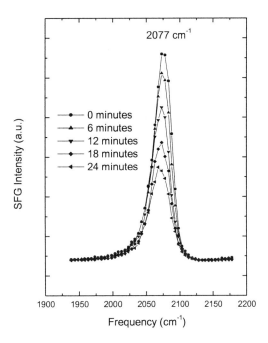

Figure 5. SFG spectra of 40 Torr of CO over Pt(557) at 548 K. The spectra evolve over time by decreasing in amplitude.

3.1.3 Study of Pt(100) crystal surface.

When a Pt(100) surface is cleaned and prepared properly, the outer most layer of Pt atoms will reconstruct to form what is known as a pseudo-hexagonal Pt(100)-(5 x 20) surface structure [34-36]. Once the surface coverage of CO increases to above 0.5 ML, this hexagonal surface reconstructs to yield the Pt(100)-(1 × 1) square surface structure [37,38]. A saturated surface exhibits a CO top site stretching frequency at 2095 cm^{-1}.

When 40 Torr of CO is exposed to Pt(100) at 300 K, a single resonance is observed at 2100 cm^{-1} (Figure 8). As the crystal is heated up to 450 K, a small shift to 2085 cm^{-1} is observed for the CO top site frequency. Once the surface is heated to 500 K, the frequency immediately shifts to 2065 cm^{-1}, and the SFG spectra evolve over time at this temperature (Figure 9). The crystal was then cooled back to 300 K, and Figure 10 shows the SFG spectra of the Pt(100) crystal in 40 Torr of CO before and after heating. As with Pt(111) and Pt(557), an irreversible change is observed (Figure 10). After the system is evacuated, the surface appears to be carbon covered indicating CO decomposition has occurred on Pt(100) at 500 K.

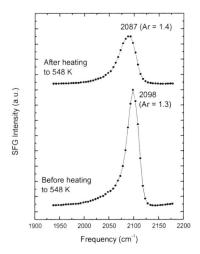

Figure 6. SFG spectra at 300 K of 40 Torr CO on Pt(557) before and after heating to 673 K. An irreversible change has occurred during heating. Ar is the value for the fitted amplitude.

To summarize the results for CO decomposition on the three single crystal surfaces, Figure 11 shows the frequency as a function of temperature for the three surfaces. As is evident from the figure, the frequency of the top site CO decreases until it reaches a critical temperature where dissociation and carbon deposition occurs. From these results, it is apparent that CO dissociation is structure sensitive as the dissociation temperatures for Pt(111), Pt(557), and Pt(100) are 673 K, 548 K, and 500 K respectively.

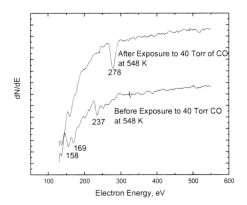

Figure 7. Auger spectrum before and after heating Pt(557) in the presence of 40 Torr of CO to 548 K.

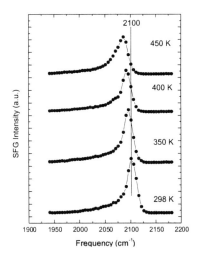

Figure 8. SFG spectra of 40 Torr of CO over Pt(100) as a function of temperature.

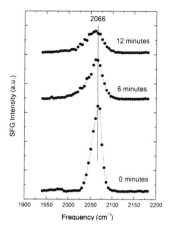 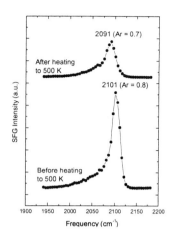

Figure 9. Time evolution of SFG spectra of 40 Torr of CO over Pt(100) at 500 K.

Figure 10. SFG spectra at 300 K of 40 Torr CO on Pt(100) before and after heating to 500 K. An irreversible change has occurred during heating. Ar is the value for the fitted amplitude.

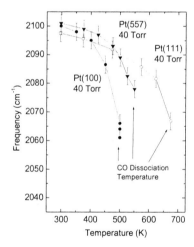

Figure 11. Comparison of the CO top site frequency as a function of temperature for Pt(111), Pt(557) and Pt(100) under 40 Torr of CO.

3.2   CO Oxidation on Pt(111), Pt(557), and Pt(100) crystal surfaces.

Under high-pressure reaction conditions, there are two activity regimes associated with CO oxidation, and an ignition temperature at which point the reaction becomes highly exothermic and self-sustaining separates these two regimes [6,7]. Below the ignition temperature, the reaction rate is slow and above the ignition temperature, the reaction rate is high. From previous studies, it has been found that the ignition temperature depends on the relative concentrations of CO and O_2 [6]. In this paper, we report the results of CO oxidation under a condition of 40 Torr CO and 100 Torr of O_2 for all three surfaces. The SFG spectra above and below the ignition temperature are considerably different as are the rates.

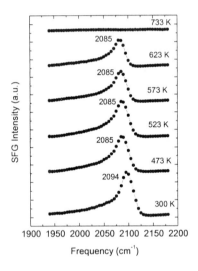

Figure 12. SFG spectra of Pt(557) under 40 Torr of CO and 100 Torr of O_2 and 630 Torr of He at temperatures below and above ignition.

For all three surfaces, the SFG spectra are essentially the same when compared to each other for both above and below the ignition temperature. Below the ignition temperature, the top site CO peak slowly shifts as a function of temperature until the ignition temperature is reached. At the ignition temperature, the crystal temperature increases rapidly, and the top site CO peak decreases rapidly. Once above the ignition temperature, no spectral feature is observed under these pressure conditions. The specifics for CO oxidation on Pt(557) will now be discussed.

To begin the CO oxidation reaction on Pt(557), the crystal was cleaned as described above and 4L of CO was exposed of the surface. Once SFG spectra were obtained under UHV conditions, the sample was introduced to 40 Torr of CO, 100 Torr of O_2, and 630 Torr of He. SFG spectra were then acquired at 300K and then as a function of temperature up to 623 K (Figure 12). At 300 K, the top site CO peak was at 2094 cm^{-1}. As the temperature was raised to 623 K the CO peak shifted to 2085 cm^{-1}. Around 640 K, the ignition temperature was reached, and the reaction became self-sustained and crystal temperature rose to 733 K. The ignition temperature was observed to be slightly dependent on the heating rate as the temperature approached the ignition point. An error of ±20 K was determined for the measured value of the ignition temperature. At the ignition point, the CO top site peak was no longer detectable, and the spectrum remained essentially free of any spectral features.

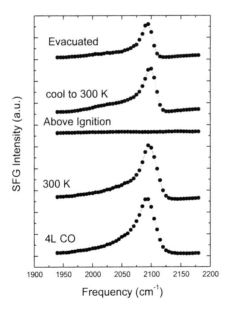

Figure 13. SFG Spectra of CO on Pt(557) before and after ignition in the presence of 40 Torr of CO, 100 Torr of O2 and 630 Torr of He.

The sample was then cooled back down to room temperature to quench the reaction. The SFG spectra at 300 K before and after ignition are compared with the spectrum above the ignition temperature in Figure 13. The spectrum at 300 K after the reaction is essentially the same, indicating no irreversible process has occurred as a consequence of the CO oxidation reaction as compared to the CO decomposition experiment. Figure 14 monitors the SFG signal of the CO top site frequency at 2085 cm^{-1} and temperature as a function of time. At 623 K, the SFG intensity is large compared to the baseline. The temperature is slowly ramped to the ignition point at which time the temperature increases rapidly, and the SFG intensity rapidly decreases to baseline. After 30 minutes, the sample was slowly cooled to quench the reaction and the SFG intensity did not recover until the temperature dropped below the ignition temperature.

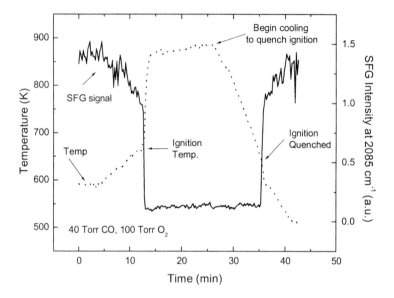

Figure 14. SFG top site CO frequency and sample temperature as a function of time during CO oxidation. 40 Torr of CO and 100 Torr of O_2.

As stated previously, the SFG spectra were very similar for all three crystals. However, the ignition temperature was considerably different for the Pt(100) surface as compared to the Pt(111) and Pt(557) surfaces. Table 1 compares the CO decomposition temperature and the CO oxidation ignition temperatures for the three surfaces. As evident from the table, the Pt(100) surface decomposes CO near the same temperature that CO oxidation ignites under these reaction conditions. The ignition temperature for Pt(111) is considerably higher than on Pt(100), but not quite as high as the CO dissociation temperature for Pt(111). A similar trend for the structure sensitivity of both CO dissociation and ignition is observed in that both ignition and CO dissociation occur at a higher temperature for the (111) surface as compared to Pt(100). It is important to note that the CO ignition temperature for Pt(111) and Pt(557) are very similar to each other within experimental error indicating that CO oxidation occurs mainly on the (111) terraces. This agrees with other studies, which indicated that the (111) terrace sites are more important for CO oxidation than step sites.

Table 1. Comparison of CO dissociation temperature and CO oxidation ignition temperature for Pt(111), Pt(557) and Pt(100)

	Pt(100)	**Pt(557)**	**Pt(111)**
CO Dissociation Temp. (40 Torr CO)	500 ±10 K	550 ±10 K	673 ± 10 K
Oxidation Ignition Temp. (40 Torr CO, 100 Torr O_2)	500 ± 20 K	640 ± 20 K	620 ± 20 K

4.   DISCUSSION

4.1   High Pressure CO Dissociation on Platinum.

Platinum is generally thought to be unable to dissociate CO as determined from extensive studies under UHV or low CO pressure conditions [41,27,28,42,43]. Most studies agree that CO does not dissociate on low Miller index faces of Pt, but there are a few studies, which have shown that CO can dissociate on Pt surfaces with step and kink sites under UHV or low CO pressure conditions [44-46]. The mechanism of this dissociation is known as the Boudouard Reaction.

$2CO \rightarrow C + CO_2$

Iwasawa and coworkers showed using X-ray photoelectron spectroscopy (XPS) that adsorbed CO dissociates on a stepped Pt(6(111) × (710)) surface [44]. Using a spherical single crystal, Li and coworkers showed CO dissociated on various surfaces of the sphere with step edges and kink sites [45]. For the Pt(210) face of the sphere, they observed CO dissociation begins around 380 K. Park and coworkers used XPS to observe CO dissociation on Pt(410) around 450 K and calculated that this surface is the most active for dissociating CO [46].

From these adsorption studies on Pt, it becomes clear that step and kink sites are needed to dissociate CO, supporting the idea that low Miller index crystal faces will not dissociate CO under UHV conditions. There are several different adsorption sites available on Pt single crystal surfaces with kinked and stepped sites. Using vibrational spectroscopy to probe the stretching frequency of CO adsorbed on these stepped and kinked surfaces, it has been found that the CO stretching frequency shifts to lower frequencies if it is adsorbed on stepped or kink sites [32]. This observation means that CO is more strongly bound to these sites, which inherently causes the C≡O bond to weaken making it easier to break the CO bond and deposit carbon on the surface. This is an important concept in understanding CO dissociation.

With the advent of surface specific techniques that can be employed at high-pressures such as sum frequency generation surface vibrational spectroscopy, it is now possible to observe the behavior of CO dissociation from UHV to near atmospheric pressures. Previously we reported, using SFG and Auger electron spectroscopy, that CO does dissociate on Pt(111) under high pressure (400 Torr) and high temperature (~673 K) conditions. This study showed that high-pressures of CO could roughen or anneal the surface depending on the temperature. At high temperatures, it is believed that platinum carbonyls are formed which then facilitate the roughening of the surface producing step and kink sites, which are then capable of dissociating CO. In this study under 400 Torr of CO on Pt(111), a red shift was observed for the top site CO peak as the sample was heated. This red shift is probably due to multiple effects caused by heating.

The first effect, which could explain the red shift, is anharmonic coupling to the frustrated translation mode [42,47]. This model explains that a temperature dependent red shift could result from vibrational dephasing which is due to a rapid energy exchange between low frequency modes of the adsorbed molecule and substrate which are anharmonically coupled to the top site vibrational mode. This model may account for the red shift down to 2075 cm^{-1}.

Surface roughening could explain the observed frequency shift below 2075 cm^{-1}. For Pt(111) and Pt(100), dissociation did not occur until the peak shifted below 2075 cm^{-1}. Brandt and coworkers showed that the CO frequency is not only dependent on the harmonic coupling between CO molecules; it is also dependent on the coordination number (n) of platinum atoms in which the top-site CO is adsorbed [48]. For n=9, which corresponds to platinum atoms on (111) terraces, and a coverage of 0.33 ML, Brandt and coworkers calculate a top-site CO frequency of 2098 cm^{-1}. For n=6, which corresponds to platinum atoms at kink sites, they calculate a top-site CO frequency of 2076 cm^{-1}. As the coordination number decreases below 6, the data extrapolates to even lower frequencies. It is possible that as the temperature is raised in the presence of high pressures of CO, the surface slowly roughens, decreasing the coordination to a point where platinum binary carbonyls can form to facilitate the dissociation of CO.

Dipole coupling does not explain the red shift as a function of temperature. The dipole coupling argument explains that the observed CO top site stretching frequency is a function of coverage [49,50]. As the coverage increases, the frequency increases. Dipole coupling was ruled out because the fitted amplitudes, which depend partly on the CO coverage, did not change by more than 10%. This indicated the coverage at high temperatures were very similar to the coverage at low temperatures.

The specifics for CO dissociation over each crystal will now be discussed.

Pt(111) Under 40 Torr of CO on Pt(111), the SFG spectrum changed as a function of temperature up to 623 K. After heating to 623 K, the crystal could be cooled back down to 300 K and the CO spectrum recovered completely without any decrease in intensity or shift in frequency. These results are very similar to results obtained on Pt(111) under 400 Torr pressure [26].

At 673 K, the SFG spectra changed as a function of time with the top site peak red shifting and becoming broader. Most likely at this temperature, the coverage of CO is decreasing, and the surface is modified by means of roughening. The frequency of the observed CO is similar to carbonyl binary complexes, which may then facilitate the modification of the surface, producing step and kink sites that are needed to dissociate CO.

When the crystal is cooled back down to 300 K, an irreversible change was observed. The frequency of the top site CO is now at 2080 cm^{-1}, a 17 cm^{-1} red shift from the frequency observed before the heating cycle at 300 K. The fitted amplitude is essentially the same but the line width is broader, so it appears that CO is now adsorbed on a modified Pt(111) surface. The system was then pumped down to 5 x 10^8 Torr, and Auger spectroscopy revealed a carbon-covered surface. The lower frequency and broader peak is due to CO co-adsorbed with carbon on the surface.

Pt(557) CO dissociation occurs at a lower temperature on Pt(557) than on Pt(111) (548 vs. 673 K). It is known that step sites are much more active during catalysis, so it is no surprise that chemistry is done at a lower temperature on a surface with introduced defects. A similar shift in the CO top site frequency is observed up to 523 K as observed on Pt(111) which is attributed to anharmonic coupling to the frustrated translational mode. The difference between Pt(557) and Pt(111) is that once the Pt(557) surface is heated to 548 K, the top site CO frequency no longer shifts below 2078 cm^{-1}, but the amplitude of the peak decreases over time. The SFG spectrum evolves differently for Pt(557) at the dissociation temperature as compared to Pt(111) and there doesn't appear to be any evidence for the production of platinum binary carbonyls as observed on Pt(111). Since it is most likely that step and kink sites are needed for CO dissociation, it requires the Pt(111) surface to be heated to 673 K before the production of platinum binary carbonyls can modify the surface to produce step or kink sites. Because the Pt(557) surface has step sites already, the surface does not need to be heated to a temperature to produce platinum carbonyls that induce surface roughening. This explains why the Pt(557) surface does not need to be heated to 673 K to dissociate CO as needed for Pt(111).

Pt(100) Pt(100) exhibits the lowest temperature to dissociate CO. Under 40 Torr of CO, the dissociation temperature is observed to be 500 K. This temperature is considerably lower than dissociation observed on Pt(111). When a Pt(100) surface is properly cleaned, the outermost surface layer of metal atoms will reconstruct to form what is known as a pseudo-hexagonal Pt(100)-(5 × 20) surface structure [34-36]. This hexagonal surface structure will remain until CO or other contaminants such as NO is introduced at which time the surface reconstruction will lift to reveal the truncated square Pt(100)-(1×1) surface [37-38]. Since Pt(100) easily reconstructs, the surface atoms are most likely very mobile as compared to Pt(111), the lowest energy surface of Pt. Because of this, the surface atoms can produce platinum binary carbonyls at a much lower temperature to induce surface roughening.

4.2   Discussion of CO Oxidation

During the CO oxidation experiments under 40 Torr of CO and 100 Torr of O_2, the reaction exhibits two different regimes, which are separated by an ignition temperature. Below the ignition temperature, the rate of CO_2 formation is slow (< 20 molec/site/sec), and CO covers the surface. As the surface is heated, the ignition temperature is reached and the reaction becomes highly exothermic causing an uncontrolled increase in the crystal temperature. The CO_2 formation rate then becomes high (> 1000 molec/site/sec) and the surface becomes oxygen covered. The transition between these two reactivity regimes is very fast.

Below the ignition temperature, the reaction probably proceeds through a Langmuir-Hinshelwood mechanism [6,7,10]

$$CO_{(a)} + O_{(a)} \rightarrow CO_{2(g)}$$

This reaction is inhibited since it depends on atomically adsorbed oxygen. Molecular oxygen requires two sites to dissociate on Pt and if the surface is CO covered, the O_2 dissociation is inhibited [7].

Above the ignition point, the surface is covered by atomic oxygen, and the rate becomes mass transport limited most likely by the approach of CO to the surface or the departure of CO_2. These results are well known throughout the literature. What remains to be unclear is the mechanism that causes this sudden transition from a CO covered surface to an atomic oxygen covered surface. One explanation describes the ignition as the point at which oxygen adsorption is no longer inhibited once the CO coverage decreases below a critical inhibition coverage. The CO coverage can decrease either by desorption or reaction. Bowker and coworkers using molecular beams, however, showed that even before ignition, a surface under CO oxidation conditions is already covered with oxygen to about one-fourth of the saturation value of oxygen [7]. They also found that ignition would only occur when the CO coverage decreases to below 0.4 ML.

Alone, the critical CO coverage explanation for ignition does not explain our observations. As a Pt(557) single crystal surface is heated in the presence of 40 Torr of CO and 100 Torr of O_2, the CO top site frequency shifts by only 10 cm^{-1} when the sample is heated to 473 K. Between 473 K and 628 K, the peak remains at 2085 cm^{-1}, and the fitted amplitudes of the spectra up to the ignition temperature are essentially the same. If the CO coverage was decreasing as a function of temperature, we would expect to observe a dipole coupling affect in which the top site CO peak frequency would red shift beyond 2085 cm^{-1}. Also, the fitted amplitudes of the peaks would decrease if the coverage decreased. However, the amplitudes observed do not decrease between 473 K and the ignition temperature.

Under the conditions discussed here, the top site CO peak remains essentially unchanged between 473 K and the ignition point indicating the CO coverage does not change until after ignition. From the CO dissociation studies presented above, a possible explanation for the ignition could be the onset of CO dissociation or the formation of platinum binary carbonyls. When the CO dissociation temperatures are compared to the CO oxidation ignition temperatures for the three crystal surfaces, we find that both the ignition and dissociation temperatures are structure sensitive. The ignition and dissociation temperatures for Pt(100) are essentially the same. The ignition temperature for Pt(111) and Pt(557) are considerably higher than the ignition temperature on Pt(100) indicating a similar trend for the structure sensitivity of ignition as compared to CO dissociation. The similar ignition temperatures on Pt(111) and Pt(557) can be explained if the (111) terrace sites, which are common to both crystal faces, are more important than step sites for CO oxidation [5,39,40].

Because structure sensitivity is observed for both CO oxidation ignition and CO dissociation, it is likely that CO dissociation is an important mechanism during ignition. In order for the ignition of CO oxidation to occur, CO first dissociates to deposit carbidic carbon on the surface, or platinum binary carbonyls are formed. Either the carbidic carbon or the platinum binary carbonyls may then be oxidized by atomic oxygen to initiate the ignition.

5.   CONCLUSIONS

Using sum frequency generation surface vibrational spectroscopy, the structure sensitivity of CO dissociation and CO oxidation ignition were investigated. The surface of platinum single crystals can either be roughened or annealed depending on the crystal temperature. It was found that CO dissociates at 673 K, 550 K, and 500 K for Pt(111), Pt(557) and Pt(100), respectively at 40 Torr of CO pressure. From UHV studies, it is known that CO will only dissociate on step or kink sites on platinum. On Pt(111) and Pt(100), the surface must be heated to a temperature at which platinum binary carbonyls are formed which then may facilitate the roughening of the surface to produce step and kink sites. Pt(557) is essentially a (111) surface with defects already introduced as steps. Because of this, the Pt(557) surface does not need to be heated to 673 K, the dissociation temperature on Pt(111), to produce defects. CO is able to dissociate on the steps of Pt(557) at a lower temperature than needed to roughen the surface. Pt(100) has the lowest CO dissociation temperature probably because the outermost surface platinum atoms are more mobile since the Pt(100) surface can easily reconstruct, therefore the surface can roughen at a lower temperature.

The CO oxidation ignition temperatures were also structure sensitive. The Pt(111), Pt(557), and Pt(100) have ignition temperatures of 620 K, 640 K, and 500 K respectively. These ignition temperatures follow the same trend in structure sensitivity as the CO dissociation temperatures on the three crystal faces. The one exception is the CO oxidation ignition temperature on Pt(557). This ignition temperature is similar to ignition found on Pt(111). Because the rate of CO_2 formation is mass transport limited above the ignition temperature, the terrace sites are more important for CO_2 formation than the step edges. Pt(557) has a similar ignition temperature as compared to Pt(111) because they share the same (111) hexagonal terrace structure. Because of the structure sensitivity of both CO dissociation and the CO oxidation ignition, it is possible that CO dissociation or platinum binary carbonyl formation may be needed to initiate the ignition mechanism. The carbon or carbonyls are oxidized in addition to the oxidation of molecular CO. The two exothermic reactions produce the temperature rise that facilitates CO desorption to allow the metal surface to become oxygen covered.

6.   ACKNOWLEDGEMENTS

This work was supported by the Director, Office of Energy Research, Office of Basic Energy Sciences, Materials Science Division, of the US Department of Energy.

7.   REFERENCES

1.   Roberts, M.W. *Surf. Sci.* 1994, 299/300, 769.
2.   Carley, A.F.; Yan, S.; Roberts, M.W. *J. Chem. Soc. Faraday Trans.* 1990, 86, 2701.
3.   Carley, A.F.; Davies, P.R.; Harikumar, K.R.; Jones, R.V.; Kulkarni, G.U.; Roberts, M.W. *Topics in Cat.* 2001, 14, 101.
4.   Hong, S.; Richardson, H. H. *J. Phys. Chem.* 1993, 97, 1258.
5.   Ohno, Y.; Sanchez, J.R.; Lesar, A.; Yamanaka, T.; Matsushima, T. *Surf. Sci.* 1997, 382, 221.
6.   Rinnemo, M.; Kulginov, D.; Johansson, S.; Wong, K.L.; Zhdanov, V.P.; Kasemo, B. *Surf. Sci.* 1997, 376, 297.
7.   Bowker, M.; Jones, I.Z.; Bennett, R.A.; Esch, F.; Baraldi, A.; Lizzit, S.; Comelli, G. *Catal. Lett.* 1998, 51, 187.
8.   Su, X.; Cremer, P.S.; Shen, Y.R.; Somorjai, G.A. *J. Am. Chem. Soc.* 1997, 119, 3994.
9.   Wartnaby, C.E.; Stuck, A.; Yeo, Y.Y.; King, D.A. *J. Chem. Phys.* 1995, 102, 1855.
10.   Berlowitz, P.J.; Peden, H.F.; Goodman, D.W. *J. Chem. Phys.* 1988, 92, 5213.
11.   Imbihl, R.; Cox, M.P.; Ertl, G. *J. Chem. Phys.* 1986, 84, 3519.
12.   Hafner, J.; Eichler, A. *Surf. Sci.* 1999, 433, 58.
13.   Imbihl, R.; Cox, M.P.; Ertl, G. *J. Chem. Phys.* 1985, 83, 1578.
14.   Somorjai, G.A. "Introduction to Surface Chemistry and Catalysis" Wiley, New York, 1994.
15.   Somorjai, G.A. *Surf. Sci.* 1994, 299/300, 849.
16.   Blakely, D.W.; Kozak, E.; Sexton, B.A.; Somorjai, G.A. *J. Vac. Sci. Technol.* 1976, 13, 1091.
17.   Cabrera, A.L.; Spencer, N.D.; Kozak, E.; Davies, P.W.; Somorjai, G.A. *Rev. Sci. Instrum.* 1982, 53, 1888.
18.   Somorjai, G.A. *Proc. Symp. on Advanced Surface Analytical Techniques, Kyoto, Japan* (1998).
19.   Somorjai, G. A.; McCrea, K. R. *Adv. Catal.* 2000, 45, 386.
20.   McCrea, K. R.; Somorjai, G. A. *J. Molec. Cat. A*, 2000, 163, 43.
21.   Shen, Y. R., *Surf. Sci.* 1994, 299/300, 551.
22.   Du, Q.; Superfine, R.; Freysz, E.; Shen, Y. R., *Phys. Rev. Lett.* 1993, 70, 2313.
23.   Shen, Y. R. *Nature* 1989, 337, 519.
24.   Shen, Y. R. "The Principles of Nonlinear Optics", John Wiley Inc., New York, 1984.
25.   Bain, C. D. *J. Chem. Soc., Faraday Trans.* 1995, 91, 1281.
26.   Kung, K.Y.; Chen, P.; Wei, F.; Shen, Y.R.; Somorjai, G.A. *Surf. Sci.* 2000, 463, L627.
27.   Klunker, C.; Balden, M.; Lehwald, S.; Daum, W. *Surf. Sci.* 1996, 360, 104.
28.   Hayden, B.E.; Bradshaw, A.M. *Surf. Sci.* 1983, 125, 787.
29.   Rupprechter, G.; Dellwig, T.; Unterhalt, H.; Freund, H.J. *Topics in Cat.* 2001, 15, 19.
30.   Lin, T.H.; Somorjai, G.A. *Surf. Sci.* 1981, 107, 573.
31.   Somorjai, G.A.; Joyner, R.W.; Lang, B.; *Proc. R. Soc. Lond. A.* 1972, 331, 335.

32. Hayden, B.E.; Kretzshmar, K.; Bradshaw, A.M.; Greenler, R.G. *Surf. Sci.* 1985, 149, 394.
33. Hopster, H.; Ibach, H. *Surf. Sci.* 1978, 77, 109.
34. Heilmann, P.; Heinz, K.; Muller, K. *Surf. Sci.* 1979, 83, 487.
35. Morgan, A.E.; Somorjai, G.A. *Surf. Sci.* 1968, 12, 405.
36. Broden, G.; Pirug, G.; Bonzel, H.P. *Surf. Sci.* 1978, 72, 45.
37. Behm, R.J.; Thiel, P.A.; Norton, P.R.; Ertl, G. *J. Chem. Phys.* 1983, 78, 7437.
38. Gardner, P.; Martin, R.; Tushaus, M.; Bradshaw, A.M. *J. Electron Spec. Relat. Phenom.* 1990, 54/55, 619.
39. Szabo, A.; Henderson, M.A.; Yates, J.T. Jr. *J. Chem. Phys.* 1992, 96, 6191.
40. Akiyama, H.; Moise, C.; Yamanaka, T.; Jacobi, K.; Matsushima, T. *Chem. Phys. Lett.* 1997, 272, 219.
41. Olsen, C.W.; Masel, R.I. *Surf. Sci.* 1988, 201, 444.
42. Harle, H.; Mendel, K.; Metka, U.; Volpp, H.R.; Willms, L.; Wolfrum, J. *Chem. Phys. Lett.* 1997, 279, 275.
43. Morikawa, Y.; Mortensen, J.J.; Hammer, B.; Norskov, J.K. *Surf. Sci.* 1997, 386, 67.
44. Iwasawa, Y.; Mason, R.; Textor, M.; Somorjai, G.A. *Chem. Phys. Lett.* 1976, 44, 468.
45. Li, X.Q.D.; Radojicic, T.; Vanselow, R. *Surf. Sci. Lett.* 1990, 225, L29.
46. Park, Y.O.; Masel, R.I.; Stolt, K. *Surf. Sci.* 1983, 131, L385.
47. Persson, B.N.J.; Ryberg, R. *Phys. Rev. B.* 1985, 32, 3586.
48. Brandt, R.K.; Sorbello, R.S.; Greenler, R.G. *Surf. Sci.* 1992, 271, 605.
49. Crossley, A.; King, D.A. *Surf. Sci.* 1977, 68, 528.
50. Crossley, A.; King, D.A. *Surf. Sci.* 1980, 95, 131.

Chapter 5

MODELING HETEROGENEOUS CATALYTIC REACTIONS

I. M. Ciobîcă, F. Frechard, C. G. M. Hermse, A. P. J. Jansen and
R. A. van Santen

Key words: DFT, Dynamic Monte Carlo, C_{Hx}, NO, Ru(0001), Rh(111), molecular
              modeling, heterogeneous catalysis, total energy calculations, Transition States,
              lateral interactions

Abstract: The determination of Transition States energies and structures has been done
              with Density Functional Theory. Several calculations are made to determine
              the lateral interaction of species on the surface. A model to use those results as
              input in Dynamic Monte Carlo is proposed. Together they form a good tool to
              help understand the kinetics of heterogenous catalytic reactions. Two
              examples are provided: CH_x on Ru(0001) and NO on Rh(111).

1.   INTRODUCTION

The use of quantum chemical data to simulate overall kinetics of a catalytic reaction will be illustrated using the Eindhoven Dynamic Monte Carlo code. A kinetic model for CH_4 decomposition and C hydrogenation on Ru(0001) surface is proposed, and a comparison will be made between experimental and theoretical Temperature Programmed Reaction spectra of NO reduction on Rh(111).

Surface Chemistry and Catalysis, Edited by Carley *et al.*
Kluwer Academic/Plenum Publishers, New York, 2002

2.   THEORY

2.1   Density Functional Theory

The Density Functional Theory developed by Kohn, Hohenberg and Sham [1, 2] is widely used nowadays and implemented in many programs performing quantum chemical calculations. A good introduction to DFT can be found in the the book of Parr and Yang [3], also for a shorter presentation of DFT one can have a look to the Nobel lecture of Walter Kohn [4]. The main advantage of DFT, compare to the the the traditional Hartree–Fock and post Hartree–Fock methods, is the relatively low computational cost for accurate results. Recent functionals allow one to compute, most of the time, relative energies within 10 kJ/mol while post Hartree–Fock techniques can be more accurate but at a prohibitive cost: with DFT the effective scaling of the calculations is n^{2-3} *(n* being the size of the system), while post HF are at least above n^4.

The program VASP [5, 6] developed by the group of Prof. J. Hafner has been used extensively by us to obtain a fundamental understanding of reactions on metal surfaces. VASP applies the DFT to periodical systems, using plane waves and ultrasoft pseudopotentials (US–PP) [7, 8]. The US–PP reduce significantly the number of plane waves needed by relaxing the norm conservation constraint on the pseudo wave function. The Kohn–Sham equations are solved self–consistently with an iterative matrix diagonalization combined with a Broyden mixing [9] method for the charge density. The combination of these two techniques makes the code very efficient, especially for transition metal systems which present a complex band structure around the Fermi level. The forces acting on the atoms are calculated and can be used to relax the geometry of the system.

The functional from the Generalized Gradient Approximation of Perdew and Wang [10] has been chosen because of its good description of chemical bond energies. The Ru, Rh and H pseudopotentials are converged with an energy cut–off for the plane–wave basis set (E_c) of 200 eV while for the C, N and O pseudopotential it is 300 eV, 350 eV and 400 eV respectively. Consequently all the calculations were performed with a E_c of 300 eV for CH_x on Ru(0001) and 400 eV for NO on Rh(111). We have employed periodic DFT calculations to study the activation of C–H bonds on a Ru(0001) surface and N–O bonds on a Rh(111) surface. Two coverages of 0.25 ML and 0.11 ML were considered, corresponding to 2 x 2 and 3 x 3 cells respectively. Also c(4 x 2) cells were needed for 0.5 ML and 0.75 ML coverages of NO on Rh(111) surface.

The supercell consists of 4 layers slab and 5 vacuum layers in the case of Ru(0001) surface and 5 layers slab and 5 vacuum layers in the case of Rh(111) surface. The k–points sampling was generated following the Monkhorst–Pack [11] procedure with a 5 x 5 x 1 (and 7 x 7 x 1) mesh for 2 x 2 supercell and 3 x 3 x 1 (and 5 x 5 x 1) for 3 x 3 supercell. For the 2 x 2 supercell the use of this mesh gives 5 to 25 irreducible k–points (in the irreducible Brillouin zone) and for the 3 x 3 supercell the use of this mesh gives 3 to 13 irreducible k–points.

Adsorption on both sides with an inversion centre avoids dipole–dipole interactions between the cells, no other symmetrical constraint was imposed. Complete geometry optimisations are performed on all models, the electronic structure analyses being used to help to rationalise the behaviour of methane on the Ru(0001) surface.

The Nudged Elastic Band (NEB) Method developed by J′onsson et al. [12] is used to determine the Transition States. This is a chain–of–states method. Two points in the hyperspace containing all the degrees of freedom are needed (initial and final state) and a linear interpolation can be made to produce the images along the elastic band. The program will run simultaneously each image and will communicate at the end of each ionic cycle in order to compute the force acting on each image.

The term "nudged" indicates that the projection of the parallel component of true force acting on the images and the perpendicular component of the spring force are cancelled. A smooth switching function is introduced that gradually turns on the perpendicular component of the spring force where the path becomes kinky at large differences in the energies between images.

The results obtained with the NEB are refined with a quasi–Newton algorithm [13]. This implies that the atoms are moved according to minimise the forces. The total energy is not taken into account for minimisation. In this way the program is searching a stationary point. Only in the very few cases when the given initial geometry is close to the geometry of the TS it is possible to reach the TS directly with the quasi–Newton technique, so the NEB is still required.

2.2 Dynamic Monte Carlo

Although kinetics plays such an important role in catalysis, its theory has for a long time mainly been restricted to the use of macroscopic deterministic rate equations. These implicitly assume a random distribution of adsorbates on the catalyst surface. Effects of lateral interactions, reactant segregation, site blocking, and defects have only been described *ad hoc*. With the advent of Dynamic Monte–Carlo simulations (DMC simulations), also called Kinetic Monte–Carlo simulations, it has become possible to follow reaction systems on an atomic scale, and thus to study these effects properly. Three parts can be distinguished in our DMC method; the model representing the catalyst and the adsorbates, the Master Equation (ME) that describes the evolution of the system, and the DMC algorithms to solve the ME. The ME and the DMC algorithms have been described extensively elsewhere [14, 15, 16].

The three parts contribute differently to making our DMC method useful. The model insures that it is easy to study a very broad range of systems and phenomena. The ME forms the link with other kinetic theories like macroscopic rate equations and reaction–diffusion equations. As the parameters in the ME can be calculated using *ab–initio* quantum chemical methods, very similar to normal rate constants, it is the ME that allows us to talk about *ab–initio* kinetics. Finally, the DMC algorithms make our DMC method extremely efficient.

2.2.1 The physical model.

For our model we assume that adsorption takes place at well–defined sites. These sites are represented by a grid of points. We assume that these points form a regular grid, a lattice, although this is not strictly necessary. One can block this grid into unit cells and we admit the case with more than one grid point per unit cell. It is important to realise that the model does not contain any information on the distance between the sites, or which sites are nearest neighbours, next–nearest neighbours, etc. This kind of information is contained implicitly in the description of the reactions.

A label is attached to each grid point. The most common use of this label is that it specifies the adsorbate at the site or that the site is vacant. Because of reactions, this implies that the labels will change during a simulation. Indeed, a simulation consists of nothing but changes of the labels according to reactions, and the determination of times when the reactions take place. The specification of a reaction consists of a set of grid points, labels attached to them corresponding to the adsorbates before the reaction has taken place (reactants), labels corresponding to the adsorbates after the reaction has taken place (products), and some rate constant. The set of grid points should be regarded as a representation for all sets of grid points where the reaction may occur. All these sets are related *via* translational symmetry and possibly (combinations with) rotations and reflections.

The use of labels need not be restricted to the specification of the adsorbate at a site. It may also tell something about the type of site. This usage of the label allows us to handle different sites reconstructions, defects, steps, etc.

2.2.2 The Master Equation.

The evolution of the adlayer and the substrate is described by the Master Equation (ME)

$$\frac{dP_\alpha}{dt} = \sum_\beta \left[W_{\alpha\beta} P_\beta - W_{\beta\alpha} P_\alpha \right] \qquad (5.1)$$

where α and β refer to the configuration of the adlayer, the P's are the probabilities of the configurations, 't' is time, and the W's are transition probabilities per unit time. These transition probabilities give the rates with which reactions change the occupations of the sites. They are very similar to reaction rate constants and we will use this term in the rest of this paper. $W_{\alpha\beta}$ corresponds to the reaction that changes α into β. A configuration can be regarded as the way the adsorbates are distributed over all sites in the system.

The ME has been used extensively in the statistical physics literature for the dynamics of all kinds of lattice–gas models. We would like to point out, however, that for reactions on surfaces this equation can be derived from first principles [16]. The derivation shows that the rate constants can be written as:

$$W_{\alpha\beta} = \frac{k_B T}{h} \frac{Q^\ddagger}{Q} \exp\left[-\frac{E_{bar}}{k_B T} \right] \qquad (5.2)$$

with k_B is the Boltzmann constant, h, Planck's constant, T temperature, and E_{bar} the energy barrier of the reaction that transforms configuration β into configuration α. The partition function $Q^\ddagger$ and Q can be interpreted as the partition functions of the transition state and the reactants, respectively, although there are some small, often negligible, differences [16]. This expression is, of course, familiar from Transition–State Theory (TST). Indeed, the derivation of the ME is very similar to Keck's derivation of the variational form of TST [17, 18, 19]. The important point is that the derivation of the ME from first principles makes an *ab–initio* approach to kinetics possible.

2.2.3 The Dynamic Monte Carlo algorithm.

The DMC simulations form a powerful numerical method to solve the ME exactly. In fact there are numerous DMC algorithms that might be used; a recent taxonomy of these algorithms contained no less than 48 [20]. Most of them are not efficient for any reaction system, however. For a general ME various algorithms have been given by Binder [21]. DMC algorithms for rate equations have even been given earlier by Gillespie [22, 23]. Our work has mainly focused on making these algorithms efficient for lattice–gas systems [14, 15, 20]. We will deal here only with the essential aspects, and we will point out the most important factors that determine the choice of the algorithm.

All DMC algorithms generate an ordered list of times at which a reaction takes place, and for each time in that list the reaction that occurs at that time. A DMC simulation starts with a chosen initial configuration. The list is traversed and changes are made to the configuration corresponding to the occurring reactions.

The method that is conceptually simplest is the First–Reaction Method (FRM) [14, 15, 21, 22, 23]. It can be applied to any system, but it may not be the most efficient method [20]. For each configuration that occurs a list of all possible reactions is computed, and for each reaction a time of occurrence is generated. If the rate constant is time independent and the current time is t, then the reaction $\beta \rightarrow \alpha$ will occur at time t+Δt with Δt=-(ln r)/$W_{\alpha\beta}$ where r is a random deviate of the unit interval [24]. The list of all reactions is ordered according to time of occurrence, and the configuration is changed corresponding to the first reaction in the list. This leads to a new configuration and a new time, and then the whole procedure is repeated. As the list of reactions does not have to be regenerated entirely after each configuration change, FRM is not as inefficient as it may seem. However, computer time per reaction does depend logarithmically on the number of sites in the system. This is because the list of reactions is implemented as a priority queue. Operations on it scale logarithmically with its size, [25] and the number of reactions, which is its size, is proportional to the size of the system. It can be shown that FRM generates configurations with probabilities that are solutions of the ME.

2.2.4 Lateral interactions.

Simulations with lateral interactions are significantly slower than simulations without lateral interactions. This is because a new reaction time has to be computed when the occupation of a site involved in the reaction changes. With lateral interactions this has to be done even each time that a change occurs in the neighbourhood contributing to the activation energy. This is simply an effect of using large neighbourhoods and cannot be avoided. Because the rate of a reaction varies significantly depending on the neighbourhood we use FRM as the simulation method. Other methods are less efficient in this case.

3. CH$_X$ ON Ru(0001)

The activation of methane and methane formation on metal surfaces are catalytically significant reactions. The sequential dehydrogenation of methane is important for CO production and the sequential hydrogenation of carbon are essential parts of the Fisher–Tropsch mechanism. Many theoretical and experimental studies have been conducted for CH$_x$ species adsorbed on transition metal surfaces.

3.1 DFT calculations

The main limitation of DFT is the description of the weak interactions (as van der Waals or hydrogen bonds). The errors due to this limitation are small if the molecules adsorbed are strongly bonded. This is the case of CH$_x$, x=0,1,2,3 and H, but it is not the case for CH$_4$ [26]. Methane in very weakly adsorbed and DFT calculations show adsorption energies around 0 kJ mol^{-1}.

Recent calculations for adsorbed CH$_x$ (x=0,1,2,3) species and adsorbed H on Ru(0001) surface have been reported by us [26] for 0.25 ML coverage. They show that the threefold sites are preferred for the adsorption of those species over top or bridge sites. CH$_3$ and H prefer the fcc hollow site (with small difference to the hcp site) and CH$_2$, CH and C prefer the hcp hollow site for adsorption.

In order to investigate the lateral interaction, calculations for 0.11 ML coverage were done for the same species in their most stable sites. We were able to show that the difference between the adsorption energy of a CH$_x$ species in 2 x 2 (0.25 ML coverage) and 3 x 3 (0.11 ML coverage) structure is 10 kJ mol^{-1} for CH$_3$, 21 kJ mol^{-1} for CH$_2$, 28 kJ mol^{-1} for CH and 45 kJ mol^{-1} for C. So, the idea that the lateral interactions are only significant for species adsorbed on a surface which share a metal atom is not true. In the 2 x 2 structure the three metal atoms which bind to the adsorbed molecule have one or two metal atom neighbours on the surface which accommodate another molecule. In the 3 x 3 structure the three metal atoms bound to the adsorbed molecule have no metal atom neighbours on the surface interacting directly to another adsorbate. At least one metal atom not bounded to an admolecule is inserted between the surface metal atoms bound to the adsorbates.

For a pair of identical species (CH$_x$, x=0,1,2,3, H), calculations were made for the adsorption energies of two species adsorbed on threefold hollow sites in a 2 x 2 supercell. They could be on the same kind of site or in a different. site. In all the cases they share one metal atom on the surface. The case when they share two metal atoms was not considered because of the large repulsion [27]. For a pair of different species A and B (CH$_x$, x=0,1,2,3, H), calculations were made for the adsorption energies for those two species adsorbed on three fold sites in a 2 x 2 supercell. The species can be on the same kind of site or in different sites, and also they share one metal atom on the surface. The results of these calculations [28] were used to estimate the lateral interactions (see next section).

We also used DFT to calculate the reaction barriers. Previously reported Transition States (TS) [27] were calculated in 2 x 2 structures which imply important lateral interaction. New calculations were performed in 3 x 3 supercells, in order to have the barriers without the lateral interaction. The geometries of the TS found for the 2 x 2 supercells were the initial guess and the TS were located by minimising only the forces.

The barriers calculated at 0.11 ML coverage differ from those obtained at 0.25 ML coverage. For the reactions $CH_{x-1} + H \rightarrow CH_x$ (x=1,2,3,4), the general trend is an increase of the barrier at low coverages compared with high coverage. This can be explained by the fact that the more H atoms are in a CH_x species the weaker the lateral interactions become. Therefore, if the species in the left–hand–side of the chemical equation have more repulsive interactions with the neighbouring species than the species from the right–hand–side of the chemical equation the barrier will go up. (If you destabilise the reactant more than the product, the barrier will increase.) For the opposite reactions $CH_x \rightarrow CH_{x-1} + H$ (x=1,2,3,4), there is not a general trend. In the case of CH and CH_4 decomposition, the barriers decrease at lower coverage.

In the case of CH_2 and CH_3 dissociation no significant modification of the barriers is found. The decrease of the barriers can be explained in a similar way as for the dehydrogenation reactions. The small change of the barrier for CH_2 and CH_3 can be due to the similar repulsive interactions with the neighbouring species for both the products and the reactants. See table 1 for values.

In the case of $CH_3 \rightarrow CH_2 + H$ reaction, due to the fact that CH_3 and CH_2 prefer different threefold hollow sites for adsorption, two different reactions were considered: $CH_{3(hcp)} \rightarrow CH_{2(hcp)} + H_{(fcc)}$ and $CH_{4(fcc)} \rightarrow CH_{3(fcc)} + H_{hcp}$ (fcc is the threefold site without a Ru atom below, in the second layer and hcp is the threefold site with a Ru atom below, in the second layer). The barriers were computed for each reaction in 2x2 and 3x3 supercells. For the 2x2 supercell there is no large difference for the activation energy, only 7 kJ mol^{-1} . With the 3 x 3 supercell appears a more important decrease for the reaction barrier if the CH_x species are in hcp sites. The difference becomes 14 kJ mol^{-1} (see table 2).

Since the H atom, CH_3 and the CH_2 present a small difference for fcc and hcp adsorption energies (respectively 3, 5 and 5 kJ mol^{-1}) it was difficult to conclude which reaction will have the smaller barrier without performing the calculations.

Table 1. The energy barriers (TS) in kJ mol^{-1} for the elementary reactions for methane dehydrogenation and C hydrogenation on Ru(0001) surface, calculated at 0.25 ML and 0.11 ML coverages, 2 x 2 and 3 x 3 supercells respectively. The third column represents the TS calculated for 3 x 3 with a Brønsted–Polanyi formula (see after) where we consider the final state the CH_x and the H already diffused away.

Reaction vs. coverage	2x2	3x3	3 x 3 (Brønsted–Polanyi formula)
$CH_4 \rightarrow CH_3 + H$	86	79	79
$CH_3 \rightarrow CH_2 + H$	49	51	52
$CH_2 \rightarrow CH + H$	16	16	22[a]
$CH \rightarrow C + H$	112	100	101
$C + H \rightarrow CH$	69	87	88
$CH + H \rightarrow CH_2$	58	74	68[b]
$CH_2 + H \rightarrow CH_3$	54	61	58
$CH_3 + H \rightarrow CH_4$	91	98	97

[a] CH_2 decomposition give the correct result if we consider an early TS.
[b] CH hydrogenation give the correct result if we consider a late TS.

Table 2. The energy barriers (TS) in kJmol^{-1} for the elementary reaction of methyl decomposition to methylene and hydrogen on Ru(0001) surface, calculated at 0.25 ML and 0.11 ML coverages (2 x 2 and 3 x 3 supercells respectively) for the two threefold sites.

reaction vs. coverage	2 x 2	3 x 3
$CH_3(hcp) \rightarrow CH_2(hcp) + H(fcc)$	49	47
$CH_3(fcc) \rightarrow CH_2(fcc) + H(hcp)$	56	61

3.2 Lateral interaction

For each A and B pairs from CH_3, CH_2, CH, C and H we calculated the energy for the following situations: A fcc + B fcc, A fcc + B hcp, A hcp + B hcp, A hcp + B fcc. If A and B are identical then only three different situations occur: fcc + fcc, fcc + hcp and hcp + hcp.

Since the Monte Carlo code used currently handles only lateral interactions in a pair wise additive fashion we correspondingly fitted the lateral interactions from the DFT calculations. It was proposed recently by King *et al.* [29] that for a coverage of about 0.1 ML the lateral interactions are minimal. Since a 3 x 3 structure leads to a 0.11 ML coverage, we assume that at larger distance the lateral interactions vanish.

If we draw a circle around a particular site with the radius of **three** times the Ru–Ru bulk distance, there are no less than 60 adsorption sites (only hollow) inside of this circle which, we believe, influence the adsorption energy of the given initial site (see figures 1 & 2). Those 60 sites are divided into two groups. They are named f or h if they are the same type as the particular site (fcc or hcp) for which the lateral interactions are calculated. The second group of sites are named d, and it includes the remaining sites which are of different type compared with the reference site (see figures 1 and 2).

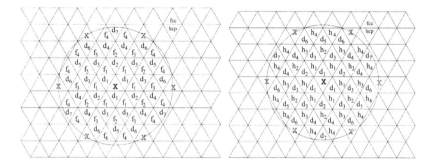

Figure 1. The lateral interactions for a "X" adsorbate in a fcc site. f_i indicates fcc–fcc interactions, and d_l indicates fcc–hcp interactions. With light gray it is shown the position of another "X" adsorbate for which the lateral interactions with the original "X" adsorbate are considered negligible.

Figure 2. The lateral interactions for a "X" adsorbate in a hcp site. h_i indicates hcp–hcp interactions and d_i indicates fcc–hcp interactions.

We also labelled those sites according to their distance to the centre of the circle. Thus we have $6f_1$ sites at the distance of 1 (on the Ru–Ru bulk distance scale), $6f_2$ sites at the distance of $\sqrt{3}$, $6f_3$ sites at the distance of 2 and $12f_4$ sites at the distance of $\sqrt{7}$ if the initial site was a fcc one. Note that they are 2 different f_2 sites. We consider them equal since they are at the same distance to the centre of the circle. Concerning the d sites, we have $3d_1$ sites at the distance of $1/\sqrt{3}$, $3d_2$ sites at the distance of $2/\sqrt{3}$, $6d_3$ sites at the distance of $\sqrt{7}/\sqrt{3}$, $6d_4$ sites at the distance of $\sqrt{13}/\sqrt{3}$, $3d_5$ sites at the distance of $4/\sqrt{3}$, $6d_6$ sites at the distance of $\sqrt{19}/\sqrt{3}$ and $3d_7$ sites at the distance of $5/\sqrt{3}$. The next sites outside the circle are f_5 at the distance of 3, d_8 at the distance of $2\sqrt{7}/\sqrt{3}$, and d_9 at the distance of $\sqrt{31}/\sqrt{3}$.

The resulting equations for the adsorption energies for an A species at different coverages are:

$$E_{Afcc}^{(2x2)} = E_{fcc}^{A} + 3\varphi f_3^{A}$$

$$E_{Ahcp}^{(2x2)} = E_{hcp}^{A} + 3\varphi h_3^{A}$$

$$E_{Afcc+Afcc}^{(2x2)} = 2E_{fcc}^{A} + 2\varphi f_1^{A} + 2\varphi f_2^{A} + 6\varphi f_3^{A} + 4\varphi f_4^{A}$$

$$E_{Ahcp+Ahcp}^{(2x2)} = 2E_{hcp}^{A} + 2\varphi h_1^{A} + 2\varphi h_2^{A} + 6\varphi h_3^{A} + 4\varphi h_4^{A} \qquad (5.3)$$

$$E_{Afcc+Ahcp}^{(2x2)} = E_{fcc}^{A} + E_{hcp}^{A} + 3\varphi f_3^{A} + 3\varphi h_3^{A} + 3\varphi d_2^{AA} + 3\varphi d_5^{AA}$$

$$E_{Ahcp}^{(3x3)} = E_{hcp}^{A} \quad \text{or} \quad E_{Afcc}^{(3x3)} = E_{fcc}^{A}$$

where φ is the change in the adsorption energy due to the lateral interactions between the two sites, f_i^{A}, h_i^{A} and d_i^{A} reflect the interactions between A fcc – A fcc, A hcp – A hcp, and A fcc – A hcp respectively (see beginning of section). E_{fcc}^{A} and E_{hcp}^{A} are the adsorption energies for the A species on three fold sites *without* any lateral interactions. The first two equations refer to one A species adsorbed on a threefold site (fcc and hcp) on a 2 x 2 structure, the next two equations refer to two A species adsorbed on the same threefold site and the next one refers to 2 A species adsorbed on different threefold site. The last equation refers to one A species adsorbed on the most stable threefold site (fcc or hcp) on a 3 x 3 structure.

If two different species (A and B) are adsorbed together, then the equations become:

$$E_{Afcc+Bfcc}^{(2x2)} = E_{fcc}^{A} + E_{fcc}^{B} + 2\varphi f_1^{AB} + 2\varphi f_2^{AB} + 3\varphi f_3^{AA} +$$
$$3\varphi f_3^{BB} + 4\varphi f_4^{AB}$$

$$E_{Ahcp+Bhcp}^{(2x2)} = E_{hcp}^{A} + E_{hcp}^{B} + 2\varphi h_1^{AB} + 2\varphi h_2^{AB} + 3\varphi h_3^{AA} +$$
$$3\varphi h_3^{BB} + 4\varphi h_4^{AB}$$

$$E_{Afcc+Bhcp}^{(2x2)} = E_{fcc}^{A} + E_{hcp}^{B} + 3\varphi f_3^{AA} + 3\varphi h_3^{BB} + 3\varphi d_2^{AB} +$$
$$3\varphi d_5^{AB} \qquad (5.4)$$

$$E^{(2x2)}_{Ahcp+Bfcc} = E^{A}_{hcp} + E^{B}_{fcc} + 3\varphi h_3^{AA} + 3\varphi f_3^{BB} + 3\varphi d_2^{BA} +$$
$$3\varphi d_5^{BA}$$

where φ is the change in the adsorption energy due to the lateral interactions between the two sites, f_i^{AB}, h_i^{AB}, d_i^{AB} and d_i^{BA} reflect the interaction between A fcc – B fcc, A hcp – B hcp, A fcc – B hcp, and A hcp – B fcc respectively (see beginning of section). E_{fcc}^{X} and E_{hcp}^{X}, (X–A, B) are the adsorption energies for the X species on the three fold sites *without* any lateral interaction. The first two equations refer to two species A and B adsorbed on the same three fold site (fcc and hcp) on a 2 x 2 structure, the next two equations refer to two species A and B adsorbed on different three fold sites. The parameters for the interactions of adsorbates of the same species are calculated before and simply used here to calculate the parameters for the interactions of two different species.

The number of lateral interactions is too large to be determined unique with the limited number of calculations available. We therefore tried various functions to describe the variations of the lateral interactions with the distance. The retained function are $d=A_d/r^x$, $f=A_f/r^y$ and $h=A_h/r^z$, where $d=(d_1, d_2, \ldots, d_7, d_8, \ldots)$, $f=(f_1, f_2, \ldots)$ and $h=(h_1, h_2, \ldots)$. A_d, A_f and A_h are the prefactors, z, x and y are the exponents and r is the distance. The parameters were computed with a least–squares procedure. Attractive terms, like $M/r^\alpha, -N/r^\beta$, were also tested but N was always very small, so it was disregarded. The solutions minimise f ≥ 4, h ≥ 4, d ≥ 4, so they can be neglected.

In a mixed case (A and B) the d term was split in two functions, because the interactions of an A molecule in a fcc site with a B molecule in a hcp site have no reason to be identical to the interactions of the same A molecule in a hcp site with the same B molecule in a fcc site.

In other words if we look at each d_i^{AB}, f_i^{AB} or h_i^{AB} then $d_i^{AB} \neq d_i^{BA}$, $f_i^{AB} = f_i^{BA}$ and $h_i^{AB} = h_i^{BA}$, $\forall i$ and $\forall A,B \in \{CH_3, CH_2, CH, C, H\}$.

The number of lateral interactions being too large, we will not display them. The large values for d_1 in the case of CH_3 are to be noted. This can be explained by the volume of the CH_3. It is not possible for any other species to approach that close. CH_3 has 2 kinds of repulsions, through the surface as for all the other CH_x species, but also steric repulsions due to its umbrella shape. CH and C present only lateral interactions through the surface. CH_2 is in between, but its repulsions due to steric hindrance are much smaller that for CH_3. A typical prefactor for the repulsions through the surface is about 150-200 kJ mol^{-1} for the 'f' and 'h' functions and around 1000 kJ mol^{-1} for the 'd' function. The exponent can be in between 2.5 and 3.5. In the case of steric repulsions, the prefactor can go from 100,000 kJ mol^{-1} to much more and the exponent is in between 4.0 and 7.5. The H atom undergoes only small lateral interactions from the very close adsorbates, all the other interactions are negligible.

All the parameters were processed using a Brønsted–Polanyi like formula:

$$\Delta E^{TS} = \frac{1}{2} \times \left(-\Delta E^R + \Delta E^P \right)$$ (5.5)

where: ΔE^{TS} is the change in the activation energy, ΔE^R is the change in the energy of the reactants and ΔE^P is the change in the energy of the products. With this relation we can determine the changes in the barrier for each reaction and for each possible configurations. If the decrease of the barrier is larger than its actual value then the barrier is fixed to zero in order to avoid negative activation energies.

We used Brønsted–Polanyi relation to check the results we have for the 2x2 and 3x3 structures. The activation energy changes compare well (see table 1, columns 2 and 3). For CH_2 decomposition $^1/_2$ has to be changed with 0 and for CH hydrogenation with 1.

3.3 DMC simulations

For comparison, Mean Field Approximation (MFA) simulations, Dynamic Monte Carlo (DMC) without lateral interactions and Dynamic Monte Carlo with lateral interaction were performed for the $C + 2H_2 \rightarrow CH_4$ and $CH_4 \rightarrow C + 2H_2$ reactions.

The MFA was carried out with Mathematica [30], by writing the differential equations and solving them at different temperatures. For the elementary reactions steps describing the dehydrogenation of methane to C and H_2 :

1. $CH_4 + 2* \rightarrow CH_3 + H$
2. $CH_3 + * \rightleftharpoons CH_2 + H$
3. $CH_2 + * \rightleftharpoons CH + H$
4. $CH + * \rightleftharpoons C + H$
5. $2H \rightarrow H_2 + 2*$

the following system of equations can be written:

$$\frac{dc_3}{dt} = k_1 p_1 \left(c_* [t] \right)^2 + \bar{k}_2 c_2 [t] c_H [t] - k_2 c_3 [t] c_* [t]$$

$$\frac{dc_2}{dt} = k_2 c_3[t]c_*[t] + \overline{k}_3 c_1[t]c_H[t] - \overline{k}_2 c_2[t]c_H[t] - k_3 c_2[t]c_*[t]$$

$$\frac{dc_1}{dt} = k_3 c_2[t]c_*[t] + \overline{k}_4 c_0[t]c_H[t] - \overline{k}_3 c_1[t]c_H[t] - k_4 c_1[t]c_*[t]$$

$$\frac{dc_0}{dt} = k_4 c_1[t]c_*[t] - \overline{k}_4 c_0[t]c_H[t] \tag{5.6}$$

$$\frac{dc_H}{dt} = k_1 p_1 (c_*[t])^2 + k_2 c_3[t]c[t] + k_3 c_2[t]c_*[t] + k_4 c_1[t]c[t]$$
$$- \overline{k}_2 c_2[t]c_H[t] - \overline{k}_3 c_1[t]c_H[t] - \overline{k}_4 c_0[t]c_H[t]$$
$$- 2k_5 (c_H[t])^2$$

$$\frac{dc_H}{dt} = \overline{k}_2 c_2[t]c_H[t] + \overline{k}_3 c_1[t]c_H[t] + \overline{k}_4 c_0[t]c_H[t] + 2k_5 (c_H[t])^2$$
$$- 2k_1 p_1 (c_*[t])^2 - k_2 c_3[t]c_*[t] - k_3 c_2[t]c_*[t] - k_4 c_1[t]c_*[t]$$

k_i and $\overline{k}_i$ are the rate constants for the forward and reverse reaction i. They are the same as the reaction rate constants as in the ME 5.1, except for a geometric factor. c_H is H coverage, c_3 is CH_3 coverage, c_2 is CH_2 coverage, c_1 is CH coverage, c_0 is C coverage, c_* is the coverage of empty sites, p_1 is methane partial pressure. In the same way, similar equations for the reactions of hydrogenation of adsorbed C on Ru(0001) surface can be obtained. The simulations were conducted for several temperatures, several partial presures for H_2 and CH_4 and different initial coverages of atomic C on Ru(0001).

The DMC simulations were done in two different ways:

- without lateral interactions, using the results for activation energies from the 2 x 2 calculations; the prefactors were assumed to be 10^{12} for reactions on the surface, (few orders of magnitude) lower if a product is in the gas phase and (few orders of magnitude) larger if a reactant is in the gas phase.
- with lateral interactions; once the values for the lateral interactions had been obtained in the way described in previous section we calculated the changes in barriers using the Brønsted–Polanyi type formula..

The simulations were performed at the following temperatures: 550 K, 650 K, 750 K and 850 K.

Figures 3, 4, 5, 6, 7 and 8 show the results at 750 K with Mean Field, Dynamic Monte Carlo simulations with and without lateral interactions for methane activation and for 10% carbon hydrogenation on Ru(0001) surface.

The results of the simulations for the methane activation process are different depending on the approach considered. MFA and DMC without lateral interactions show exactly the same results. However DMC with lateral interactions gives a different picture: The reactions are much faster. The effect of the lateral interactions on this reactions is beneficent as the reactant are destabilised by the lateral interactions.

For the carbon hydrogenation, MFA and DMC without lateral interaction shows results which are slightly different, but the overall trends are identical. The coverage on the surface increases rapidly because of the hydrogen dissociation and therefore the hypothesis on which MFA is based are not anymore valid. DMC with lateral interactions shows again a different reaction as the hydrogenation of carbon is much slower than in the previous cases.

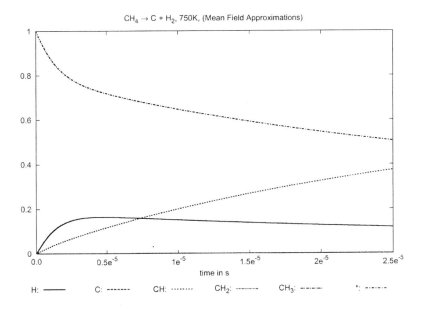

Figure 3. Mean Field Approximation results for methane decomposition at 750 K.

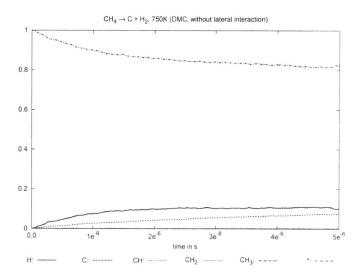

Figure 4. Dynamic Monte Carlo results for methane decomposition at 750 K without lateral interaction.

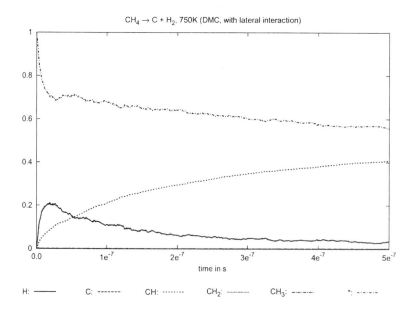

Figure 5. Dynamic Monte Carlo results for methane decomposition at 750 K with lateral interactions included.

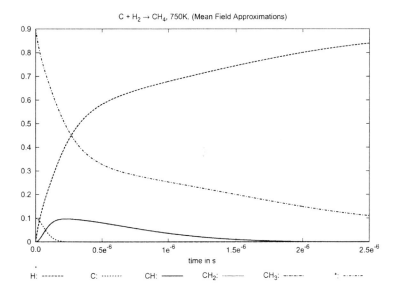

Figure 6. Mean Field Approximation results for 0.1 ML C hydrogenation at 750 K.

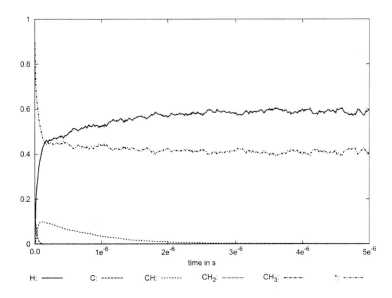

Figure 7. Dynamic Monte Carlo results for 0.1 ML C hydrogenation at 750 K without lateral interaction.

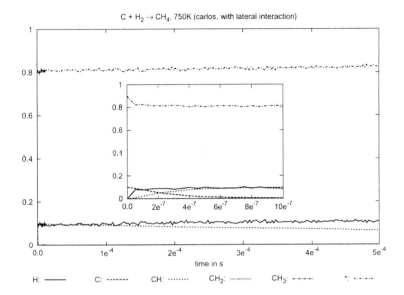

Figure 8. Dynamic Monte Carlo results for 0.1 ML C hydrogenation at 750 K with lateral interactions included.

A common feature of all the simulations is the absence of the CH_2 intermediate. This species is very reactive and its life–time is extremely short. CH_3 has a very similar behavior, with a life–time a bit enhanced. CH is by far the most abundant species on the surface, as expected from theoretical predictions [26] and in good agreement with experimental results [31, 32].

The hydrogenation of C to CH and the dehydrogenation of CH to C are the most difficult reactions. The whole system behaves as if the methane decomposes directly to CH followed by the dissociation of CH to C and H.

4.   NO ON Rh(111)

The behaviour of NO on rhodium single crystals under ultra–high vacuum has been well studied. There are various reasons for this interest. From the environmental point of view, nitrogen oxides are known to be a greenhouse gases, causing smog and acid rain, and rhodium is one of the components used in commercial three–way catalysts to remove NO. From the scientific point of view, it is still difficult to predict catalyst performance, and the elementary reaction steps are often not fully understood.

NO on Rh(111) is known to occupy multiple sites and arranges into ordered structures at certain coverages and temperatures. Upon heating NO can desorb or dissociate depending on the coverage; the fragments can recombine to form dinitrogen and dioxygen. Studying the behavior of this system can therefore yield valuable insights in many interesting phenomena, that are also known for other catalysts.

This system has been studied with DMC previously [33, 34]. Although experimental TPD spectra could be reproduced, we felt that these studies were rather unsatisfactory. First, there is the problem of the large number of kinetic parameters. Some of these parameters could be determined from experiments in certain limits (*e.g.* NO dissociation prefactor and activation energy in the absence of lateral interactions from TPSIMS in the low–coverage limit), but most could not. This made it necessary to determine many parameters simultaneously, which is very difficult. Second, the experimental data need not suffice to fix all parameters. We found various sets of parameters that reproduced the experimental results more or less equally well. It was even possible to reproduce these results with a one–site model, whereas three different sites are known to be occupied at the saturation coverage.

We were stimulated to do new simulations by seemingly high discrepancies between results of DFT calculations [35] and DMC simulations for the lateral interactions: DFT calculations showed lateral interactions that seemed to be about an order of magnitude larger than we had to use in the DMC simulations. This was caused by the neglect of other sites in the DMC simulations, and DFT calculations of high–coverage adlayers which are not formed in reality because of unfavorable lateral interactions. DMC simulations with a three–site model resolved the discrepancies.

4.1 DFT calculations

The reconstruction of the surface has not been explored as there are no experimental nor theoretical indications for its occurrence. The lattice parameter a=3.85 Å (Rh–Rh=2.72 Å) for bulk Rh has been taken from a previous study [36]. The reference energies for the N_2, O_2 and NO molecules in gas phase have been calculated with two molecules in the (3 x 3) cell with an inversion centre. The (3 x 3) cell is large enough for the energy of an isolated molecule to be computed with an accuracy around 1 kJ mol^{-1}.

Except for O_2 and NO, the spin polarisation has not been taken into account as test calculations for a bare surface and a surface with adsorbed NO give systems without any spin polarisation, once NO interacts with the surface it looses its magnetic moment. The NO molecule has been placed on the different sites (top, hollow fcc and hcp or bridge) of the (111) surface with the nitrogen atom pointing toward the surface.

The NO adsorbed atop or hollow has been optimised with a slightly displaced or tilted starting geometries. It results in identical final geometries and energies within numerical error after optimisation. The O and N atoms were placed in the hollow and bridge sites, the top site was discarded as it is unstable. The bridge site is also unstable but it gives the activation barrier for the diffusion of the atomic species..

The most stable sites for NO are the hollow sites, followed by the bridge site and finally the top site. If the coverage is increased up to 0.75 ML (maximum value) two NO will be hollow (fcc and hcp) but the third NO will be atop as it minimises the repulsions with the other two NO. The remaining free hollow or bridge sites are less stable because the NO are bound to the same Rh atoms.

4.2   DMC simulations

Three adsorption sites are known and simulated for NO on Rh(111): fcc, hcp and at high coverages also top. These sites form a regular grid with a spacing of $1/\sqrt{3}d_{Rh-Rh}$. Our model includes all these sites. Adsorbates are not allowed on sites directly next to an adsorbate. Adsorbates d_{Rh-Rh} and $2d_{Rh-Rh}/\sqrt{3}$ apart experience a strong (>20 kJ mol^{-1}) and a small (<10 kJ mol^{-1}) repulsion respectively.

We have calculated the lateral interactions by assuming pairwise additivity. Adsorption energies were calculated for, for example, a (2x2) 0.25ML and a (2x1) 0.50ML NO adsorption structure where NO was bound to threefold sites. Comparison of the NO adsorption energy on the (2x2)0.25ML and on a (3x3)0.11ML structure yielded no difference, indicating no long–range interaction. No noticeable difference in adsorption energy between hcp and fcc site was found. The difference in adsorption energy between the (2x2)0.25ML and the (2x1)0.50ML structure was about 25 kJ mol^{-1}, which we took as a measure for the lateral interaction between two NO molecules one lattice vector apart. In the same way we have calculated the lateral interaction between two NO molecules $2d_{Rh-Rh}/\sqrt{3}$ apart: this time by comparing the NO adsorption energy on a c(4x2)0.50ML and a (2x2)0.25ML structure. Similar calculations have been performed for NO, nitrogen and oxygen adatoms coadsorbed in order to get all the other interactions.

The pair repulsions were derived from DFT calculations of adsorption energies using adlayers with varying structures and compositions. The reactions include adsorption, desorption, diffusion, dissociation and recombination; prefactors and activation energies are based on experimental results.

The threefold sites are filled first in our simulations upon adsorption of NO. At 0.50 ML a c(4 x 2)2NO structure is formed for low temperatures, with NO adsorbed alternatively in fcc and hcp sites (see figure 9).

This structure has been experimentally observed for this and other systems, like NO on Ni(111), by means of LEED measurements. It consists of domains which are identical upon rotation over 120°. After continued adsorption at higher temperatures a (2 x 2)3NO 0.75 ML saturation structure forms, with one NO fcc, one hcp and one on top in each unit cell. This is the second known ordered structure, with similar structures known for CO on Rh(111) and NO on Ru(0001). Reconstruction of the c(4 x 2)2NO structure into the (2 x 2)3NO structure is strongly influenced by the presence of defects and precursor adsorption.

bridge hcp fcc top

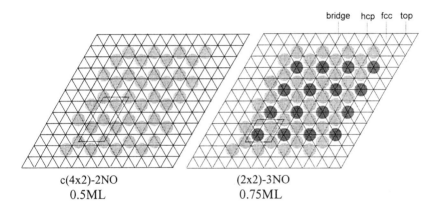

c(4x2)-2NO   (2x2)-3NO
0.5ML   0.75ML

Figure 9. Ordered structures for NO on Rh(111).

Three coverage regions have been identified for the temperature programmed desorption of NO on Rh(111) by Borg et al [37]. Our simulations reproduce the behavior of the system within these coverage regions (see figure 10).

At coverages below 0.25 ML (left panel) NO dissociation starts at 275 K. Dissociation is complete at 350 K, and the nitrogen formed recombines in a second–order peak between 450 and 650 K. Oxygen recombines above 900 K (not shown). At coverages between 0.25 and 0.50 ML (middle panel), NO desorption starts before the dissociation is complete. The lower dissociation rate is caused by a lack of free sites.

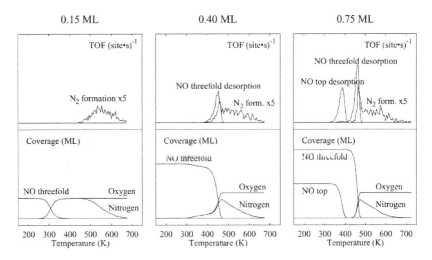

Figure 10. NO and N_2 desorption rates (top), and NO, nitrogen and oxygen coverages (bottom), during temperature programmed desorption. Starting coverages are (from left panel to right) 0.15, 0.40 and 0.75 ML. N_2 desorption rates have been multiplied by 5; the heating rate was 10 K/s.

The threefold bound NO starts desorbing at 400 K, the nitrogen formation is through a first order peak which coincides with the tail of the NO desorption peak, and a second order peak at higher temperatures. Above 0.50 ML (right panel) NO starts to occupy top sites, and dissociation is totally inhibited till the desorption of threefold NO. Top bound NO does not dissociate, it desorbs around 380 K. Threefold bound NO starts desorbing around 400 K. Nitrogen formation occurs again through both a first and a second order peak.

5.   CONCLUSIONS

Lateral interactions are primordial as demonstrated by the simulations for two different systems. We successfully developed a model to extract parameters from DFT calculations and used as input in DMC.

Including lateral interaction in DMC slows down hydrogenation of C atom and speed up the decomposition of CH_4 in the system $CH_4 \rightleftharpoons C + 2H_2$. In the case of NO on Rh(111) we have a good agreement with experimental results.

The combination of *ab–initio* calculations for kinetic parameters and DMC simulations allows one to get an insight into heterogenous catalytic reactions at the atomic scale.

6. ACKNOWLEDGEMENTS

This work is part of the research program of the "Stichting voor Fundamenteel Onderzoek der Materie (FOM)", which is financially supported by the "Nederlandse organisatie voor Wetenschappelijke Onderzoek (NWO)". This work has been accomplished under the auspices of NIOK, the Nederlands Institute for Catalysis Research, Lab Report No. TUE–2000–5–6. The DFT calculations have been performed partially with NCF support (MP043b). We gratefully acknowledge helpful discussions with dr. R. Pino.

7. REFERENCES

1. P. Hohenberg and W. Kohn, *Phys. Rev.*, 136B, 1965, 864
2. W. Kohn and L. J. Sham, *Phys. Rev.*, 140A, 1965, 1133
3. R. G. Parr and W. Yang *Density–Functional Theory of Atoms and Molecules* Oxford University Press, New York, 1989
4. W. Kohn, *Rev. Mod. Phys.* 71, 1999, 1253
5. G. Kresse, J. Furthmüller, *Comp. Mat. Sci.*, 6, 1996, 15
6. G. Kresse, J. Furthmüller, *Phys. Rev. B*, 54, 1996, 169
7. D. Vanderbilt, *Phys. Rev. B*, 41, 1990, 7892
8. G. Kresse, J. Hafner, *J. Phys.: Condens. Matter*, 6, 1994, 8245
9. D. D. Jonsson, *Phys. Rev. B*, 38, 1980, 393
10. J. P. Perdew, *Electronic Structure of Solids '91* Akademie Verlag, Berlin, Germany, 1991
11. H. J. Monkhorst J .D. Pack, *Phys. Rev. B*, 13, 1976, 5188
12. H. J´onsson, G. Mills, W. Jacobsen, Enrico Fermi Summer School Proceedings (Lenci'97), 1997
13. P. Pulay, *Chem. Phys. Lett.*, 73, 1980, 393
14. A. P. J. Jansen, *Comput. Phys. Comm.*, 86, 1995, 1
15. J. J. Lukkien, J. P. L. Segers, P. A. J. Hilbers, R. J. Gelten and A. P. J. Jansen, *Phys. Rev. E*, 58, 1998, 2598
16. R. J. Gelten, R. A. van Santen and A. P. J. Jansen, Dynamic Monte Carlo Simulations of Oscillatory ʜeterogeneous Catalytic Reactions in 57. Molecular Dynamics: From Classical to Quantum Methods, Elsevier Amsterdam, 1999
17. J. C. Keck, *J. Chem. Phys.*, 32, 1960, 1035
18. J. C. Keck, *Discuss. Faraday Soc.*, 33, 1962, 173
19. J. C. Keck, *Adv. Chem. Phys.*, 13, 1967, 85
20. J. P. L. Segers, *Algorithms for the Simulation of Surface Processes*, PhD thesis, Eindhoven University of Technology, 1999

21. K. Binder, Monte Carlo Methods in Statistical Physics, Springer Berlin, Germany, 1986
22. D. T. Gillespie, *J. Comput. Phys.*, 22, 1976, 403
23. D. T. Gillespie, *J. Phys. Chem.*, 81, 1977, 2340
24. W. H. Press, A. Teukolsky, W. T. Vetterling and B. P. Flannery, *Numerical Recipes in C: The Art of Scientific Computing*, Cambride University Press, 1992
25. D. E. Knuth, The Art of Computer Programming, Volume 3: Sorting and Searching, Addison–Wesley, 1973
26. I. M. Ciobîcă, F. Frechard, R. A. van Santen, A. W. Kleyn, and J. Hafner, *Chem. Phys. Lett.*, 311, 1999, 185
27. I. M. Ciobîcă, F. Frechard, R. A. van Santen, A. W. Kleyn, and J. Hafner, *J. Phys. Chem. B*, 104, 2000, 3364
28. I. M. Ciobîcă, R. A. van Santen, to be published
29. Q. Ge, R. Kose and D. A. King, *Adv. Catal.*, 45, 2000, 207
30. *Mathematica 4.0*, Wolfram Research Inc., 1999
31. M. C. Wu and D. W. Goodman *J. Am. Chem. Soc.*, 116, 1994, 1364
32. M. C. Wu and D. W. Goodman *Surf. Sci. Lett.,* 306, 1994, L529
33. R. M. Nieminen and A. P. J. Jansen, *Appl. Catal. A*, 160, 1997, 99
34. A. P. J. Jansen and J. J. Lukkien, *Catal. Today*, 53, 1999, 259
35. D. Loffreda, D. Simon and P. Sautet, *J. Chem. Phys.*, 108, 1998, 15, 6447
36. F. Frechard, R. A. van Santen, A. Siokou, J. W. Niemantsverdriet, *J Chem. Phys.*, 111, 1999, 8124.
37. H. J. Borg, J. F. C.J. M. Reijerse, R. A. van Santen and J. W. Niemantsverdriet, *J. Chem. Phys.*, 101, 1994, 10052

Chapter 6

MODEL SYSTEMS FOR HETEROGENEOUS CATALYSIS: QUO VADIS SURFACE SCIENCE?

H.-J. Freund, N. Ernst, M. Bäumer, G. Rupprechter, J. Libuda, H. Kuhlenbeck, T. Risse, W. Drachsel, K. Al-Shamery, H. Hamann

Key words: Model catalysts, heterogeneous catalysts, thin oxide films, vanadium oxide, V_2O_5, VO_2, V_2O_3, photoemission, $Cr_2O_3(0001)$, field ion microscopy, FIM, surface diffusion, nanocrystals, electron spin resonance, ferromagnetic resonance, molecular beam, infrared-visible sum frequency generation, SFG, photochemistry.

Abstract: The preparation of model catalyst systems based on thin epitaxial oxide films and oxide single crystals is discussed. The effect of surface pre-treatment on the distribution of metal aggregates deposited by vapour deposition is investigated, and a range of surface sensitive techniques applied to study the geometric, optical and electronic properties of these systems. The findings are correlated with the adsorption and reaction of probe molecules on the surfaces studied using multiple molecular beams. While mainly performed under UHV conditions, adsorption measurements can be made under ambient conditions by means of non-linear optical techniques such as sum frequency generation. Particle size and shape are shown to have a profound effect on the thermal and photochemistry of the systems.

1.   Introduction

Eighty to ninety percent of heterogeneous catalysts in use are either oxides or metals dispersed onto oxide supports [1]. Therefore, models must reflect this constitution. Our group has been active in preparing model systems for heterogeneous catalysts, including clean oxide surfaces [2] as well as deposited metal aggregate systems [3, 4]. We are using thin oxide films on metal substrates as samples to be able to apply the full scope of surface science methods to the study of model systems. Methods include those which are based on the analysis of charge carriers, as for example photoelectron spectroscopy, electron energy loss spectroscopy or scanning tunneling microscopy. However, also in connection with optical spectroscopy thin film systems are advantageous due to the high reflectance of the supporting metal. We restrict ourselves here to the discussion of results from our own laboratory and refer for a more complete overview of this field to review articles, which appeared in the literature recently [3-9]. We will cover a variety of different subjects including structural and morphological determinations in model catalyst systems as well as investigations of electronic properties, adsorption of molecules and reactivity studies.

While most of these results have been obtained under ultrahigh vacuum (UHV) conditions, we have recently made an attempt to extend these studies to higher pressures [10]. Thus, we are trying to close both the so called "materials gap" by preparing model systems capturing essential features of real catalysts and the "pressure gap" by moving from UHV to ambient conditions. Nevertheless, our basis are well defined UHV studies, and most of our conclusions heavily depend on a clear understanding of processes at clean surfaces.

One important aspect in using thin oxide film systems is the formation of the oxide overlayer on the metal substrate. Oxide formation and growth is a subject by itself, and not the central theme of this chapter. However, Wyn Roberts, to whom this chapter is dedicated, and his group have been interested in the initial stages of oxidation and oxygen induced chemistry on metal surfaces [11, 12], long before we became involved [13, 14]. Two quotations from one of his papers [15] document this: "Interest in metal oxidation has led us to explore in 1963 how studies of the energy distributions of photoelectrons might provide definitive information on the transition of chemisorbed oxygen to an oxide overlayer with a discrete band structure." He then goes on in the conclusions of the same paper: "The significance of oxygen transients in providing low energy pathways in surface oxygenation reactions was first established using surface sensitive spectroscopies in conjunction with the probe-molecule approach."

Roberts and his group have investigated a series of oxide overlayers with photoelectron spectroscopy, a technique that has been pioneered in its applications to surface studies in Cardiff [16-18]. NiO [19-21] is a prominent example for studies on oxides from the Roberts group. We will come back to other activities of this group concerning the chemistry of carbon dioxide in the course of the paper.

The present paper is organized as follows: We first present some results on clean oxide surfaces, and then proceed to deposited metal aggregates, discussing various specific aspects of nucleation and growth, diffusion, magnetic and electronic properties, adsorption and reactivity.

2.   OXIDE SURFACES

The preparation of a clean oxide surface in ultrahigh vacuum is a rather difficult task. Strictly speaking, a certain oxygen activity is necessary in the gas phase to establish true equilibrium and then the stoichiometry is defined according to the chosen conditions [22]. In this respect, the oxide stoichiometry is not well-defined under dynamical UHV conditions, and the system is only kinetically stabilized. It is therefore believed that defects determine the physical and chemical properties of oxide surfaces. Particularly interesting are vanadium oxides and vanadyl pyrophosphate compounds [23, 24]. Activation of hydrocarbons is thought to take place through abstraction of hydrogen atoms and the formation of surface hydroxyl groups involving defects and isolated transition metal oxide cluster sites [25].

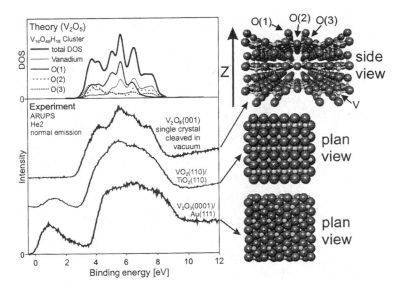

Figure 1. Photoelectron spectra and schematic representations of structures of various vanadium oxides [26, 27]. For comparison a computed density of states [28] is shown.

We have started to study various vanadium oxide surfaces, V_2O_5, VO_2 [26] and V_2O_3 [27]. Figure 1 shows a photoemission spectrum of $V_2O_5(001)$ at the top which may be interpreted on the basis of calculations by Klaus Hermann and his group [28]. Briefly, the spectrum is dominated by O2p/O2s emissions and there is a minor admixture of the V3d wave functions as indicated in the panel showing the results of the calculations. There are no features near the Fermi energy which would, if they were present, be characteristic for defects leaving V atoms in lower oxidation states. The oxygen derived part of the valence band can be separated into contributions from the various types of oxygens constituting the structure as shown in the right part of Fig. 1. The terminal vanadyl oxygen [O(1)] gives rise to the central features whereas the bridging oxygens [O(2) and O(3)] connecting the vanadyl groups are contributing to the wings of the valence band. We will use this fact to identify oxygen specific reactivity. The spectrum in the panel below V_2O_5 is that of VO_2. $VO_2(110)$ has been grown as a thin film on the isostructural rutile(110) surface. The oxygen derived valence band is slightly different from that of V_2O_5. A characteristic difference is the appearance of a feature close to the Fermi edge in VO_2 indicating the presence of vanadium 3d electrons. Its intensity represents to some extent the population of the V3d orbitals. When we turn from VO_2 to V_2O_3 an even stronger increase in the V3d intensity is observed. In Fig. 1, in the lower panel, the valence band photoemission spectrum of $V_2O_3(0001)/Au(111)$ is shown. The V_2O_3 film shows a sharp hexagonal LEED pattern representing the corundum type structure similar to Al_2O_3, Cr_2O_3 and Fe_2O_3 (see below). Again, the V_2O_3 oxygen valence band emission is not very characteristic, as compared with VO_2 and V_2O_5. It is only the change in the near Fermi edge structures that show a characteristic variation from V^{5+} to V^{3+}.

Vanadium pentoxide is rather inactive with respect to adsorbates. For example, molecular hydrogen only leads to observable effects in the photoelectron spectrum after 10^4 L exposure. On the other hand, atomic hydrogen leads already at low exposure to recognizable changes in the photoelectron spectrum. The spectrum shown in Fig. 2 in the middle has been obtained after exposing the surface to atomic hydrogen. Opposite to the finding with molecular hydrogen the surface is chemically corroded. Defects form, as we also know from electron energy loss spectroscopy, and concomitantly reduced metal centers show up near the Fermi edge. In fact, the spectrum looks very similar to the one of VO_2 which we show for comparison. We have investigated the surface with vibrational spectroscopy and found <u>no</u> indication for the formation of hydroxyls. It is very likely, that during exposure water evolves from the surface. Still the complete absence of OH on the surface is surprising and might point to the formation of hydrogen vanadium oxide bronzes which are well-known to exist and which also have been used as precursors in the preparation of catalysts [29]. We are now starting to investigate reactivity of these surfaces to more complex molecules.

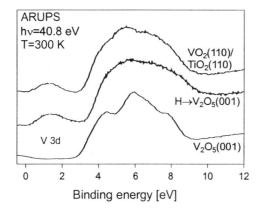

Figure 2. Photoelectron spectra of $V_2O_5(001)$ and $VO_2(110)/TiO_2(110)$ in comparison with a photoelectron spectrum obtained after dosing atomic hydrogen to a $V_2O_5(001)$ surface.

The best studied oxide surfaces are those of TiO_2, the rocksalt structures MgO and NiO as well as the corundum structured Al_2O_3, Cr_2O_3 and Fe_2O_3. Figure 3 shows the results of structural determinations for the three related systems $Al_2O_3(0001)$ [30], $Cr_2O_3(0001)$ [31, 32] and $Fe_2O_3(0001)$ [33]. In all cases a stable structure in UHV is the metal ion terminated surface retaining only half of the number of metal ions in the surface as compared to a full buckled layer of metal ions within the bulk [34]. The interlayer distances are very strongly relaxed down to several layers below the surface [2]. The perturbation of the structure of oxides due to the presence of the surface is considerably more pronounced than in metals, where the interlayer relaxations are typically of the order of a few percent. The absence of the screening charge in a dielectric material such as an oxide contributes to this effect considerably. It has recently been pointed out that oxide structures may not be as rigid as one might think judged on the relatively stiff phonon spectrum in the bulk [35, 36]. In fact, at the surface the phonon spectrum may become soft so that the geometric structure becomes rather flexible, and thus also very much dependent on the presence of adsorbed species [36].

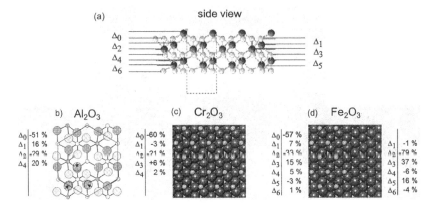

Figure 3. Experimental data on the structure of corundum-type depolarized (0001) surfaces (side and top views). Adapted from b) ref. [30], c) ref. [31, 32], d) ref. [33].

The vibrational modes of a clean $Cr_2O_3(0001)$ surface have been studied under ultra high vacuum conditions with high resolution electron energy loss spectroscopy [36]. Figure 4 shows the spectrum of the clean surface at the bottom. A mode at 21.4 meV, which is confined to the first few atomic layers of the oxide, was detected, in addition to the Fuchs-Kliewer phonons. This mode shows only a very small isotopic shift when the $Cr_2O_3(0001)$ film is prepared from $^{18}O_2$ instead of $^{16}O_2$. In contrast to the Fuchs-Kliewer phonons which extend deeply into the bulk the 21.4 meV mode is very sensitive towards the presence of adsorbates, as is shown in Fig. 4. This can be seen by its attenuation upon exposure of the surface to CO, which in this case is only weakly adsorbed. Full-potential linearized augmented plane wave calculations [36] show that this mode is an in-phase oscillation of the second-layer oxygen atoms and the Cr atoms of the two layers below. A schematic representation of the mode is shown in the inset of Fig. 4.

Bulk oxide stoichiometries depend strongly on oxygen pressure, a fact that has been recognized for a long time and we have alluded to above [2]. So do oxide surfaces, structures and stoichiometries, a fact that has been shown again in a recent study on the $Fe_2O_3(0001)$ surface by the Scheffler and Schlögl groups [34]. In fact, if a Fe_2O_3 single crystalline film is grown in low oxygen pressure, the surface is metal terminated while growth under higher oxygen pressures leads to oxygen termination. This surface would be formally unstable on the basis of electrostatic arguments [2]. However, calculations by the Scheffler group have shown that a strong rearrangement of the electron distribution as well as relaxation between the layers leads to a stabilization of the system. STM images by Weiss and co-workers corroborate the coexistence of oxygen and iron terminated layers and thus indicate that stabilization must occur. Of course, there is need for further structural characterization. For the surface of $Cr_2O_3(0001)$ infrared spectroscopic data suggest a different termination when the clean surface is exposed to oxygen [37]. Figure 5 shows the appearance of a sharp line which develops out of a broad feature after heating the surface to room temperature. As has been discussed earlier and also supported by isotopic labeling experiments by Dillmann et al. [37] chromyl groups are formed on the surface and it was suggested that the chromium ions in the surface layer of the clean surface bind oxygen to form such chromium-oxygen double bonds. As is obvious from Fig. 5, there are adsorbed oxygen precursors to the chromyl formation. The width as well as frequency is indicative of an O_2^- species being adsorbed on the surface in a more or less upright position [37]. The chemistry of this species has yet to be investigated.

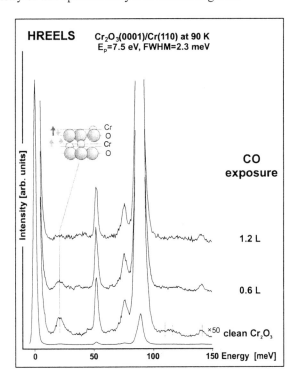

Figure 4. Electron energy loss spectra in the vibrational regime of clean and CO-covered $Cr_2O_3(0001)/Cr(110)$ surface. [E_p: 7.5 eV, specular scattering] The inset shows a schematic representation of the calculated normal mode of a $Cr_2O_3(0001)$ surface at 21.4 meV, according to ref. [36].

CO adsorption on corundum type (0001) surfaces is interesting because, opposite to checker board rocksalt type surfaces, on the (0001) surface of Cr_2O_3 CO has been shown to assume a strongly inclined almost flat adsorption geometry [38]. Recent cluster calculations by Staemmler's group have revealed the reason for this [39]. The CO molecule is situated between two adjacent Cr ions. This allows the carbon lone pair to interact with the positively charged Cr ions and to reduce, simultaneously, the Pauli repulsion between the CO 1π electrons and the oxygen ions by placing the molecules atop an open oxygen triangle. This leads to a more stable adsorption than by placing the molecule vertically atop a chromium ion. However, the energy differences are only of the order of a few kJ so that it is fair to assume that the molecule is rather dynamic on the surface. It is very interesting to note that CO adsorption on $V_2O_3(0001)$ seems to show very much related behavior although the study of orientation of the molecule is not complete yet [27].

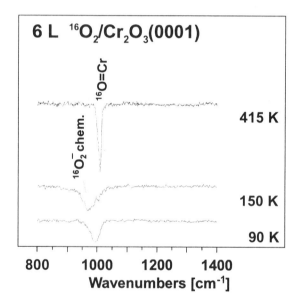

Figure 5. IRAS spectra after adsorption of a saturation coverage of O_2 on $Cr_2O_3(0001)$ /Cr(110). The spectra are taken at the temperatures indicated.

Another interesting case is CO_2 adsorption on solid surfaces. Wyn Roberts and his group have pioneered this field on metal surfaces [18, 40, 41]. Together with one of the authors he reviewed the field recently [42]. Here we report on the interesting case of CO_2 adsorption on $Cr_2O_3(0001)$ [43]. The adsorption can be almost completely suppressed when the $Cr_2O_3(0001)$ surface is oxygen terminated with chromyl groups. On the metal terminated surface, on the other hand, CO_2 is relatively strongly chemisorbed. The CO_2 molecule forms a carboxyl species bound through the carbon atom to a chromium ion. Interestingly, carbonates are only easily formed when promoters such as alkali metals are present.

3. METAL AGGREGATES ON OXIDE SURFACES: MORPHOLOGY, GROWTH AND ELECTRONIC STRUCTURE

So far, we have considered the clean oxide surface and its reactivity. In the following we will discuss the modification of the oxide surface by deposition of metal [3, 4]. The systems prepared in this way can be regarded as model systems in heterogeneous catalysis [4, 9]. They aim at bridging the so-called materials gap. Several groups have started to prepare such systems including Goodman [5], Henry [6], Madey [7], Møller [44] and our own group [2-4, 9]. The pioneer in this field was Helmut Poppa [45] who undertook the first systematic study applying transmission electron microscopy.

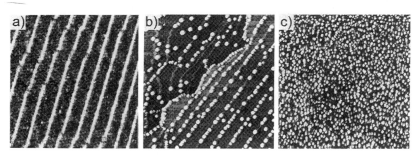

Figure 6. Scanning tunneling images (1000 × 1000 Å). a) Clean Al_2O_3/NiAl(100) film, U = 4.2 V, I = 0.5 nA. b) After deposition of Pd(0.2 Å) at 300 K, U = 3.1 V, I = 0.5 nA. c) After deposition of Pd (0.2 Å) at 300 K on a hydroxylated alumina film, U = 3.1 V, I = 0.4 nA.

Figure 6 shows STM topographs of a clean alumina film (left), after deposition of Pd at room temperature on the clean film (middle) and onto a chemically modified film which is terminated by hydroxyl groups (right) [4] which can be identified *via* vibrational spectroscopy [46].

The surface of the clean film (left) is well ordered and there are several kinds of defects present: The lines represent line defects due to anti-phase domain boundaries [47]. We know from nucleation studies that there are about 1×10^{13} point defects per cm^2 present as well [3]. Pd deposited at room temperature leads to aggregates residing on the domain boundaries and steps (middle). The surface of these aggregates can be imaged with atomic resolution, showing that the aggregates are crystalline and terminated by (111) and much smaller (100) facets [48, 49]. The mobility of Pd under these conditions is high enough to allow nucleation at the stronger interacting line defects, whereas at 90 K the point defects determine the growth behavior. At this temperature we find aggregates statistically distributed across the entire surface of the sample.

If we deposit the same amount of Pd at room temperature onto the hydroxylated film as onto the pristine film, the dispersion of aggregates is by about an order of magnitude higher as shown in the right panel of Fig. 6 [46]. This is due to stronger interaction with the substrate which also leads to higher thermal stability before agglomeration starts.

Studying this agglomeration process is an interesting subject in itself and research in this direction is only starting. A more basic aspect, of course, would be a study of metal atom diffusion on oxide surfaces. The obvious method to perform such a study is scanning tunneling microscopy. However, in contrast to diffusion studies on metal surfaces, similar studies on oxide surfaces have not been reported. On the other hand, field ion microscopy is being used in our group to study such processes. Applying neon-FIM at 79 K the feasibility to image platinum adatoms, supported on a thin alumina film grown on a [110]-oriented NiAl tip, had been demonstrated in our group [48]. At these conditions, the imaging of an *ensemble* of Pt adatoms is characterized by strong fluctuations and non-optimum contrast, partly compensated by computer aided image processing. Intending to observe cluster formation, the oxide surface had been exposed to a relative high platinum dose followed by several heating cycles. As a result, a local ordering of platinum adatoms was obtained in the vicinity of a point defect; the arrangement of adatoms was compatible with the surface unit cell [48]. From a displacement analysis of the platinum adatom ensemble only a rough estimation of the diffusion activation energy was obtained. Recently, we have prepared an *individual* Pt adatom on the apex plane of an oxidized [110]-oriented NiAl tip using neon FIM at 35 K [50]. A sequence of FIM patterns obtained in a first, exploratory experiment is shown in Fig. 7. Circular sections of FIM images show the specimen surface after oxidation. Before deposition of platinum the Al$_2$O$_3$ film was imaged allowing to visualize sites at the boundary and point defects as in Fig. 7(a). After exposing the surface (at 35 K) to a platinum beam, a new emission site appeared which was interpreted as the image spot of a single platinum adatom.

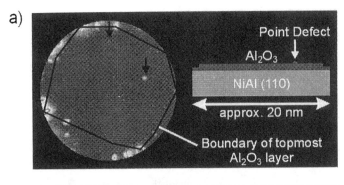

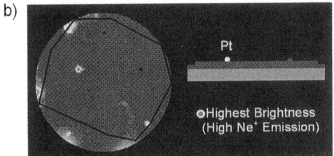

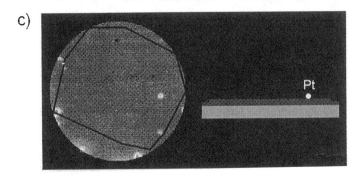

Figure 7. Surface diffusion of Pt/Al₂O₃/NiAl(110). The polygon curves had been drawn in the circular sections of the FIM-pattern to indicate the boundary of the topmost Al₂O₃ layer: a) before deposition; b) After deposition, an individual Pt adatom was imaged at 35 K using neon-FIM at 7.7 kV (black dots: positions of defects); c) After heating the surface to 170 K during 20 s at zero field.

Chapter 6

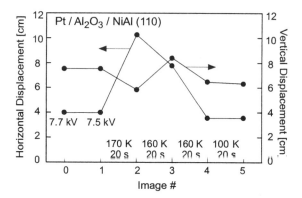

Figure 8. Positions of image spots of Pt adatoms diffusing on Al₂O₃/NiAl(110): Between the
imaging cycles the temperature was varied from 170 K (Fig. 7c) to 100 K.

Figure 8 gives a survey of exploratory examinations of the surface
diffusion behavior for alumina-supported, individual platinum adatoms,
based on FIM patterns such as in Fig. 7(b) and (c). At 160 K the Pt adatom is
in fact more mobile on the well ordered oxide film than on NiAl(110) [50].

The onset temperature for surface diffusion was especially examined in
another experiment (Fig. 9), allowing an estimation of a value for the
activation energy [51]. The identity of the adatom, prepared in the center of
the plane was ascertained before and after the diffusion steps. In particular,
upon the first heating step to 100 K for 15 s, the adatom changed its position
slightly, but noticeably. In this case, our analysis of the jump direction is
compatible with the model visualized in Fig. 10 [52].

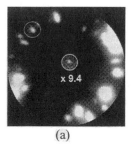
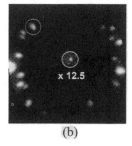

(a)   (b)

Figure 9. Onset of surface diffusion of a Pt adatom adsorbed in the center of an alumina film
on NiAl(110): To improve the visibility, the intensity around the image spot of the adatom
was multiplied by a factor 9.4 (a) and 12.5 (b). Image b) was taken after 15 s heating at 100 K
without applied field. The circle at eleven o'clock marks the position of a step site serving as
a reference at the boundary.

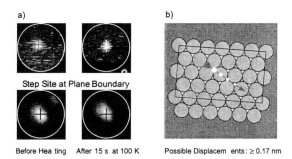

Step Site at Plane Boundary

Before Heating   After 15 s at 100 K   Possible Displacements: ≥ 0.17 nm

Figure 10. Onset of surface diffusion: a) experiment and b) model. Upper two enlarged FIM images in a) show positions of Pt adsorbed in the center of the alumina film prepared on NiAl(110) (Fig. 9). The lower two patterns display step sites whose image positions do not change upon heating. The model b) shows a simplified, hexagonal arrangement of the top layer of oxygen ions and the surface unit cell. The darkest dot marks the position of the Pt adatom prior to heating. The pale dots show possible positions after the diffusion event. The maximum displacement is estimated to approximately 0.5 nm considered through decreasing diameter of the pale dots.

In Table 1, we have compiled experimental and theoretical results of surface diffusion parameter for individual Pt adatoms. Those deposited on a thin alumina film supported on NiAl(110) showed an onset temperature for approaching 100 K. We note that our experimental estimate is probably close to 0.3 eV and thus significantly smaller than the calculated value, 0.7 eV obtained for Pt/Al$_2$O$_3$/Al(111) [53]. Though further experiments are still required, we conclude that the activation energy of Pt on a defect free surface area of Al$_2$O$_3$ is comparable with results of Pt surface diffusion on closely packed metal surfaces such as Ni(111) and Pt(111) [51]. The relatively high mobility at temperatures approaching 100 K is also in general agreement with previous STM investigations on Pt cluster formation occurring at low temperatures on alumina [3]. It is clear, however, that the characterization of the surface diffusion behavior of alumina-supported platinum requires more displacement observations.

Table 1. Surface diffusion parameter for individual Pt adatoms [50].

System	Onset Temperature (K)	Activation Energy (eV)
Pt/NiAl(110)	165	0.48
Pt/Ni(111)		<0.22 [51]
Pt/Al$_2$O$_3$/NiAl(110)	>100	>0.29
	<160	<0.47
Pt/Al$_2$O$_3$/Al(111)		0.7 [53]

Figure 11. Scanning tunneling images at atomic resolution of Pd aggregates grown on an alumina film (left 500 × 500 Å, right 50 × 50 Å) [49].

It is obvious that the area of diffusion studies will considerably profit from atomic resolution, once it is obtained routinely for deposited aggregates on oxide surfaces. While for TiO_2 and very few other oxide substrates atomic resolution may be obtained routinely, there are very few studies on deposited metal particles where atomic resolution has been reported [49], [54]. The first report for an atomically resolved image of a Pd metal cluster on MoS_2 was reported by Claude Henry and his group [54]. A joint effort between Fleming Besenbacher and our group [49] has led to atomically resolved images of Pd aggregates deposited on a thin alumina film. Figure 11 shows such an image of an aggregate of about 80 Å in width. The particle is obviously crystalline and exposes on its top facet the (111) Pd surface. Also, the (111) facets on the side, typical for a cuboctahedral particle, can be discerned. The small (100) facets predicted *via* equilibrium shape considerations on the basis of the Wulff-construction could not be atomically resolved. If we, however, apply the concept of the Wulff-construction, we may deduce the metal surface interaction energy [49]. The basic equation is

eq. (1) $\quad W_{adh} = \gamma_{oxide} + \gamma_{metal} - \gamma_{interface}$

Provided the surface energies (γ_{metal}) of the various crystallographic planes of the metal are known [55], a relative work of adhesion (W_{adh}) may be defined [49]. We find 2.9 ± 0.2 J/m^2 which is still rather different with respect to recent calculations by Jennison et al. [53] where metal adsorption energies (1.05 J/m^2) have been calculated on a defect free thin alumina film. It is not unlikely that this discrepancy is connected with the rather complicated nucleation and growth behavior of the aggregates involving defects in the substrate [56].

It is clear that these studies only represent a beginning, but we think they are promising. Parallel to studying morphology, structure and processes of cluster formation, we have investigated the electronic structure of deposited metal aggregates. In earlier studies we have used photoelectron spectroscopy and X-ray absorption to investigate deposits with finite but narrow size distributions to probe the size dependent electronic structure of these systems [56]. With the advent of scanning tunneling microscopy it has become possible to investigate individual objects, and therefore the spectroscopy of optical and electronic properties of single, selected objects on a surface has been in the focus of interest in recent years. It opens the unique possibility to detect local variations in the electronic structure which are normally hidden in the inhomogeneous broadening of the spectra due to statistical disorder on the surface. A local measurement allows the direct connection between a distinct geometric structure and its optical, electronic or chemical properties. In this way, the opening of a size dependent band gap has been observed in small Pt clusters with scanning tunneling spectroscopy [57]. The catalytic activity of small Au clusters has been correlated with the appearance of the metal non-metal transition as cluster sizes decrease [58]. Spectroscopy of single molecules in the STM has demonstrated the great potential of investigating molecular vibrations as a function of adsorption site [59]. With the first observation of light emission from the tunnel junction of a STM, the detection of optical properties of a surface with nanometer resolution has become feasible [60]. The method has been employed to measure photon maps of metal and semiconductor surfaces, nanocrystals and individual molecules [61]. The spectroscopy of emitted photons provides an insight in the underlying elementary processes, whereby tip-induced plasmon resonances have been identified as the source of the observed photon emission [61]. These special interfacial modes are connected with the strong electromagnetic interaction between the tip and the sample in a STM and are not visible in conventional optical spectroscopy. On the other side, typical features accessible to classical absorption spectroscopy, like Mie resonances in small metal particles or electronic transitions in molecules, have not been observed in the photon emission spectra from a tunnel junction of a STM. Therefore, there is still a gap to be bridged between conventional optical spectroscopy averaging over a macroscopic area of the surface and the emission spectroscopy in the STM at high spatial resolution. Here we report on photon emission from individual, alumina-supported Ag clusters [62]. The emission can clearly be distinguished from tip-induced plasmon modes and is interpreted as a decay of Mie-plasmon excitations in small metal particles, well known from absorption spectroscopy [63]. We have determined the Mie-plasmon energy and its homogeneous line width for single Ag particles as a function of cluster diameter between 1.5-12 nm.

In the experiment the tunnel tip of a beetle type STM, mounted in an UHV chamber, has been used as a local electron emitter. The electron energy could be adjusted in a wide range between 1-100 eV, the typical electron current was set to 1-10 nA. The electron injection from the tip caused electronic excitations in the sample followed by optical de-excitation processes. The emitted photons have been collected by a parabolic mirror surrounding the STM. Outside the vacuum chamber the light was focused onto the entrance slit of a grating spectrograph and detected with a liquid-nitrogen cooled CCD-camera. The spatial resolution of the emission spectroscopy depends on the tip-sample separation and therefore on the bias voltage applied. For tunnel voltages below 15 V the exciting electron current is restricted to an area smaller than 1.5 nm in diameter. Ag clusters were grown by vapor deposition at 300 K onto a thin, well-ordered alumina film prepared on a NiAl(110) single crystal [3]. The density of nucleation centers for the particles was controlled in a sputter-assisted deposition process. Depending on the amount of evaporated material the mean cluster size could be adjusted between 1-12 nm. The cluster diameter was determined after separating the effect of tip convolution. This was achieved by extracting a mean tip radius from the apparent broadening of step edges of the substrate.

The photon emission characteristics of a cluster-covered alumina film on NiAl(110) can be divided in two different excitation regimes depending on the tip voltage:

1. Spectra measured at low tunnel voltages ($U_{tip} \leq 5$ V) are dominated by two emission peaks at 1.3 eV and 2.4 eV, which are almost independent of polarity (Fig.12, curve (i)). The emission is caused by collective electron oscillations in the coupled electron gases of tip and sample induced by a strong dynamic electromagnetic field in the tunnel cavity. Inelastically tunneling electrons are the driving force for these tip-induced plasmons [61]. The shape of the spectra is mainly determined by the dielectric function of NiAl. It can be reproduced within a theoretical model, which allows to calculate the electromagnetic response function of a simplified tip-sample geometry, using the dielectric function of the W tip and the NiAl sample as material properties [64]. The thin alumina film and the presence of small metal clusters below the tunnel tip slightly modify the shape and intensity of the spectra emitted from the clean NiAl/W tunnel junction.

2. An increasing tunnel bias gradually decreases the tip-sample interaction and consequently the photon yield due to inelastic tunneling processes.

At voltages above 10 V the characteristic lines of the tip-induced plasmon vanish. In contrast to the clean alumina surface, a new, intense emission line appears in the spectra taken on top of an Ag cluster (Fig.12, curve (ii)). The peak is centered around 3.7 eV and visible only for the injection of electrons from the tip into the cluster (see curve (iii) for comparison). The energetic position of the emission line corresponds to the 1,0-mode of the Mie plasmon, well known from extinction cross section measurements on Ag cluster ensembles [63]. The resonance is interpreted as an oscillation of the free silver electron-gas perpendicular to the substrate plane. The corresponding in plane oscillation (1,1-mode) is not accessible to the experiment because the incoming electrons from the tip induce preferentially a dipole or impact excitation along the surface normal. Additionally, the 1,1-mode is strongly damped by its image dipole induced in the NiAl substrate. For a detailed discussion see ref. [62].

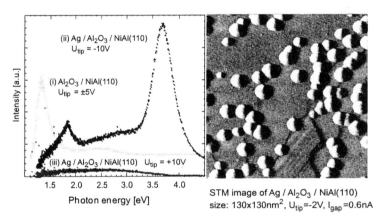

Figure 12. Photon emission spectra at different tip voltages and STM image of Ag/Al$_2$O$_3$/NiAl(110).

The energy position and the line width of the Mie plasmon show a characteristic dependence on cluster size [62](see Fig.13). For decreasing cluster diameter from 12.0 to 1.5 nm a blue shift of the plasmon energy from 3.6 eV to almost 4.0 eV is observed. In the same size range the homogeneous line width of the emission increases from 160 meV to 300 meV. Both size dependencies show an 1/d (d = cluster diameter) behavior reflecting the growing surface contribution with respect to bulk effects in small particles. For the blue shift of the peak position changes of the intrinsic electronic properties in a spatially confined system are considered. In the case of Ag particles the plasmon energy is determined by the classical Drude frequency of the free 5s electrons drastically reduced by a strong screening effect of the 4d electrons [65]. The 4d depolarization field vanishes at the cluster surface because the 4d electrons are more localized at the Ag atoms and have a low residence probability outside the classical cluster volume. In contrast, a fraction of the Ag 5s electrons spills out into the vacuum, where they are unaffected by the 4d screening and increase their plasmon frequency. With decreasing cluster size this fraction of free electrons residing outside the classical cluster volume increases with respect to the total electron number and the Mie plasmon energy shifts to higher energies. This effect is partly cancelled, because a considerable electron spill out results also in a reduction of the electron density inside the cluster volume and thus a red shift of the plasmon energy with decreasing cluster size ($\omega_{plas}^2 \sim n_{electron}$). However, in the present experiment the electron spill out is strongly quenched at the cluster-alumina interface and the blue shift dominates the size dependence of the Mie plasmon. A quantitative description of the experimental results has to treat the 5s-4d interaction on a fully quantum-mechanical level and has to include the restructuring of the bulk electronic structure in small Ag particles.

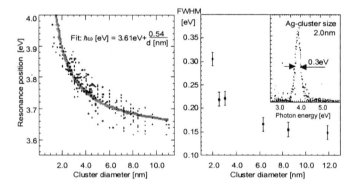

Figure 13. Energetic position and line width of the Mie resonance as a function of cluster diameter.

For the increase of the homogeneous line width of the Mie resonance with decreasing cluster size, only an empirical model can be presented as well. The behavior reflects the reduced lifetime of the collective electron oscillation in small particles due to an enhanced electron-surface scattering rate. This additional decay mechanism (apart from electron-electron, electron-phonon scattering and Landau damping) becomes highly efficient, when the electron mean free path exceeds the cluster diameter and causes a dephasing process of the collective oscillation [63]. The surface mediated character of the damping with respect to the cluster volume can explain the observed 1/d behavior of the line width with decreasing cluster size.

4.   MAGNETIC PROPERTIES

In addition to investigations on electronic properties it is interesting to develop tools which allow us to study magnetic properties. Based upon the experience we have developed using electron spin resonance (ESR) on radicals adsorbed on single crystal surfaces [66-68] we have started measurements of the ferromagnetic resonance (FMR) [69, 70] of deposited aggregates. To this end we prepare either a bulk single crystal oxide surface or an epitaxial thin oxide film under ultrahigh vacuum conditions and grow the metal aggregates on it. Such a sample is brought into a microwave cavity and the FMR is recorded. The sample, which is attached to a manipulator may be oriented with respect to the external field, and therefore the orientation of the direction of the magnetization is accessible.

Figure 14a shows such a measurement for Co particles on $Al_2O_3(0001)$, deposited at room temperature. An uniaxial orientation is found with a single minimum at orientation of the field parallel to the surface plane. This means that the magnetization is also oriented in this way [70]. A very similar behavior is found for iron as plotted in Fig. 14b. The smaller asymmetry is a property of the specific metal. While upon heating the behavior does not change for Co, it becomes more complex for Fe (see Fig. 15) [71]. This is indicative of the survival of uniaxial magnetism in the hexagonal cobalt and its breakdown for the body centered cubic iron. Fe(bcc) has three easy axes of magnetization and the formation of more crystalline aggregates is likely to be the reason for the observation.

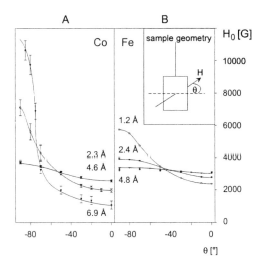

Figure 14. Angular dependence of the resonance field for various amounts of Co (A) and Fe (B) deposited at room temperature on an $Al_2O_3(0001)$-single crystal surface. θ denotes the angle between the crystal surface and the static magnetic field. The deposited amounts are given in terms of the effective layer thickness.

Since atomic resolution is very hard to achieve on the small aggregate at present, these experimental observations in the ferromagnetic resonance are very useful. FMR can also be used to follow adsorption of the aggregates. We find that chemisorption of CO quenches the surface magnetism of the small particles. Oxidation of the small particles leads to the formation of an oxide skin on a ferromagnetic kernel. Since the oxide signal occurs at very different fields only the FMR of the kernel has been recorded. Such measurements may be used to follow the formation of oxide aggregates deposited on oxide supports which would have interesting catalytic properties.

5.   ADSORPTION AND REACTION

It is the next step to investigate adsorption and reaction of molecules on the surfaces of the deposited metal aggregates. Several groups have used Fourier Transform Infrared (FTIR) spectroscopy to study adsorption on such systems [72]. The results have been published and discussed in several review articles [3, 72]. Here we describe and discuss possibilities that arise from a combination of molecular beam studies in direct combination with a fast FTIR spectrometer where in-situ studies are possible [73, 74].

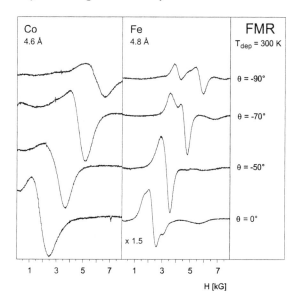

Figure 15. Comparison of the angular dependent FMR-spectra of Co and Fe on $Al_2O_3(0001)$ after heating the deposits (prepared at 300 K) to 870 K. The spectra were recorded at 300 K. θ denotes the angle between the static magnetic field and the crystal surface. The deposited amounts are given in terms of the effective layer thickness.

The experiments have been performed in a new UHV molecular beam / surface spectroscopy apparatus. It has been specifically designed for kinetic studies on complex model systems.

After preparation the sample is transferred to the scattering chamber, which contains an experimental setup as depicted schematically in Fig. 16. Up to three molecular beams can be crossed on the sample surface providing the reactants. In kinetic studies requiring high reactant fluxes two modulated effusive sources can be used, for sticking coefficient measurements or angular resolved scattering a third beam with narrow velocity distribution and well-shaped profile is generated from a supersonic source. Angular-integrated gas-phase measurements are performed with a quadrupole mass spectrometer, angular-resolved measurements with a second rotatable and doubly differentially pumped QMS. Simultaneously, time-resolved IR absorption spectra can be obtained *via* a vacuum FT-IR spectrometer. A detailed description of the experimental setup and data acquisition procedure can be found elsewhere [74].

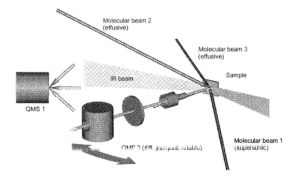

Figure 16. Schematic representation of the molecular beam/IR spectroscopy experiment.

One general effect which has to be taken into account in a description of the adsorption or reaction kinetics on supported metal catalysts is *support mediated adsorption* (sometimes also denoted as *reverse spillover*). The process is schematically illustrated in Fig. 17d: An impinging gas molecule may directly collide with the Pd particle and adsorb with a probability given by the sticking coefficient on Pd. Alternatively, the molecule may impinge on the substrate, be trapped in a physisorbed state and reach the metal particle *via* surface diffusion (this established a "capture zone" for the particle). Depending on the diffusion length, this effect may considerably enhance the adsorbate flux to the metal particle and has to be taken into consideration when comparing catalytic activities. Gillet et al. [75] have formulated the effect for CO adsorption on Pd particles. Later it was taken into account in several adsorption and reaction studies on the same system [76-78]. Models have been derived, which allow support diffusion to be taken into account quantitatively in kinetic studies [79-82].

Combining structural information with molecular beam experiments, a possible contribution to adsorption due to support trapping can be easily quantified. The procedure is illustrated in the example shown in Fig. 17. First, we determine the net sticking coefficient for the model system (Fig. 17c) *via* a King and Wells type sticking coefficient measurement [83, 84]. Taking into account the fraction of the support, covered by the active metal (from structure studies, [3, 85, 86]) and the sticking coefficient on the clean metal surface (from single crystal data, [87]) the support trapping contribution can be directly estimated. In order to quantify a 'capture zone' (an average area from which an adsorbate diffuses to the metal particle), however, we have to take into account that only a certain fraction of the adsorbate molecules is initially trapped in a physisorption well on the oxide surface (trapping-desorption channel, TD), while the remaining molecules are directly scattered back into the vacuum (direct inelastic scattering, DI).

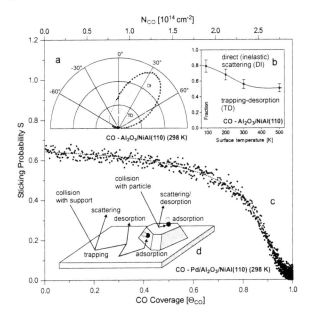

Figure 17. (a) Angular distribution of CO (kinetic energy 86 meV, incidence angle: -35°) scattered from Al_2O_3/NiAl(110); (b) Trapping-desorption and direct inelastic scattering components for CO scattered from Al_2O_3/NiAl(110) as a function of surface temperature; (c) Sticking probability for CO on Pd/Al_2O_3/NiAl(110); (d) Schematic representation of the process.

The probabilities for both processes can be derived from angular and/or time resolved scattering studies from the free support surface (Fig. 17a): The directly scattered molecules are characterized by an angular distribution centered around the specular direction and a velocity distribution dependent on the incident beam. The trapped adsorbates exhibit a symmetric distribution with respect to the surface normal (the information on the initial impulse is lost upon accommodation) and a velocity which depends on the surface temperature (in practice the situation might be complicated by partial accommodation, inefficient coupling of parallel and perpendicular momentum [88, 89] or due to diffuse scattering from defects, see *e.g.* [6]). As an example we show the corresponding measurements for CO adsorption at 300 K on the alumina-supported Pd model catalyst (Fig. 17). Under these conditions it is found that the support-trapping represents the dominating adsorption channel, as the 'capture zone' of the Pd particles exceeds the particle distance [90]. With increasing surface temperature, the diffusion length on the support decreases and eventually, depending on the strength of the adsorbate-support interaction, the support-trapping effect is expected to become less important.

The interaction of oxygen with Pd surfaces as well as the formation and influence of subsurface and bulk oxygen is a controversially discussed topic (see *e.g.* [76, 87, 91-111] and refs. therein). The remaining ambiguities, which exist in spite of the numerous studies published, are mainly related to the difficulties in the identification of bulk and subsurface oxygen species, the lack of quantitative adsorption and reactivity studies and the limited knowledge about the surface defect structure. Here, we may again take advantage of the quantitative character of molecular beam experiments, as will briefly be illustrated in this section. A more detailed discussion of the topic can be found elsewhere [73].

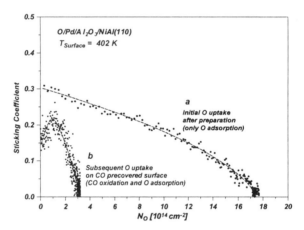

Figure 18. Oxygen sticking coefficient as a function of oxygen uptake (in O atoms/cm^2) at 402 K for Pd particles supported on Al$_2$O$_3$/NiAl(110) directly after preparation under UHV conditions and after several cycles of O$_2$ and CO exposure.

In CO oxidation experiments, there have been indications that even at low sample temperatures and oxygen exposures, rapid subsurface or bulk diffusion of adsorbed oxygen may occur [73]. In order to test this hypothesis, we have performed O$_2$ sticking coefficient measurements in a temperature range from 100 to 400 K. An example is depicted in Fig. 18, where we compare the sticking coefficient displayed as a function of oxygen uptake for two situations, (a) the freshly prepared Pd sample (Fig. 18a) and (b) after repeated cycles of O$_2$ and CO exposure (Fig.18b). Considering the situation after repeated O$_2$ and CO exposure, the total oxygen uptake can be subdivided into a reactive contribution due to oxidation of the preadsorbed CO and adsorptive contribution. Taking both into account, the adsorption uptake turns out to be consistent with what is expected for simple Pd(111)-like surface adsorption behavior based on the Pd surface area estimate from STM and high-resolution low energy electron diffraction LEED [73].

If we compare this result with the oxygen sticking diagram for the Pd particles initially after preparation (Fig. 18a), we find a much larger oxygen uptake. These values are by far too large to be compatible with pure surface adsorption. Therefore, we have indeed to take into account rapid surface and bulk diffusion on the time-scale of the experiment. Moreover, the reduced oxygen uptake after subsequent CO and O_2 exposure shows that the bulk oxygen species are not accessible to the CO oxidation in this temperature region. This observation is in agreement with a recent study on Pd(111) [106].

Next, we may consider the total oxygen uptake detectable in sticking coefficient measurements as a function of the substrate temperature (Fig. 19). Initially, we find a low and slowly decreasing uptake in the temperature range between 100 K and 250 K, followed by a step-like increase in the temperature regime $\geq$ 300 K. Based upon previous adsorption studies, we may differentiate between three adsorption regimes: at temperatures around 100 K the adsorption process is molecular [96], before at higher sample temperatures O_2 will dissociate [96]. This process is connected to a decreasing saturation coverage [99]. The sudden increase in the uptake above 250 K marks the onset of bulk diffusion. Note that in contrast to adsorption experiment on bulk metal samples, the bulk reservoir saturates rapidly and the uptake becomes independent of the sample temperature. A simple estimate reveals that the total oxygen uptake upon saturation corresponds to a total stoichiometry of $PdO_{0.5-0.8}$. Beyond this saturation uptake, the net oxygen adsorption decreases below the detection level of the experiment.

Previously, it has been reported that supported Pd particles may undergo morphological changes induced by strongly interacting adsorbates [108]. Although the conditions are rather moderate in this work, we have performed STM measurements after bulk-O-saturation (*i.e.* after a repeated O_2 and CO treatment) in order to exclude such effects as a possible reason for the changes in adsorption behavior [73]. A comparison with the STM images taken before oxygen adsorption shows that neither the particle density nor the overall morphology of the Pd aggregates is affected by the large oxygen uptake. Only slight distortions of the particle edges appear, which might be interpreted as an indication for a distortion of the Pd lattice structure due to the incorporated oxygen.

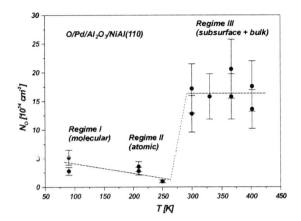

Figure 19. Oxygen uptake (in O atoms/cm^2) of a sample of Pd particles supported on Al$_2$O$_3$/NiAl(110) directly after preparation under UHV conditions as a function of the sample temperature.

Thus, we may conclude that, starting at temperatures above 250 to 300 K, the Pd particles incorporate large amounts of oxygen on the time-scale of the sticking or reactivity experiments. In comparison, on single crystal surfaces this process is typically found to proceed significantly slower and is observed at higher temperatures ([106] and refs. therein). It has been suggested, however, that the kinetics of bulk diffusion may strongly depend on the defect density of the system [92, 94, 95]. This appears consistent with the present study, as in comparison with most single crystals, the supported Pd nanocrystallites in this work represent a high defect-density system. In addition to step and edge sites, we have to take into account the particle-oxide interface, where the crystallite structure may be additionally distorted due to interactions with the support. These effects may facilitate a drastically enhanced bulk diffusion for small particles, even in comparison with defective single crystal samples. Indeed, an anomalous large oxygen uptake at low temperature was previously observed for Pd supported on γ-alumina [76, 91], although in this study no exact morphological characterization of the particles was available and no systematic temperature dependence was reported.

In view of these results, it becomes evident that studying the kinetics of test reactions such as the CO oxidation on supported nanoparticles, we have to explicitly take into account the influence of bulk diffusion, even at low reaction temperatures.

After the discussion of the adsorption and interaction of the model systems with simple adsorbates we will now proceed to the CO oxidation as a simple model reaction. Using molecular beam methods, the reaction has been studied both on single crystal Pd surfaces [112-114] and on supported Pd particles [6, 76-78, 91].

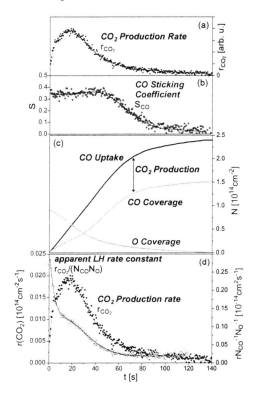

Figure 20. (a) Relative CO_2 production for oxygen-precovered Pd particles on Al_2O_3/NiAl(110) at a sample temperature of 402 K; (b) CO sticking coefficient as a function of experiment time; (c) O and CO surface coverages; (d) absolute CO_2 production rate and apparent Langmuir-Hinshelwood rate constant.

As discussed elsewhere [73], stable oxidation kinetics on the Pd particles is only obtained after saturation of the bulk oxygen reservoir. Therefore, we will consider the reaction kinetics on the fully oxygen saturated system, only. Again, we may take advantage of the single scattering conditions of the molecular beam experiment, which allow us to perform a simultaneous measurement of the reactant sticking coefficients and the CO_2 production rate. An example is displayed in Fig. 20, where the oxygen precovered surface was exposed to the CO beam. From a simultaneous measurement of the CO sticking coefficient (Fig. 20b) and the CO_2 production rate (Fig. 20a), the CO and oxygen coverages (Fig. 20c) as well as the absolute reaction rate (Fig. 20d) and eventually the apparent Langmuir-Hinshelwood (LH) rate constant (Fig. 20d) can be determined in a straightforward manner. A detailed description of the procedure can be found in the literature [76, 90, 91, 112]. From a series of similar measurements at different temperatures between 350 and 450 K, the activation barrier for the LH step as a function of oxygen or CO coverage is determined. In the limiting case of high oxygen and low CO coverage we derive an activation barrier of 57 ± 8 kJ mol^{-1}. For the reverse experiment, *i.e.* exposing a CO precovered sample to an oxygen beam, we obtain an activation barrier of 62 ± 9 kJ mol^{-1} (limiting case of high CO and low oxygen coverage).

It has to be concluded that the present results do not confirm a reduced activation barrier for the small Pd particles as has been suggested previously [76, 91]. Moreover, it is noteworthy that the activation energies are in excellent agreement with the activation barrier determined in the low temperature and high oxygen coverage regime for Pd(111) [112, 113]. Although this may not have been explicitly checked in the mentioned study [112], we may anticipate that subsurface and bulk oxygen formation play a less important role in kinetic studies on the close packed Pd(111) surface. This, however, would mean that the large oxygen uptake of the Pd nanoparticles may only have a minor influence on the kinetics of the CO oxidation reaction. In future studies this conclusion will have to be tested employing experimental techniques which allow a direct and quantitative detection of subsurface oxygen species in combination with a well controlled variation of the structural parameters of the metal deposits.

Among the different kinetic effects, which may influence the reaction kinetics on supported catalysts, surface diffusion mediated coupling between different types of adsorption sites has recently attracted considerable attention [79-81]. The presence of different types of adsorption sites in close proximity is an intrinsic property of complex catalytic systems. Such types of adsorption sites may be *e.g.* different facets on nanocrystallites, step, edge or other defect sites, interface sites located at the particle boundary or adsorption sites on the support material itself. Recently, Zhdanov and Kasemo [115] have extensively reviewed simulations of such coupling processes. They have shown that relatively moderate changes in the sticking coefficients or reaction/adsorption energetics may drastically change the reaction rate under certain conditions. Still, the experimental verification of these effects is not straightforward and very few examples exist in the literature.

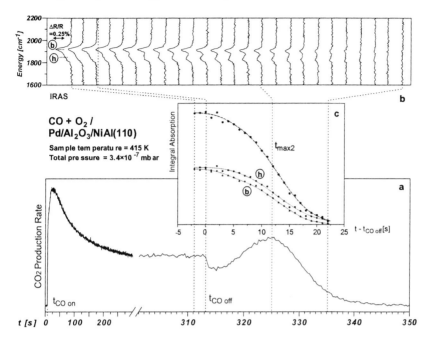

Figure 21. Simultaneous reactivity (a) and time-resolved IR reflection absorption spectroscopy experiment (b) during the transient region upon termination of the CO beam; (c) Integral absorption for the IRA spectra presented in (b) and partial absorption in the bridge (b) and hollow (h) regions.

Recently, Becker and Henry have observed a transient behavior in a beam experiment, which was related to an such heterogeneity effect [77, 78]: Upon termination of the CO flux for a system under steady-state reaction conditions, they observed a sudden decrease of the CO_2 production rate, followed by a smaller CO_2 production peak. Becker and Henry originally suggested an explanation which involves rapid desorption of CO from regular sites. CO adsorbed at defect sites, on the other hand, was assumed to be bound more strongly. Thus, these sites may remain fully saturated until the CO reservoir on the facets is largely depleted. Only after vacancies among these defect sites are generated, the reaction rate increases again, giving rise to the observed smaller maximum in the CO_2 production rate.

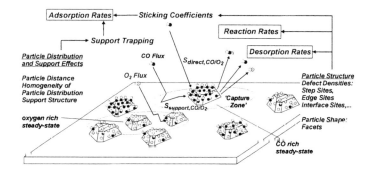

Figure 22. Schematic representation of the factors which may influence the individual adsorption and reaction behavior of the particles. Depending on their structure the steady-state adsorbate coverages of the particles may differ considerably, depending on the individual particle structure and surrounding.

Since the Pd crystallites in this study are composed of similar facets ((111) and (100)), the effect is also expected for the present model system. Indeed, we find that under certain flux conditions we observe a transient CO_2 peak (Fig. 21a), which appears very similar to the effect originally reported [77, 78]. By performing a time-resolved FTIR experiment simultaneously with the reactivity measurement, however, it can be shown that the sudden decrease in the reactions rate is not related to spontaneous desorption of CO (Fig 21b). Moreover, the CO_2 peak intensity shows a continuous dependence on the CO flux, indicating that it is not likely to be related to defect sites. A more detailed discussion of this issue can be found elsewhere [116].

In contrast to the previously suggested explanation, experimental observations indicate that the effect, at least for the present supported model catalyst, is not related to a specific type of defect sites present on the Pd particles. A detailed discussion of the effects which instead may give rise to the observed behavior can be found in the literature [116]. Here, we would like to further discuss one particular contribution, which is related to the intrinsic heterogeneity of the system and thus is likely to be characteristic for most supported model catalysts. We have to take into account that every metal particle represents a largely isolated reaction system, in a sense that adsorbate diffusion from one particle to another is strongly suppressed (due to the weak adsorption on the support). Still, the reactant flux to the particle, the adsorption and the reaction rate for each individual particle may differ considerably. Possible reasons are variations in the adsorption rate due to a different overlapping of the 'capture zones' discussed above, as well as variations in the sticking coefficients, the desorption or reaction rate induced by differences in the individual particle structure (see schematic representation in Fig. 22).

As for the CO oxidation reaction the adsorbate surface coverage may sensitively depend on the adsorbate fluxes and reaction rates, a supported particle system may under steady-state conditions be in a 'mixed' state, *i.e.* with a fraction of the metal particles may be in an oxygen-rich and a fraction in a CO-rich reaction state. A superimposition of the transient behavior of both states may explain the observed effects [116].

It remains the question in how far these contributions can be included in further microkinetic simulations to finally yield a more complete picture of the reaction kinetics of the model system. On complex catalytic surfaces, on which naturally a certain degree of heterogeneity exists, such simulations will be a necessity in order to identify and quantify the kinetic effects under discussion. As a basis, this task, however, will require a detailed knowledge of the energetic and structural properties of the model system. Here, the available data are still rather limited. Whereas some detailed and representative structural information is available from STM and TEM (transmission electron microscopy), the data basis on gas-surface interaction and surface adsorption/reactions is largely limited to single-component systems and to ideal single crystal surfaces. Starting from these simple cases, studies on supported model catalysts with a reduced and controllable level of complexity may provide a way to address such questions.

In the preceding examples we have shown how metal nanoparticles adopt themselves to reaction conditions, induced by gases and/or high temperature. It is obvious that this effect has a strong influence on the adsorption and turnover of reactants. However, even if the catalyst surface is unaffected and behaves rigid and simply "waits" for the reaction to occur, the structure of an adsorbate under reaction conditions is probably still different from UHV experiments. At the high pressure of a catalytic reaction surface coverages may be obtained that can not be reproduced by the small exposures of surface science studies (typically on the order of Langmuirs, 10^{-6} Torr sec). If the "saturation" coverage of UHV experiments is exceeded, new adsorbate structures may form, as shown *e.g.* by a high pressure STM study of CO on Pt(111) [117]. Furthermore, at high pressure weakly adsorbed species are present with much higher concentration (resulting from a higher rate of adsorption) than under low pressure when they quickly desorb [118]. In the most unfavorable case, prominent species of low pressure studies may simply be spectators under high pressure conditions.

In order to tackle this problem a surface sensitive technique is needed that allows one to monitor adsorbates under reaction conditions (~ 1 bar). A high pressure environment prevents the use of electron spectroscopies but is compatible with photon-based techniques such as infrared-visible sum frequency generation (SFG) spectroscopy. IR-vis SFG is a type of laser spectroscopy that is able to acquire vibrational spectra of adsorbates from UHV to ambient conditions. Due to its inherent surface sensitivity, surface vibrational spectra can be recorded even in the presence of a gas phase, in contrast to infrared reflection absorption spectroscopy (IRAS) that encounters problems with the excitation of rotational bands in the gas phase at pressures > 1 mbar, obstructing the surface species information. A detailed description of the SFG process can be found in the literature [118-121]. SFG is a second-order nonlinear optical process which involves the mixing of tunable infrared (ω_{IR}) and visible light (ω_{vis}) to produce a sum frequency output ($\omega_{SFG} = \omega_{IR} + \omega_{vis}$). The process is only allowed in a medium without inversion symmetry (in the electric dipole approximation), *e.g.* at surfaces where the inversion symmetry is broken. The dominant SFG signal is hence generated by the modes of the adsorbate, while the centrosymmetric bulk of face-centered cubic metals and an isotropic gas phase give nearly zero contribution to the SFG signal.

It was already mentioned in the introduction that single crystals can not fully represent supported metals. In order to include size and electronic effects, surface rearrangements, etc. in a pressure-dependent study of gas adsorption, SFG spectroscopy should be ideally carried out on supported nanoparticles. Although the applicability of SFG spectroscopy to nano-structured supported catalysts has been questioned for several reasons (scattering of laser beams on rough surfaces, disordered adsorbates, small total coverages), we recently succeeded to obtain SFG spectra from CO adsorbed on supported Pd nanoclusters [10]. Pd/Al$_2$O$_3$/NiAl(110) model catalysts were prepared as described above, and transferred under vacuum to a SFG-compatible UHV high pressure cell. Details about the sample preparation and cell design are published elsewhere [10, 119].

Figure 23 shows SFG spectra of CO adsorption on alumina supported Pd particles of 3 nm mean size (about 300 atoms per particle), grown at 90 K (particle density 10^{13}/cm^2). According to LEED and STM measurements [3, 122], the particle surface exhibits a high defect density as a result of the low growth temperature. Figure 23a displays SFG spectra taken at 190 K. Two peaks are clearly identified at 10^{-7} mbar CO and, according to IRAS results on Pd single crystals [123-125], they originate from bridge bonded CO at 1976 cm^{-1} and from terminal (on-top) CO at 2106 cm^{-1}. If one takes the integrated SFG signal intensity as a measure of the ratio of on-top to bridge-bonded CO, a value of about 0.5 is obtained. However, this value is only taken as an estimate here, since the SFG intensity can not be easily correlated with the concentration of a particular surface species. The coexistence of bridge and on-top bonded CO, a situation which is not found on Pd single crystals, reflects the defective structure of the Pd nanoparticles.

In fact, the site occupancy on the Pd aggregates compares best with defect-rich single crystal Pd(111) or rough Pd thin films [123, 124, 126]. The difference between supported Pd aggregates and Pd(111) could result from a reduced lateral CO interaction on the nanoparticles or from the presence of additional crystal planes.

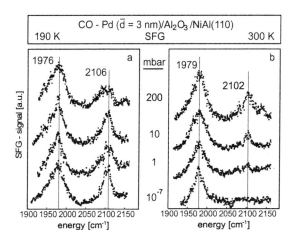

Figure 23. SFG spectra of CO adsorption on a Pd/Al$_2$O$_3$/NiAl(110) model catalyst at 190 K (a) and 300 K (b) (3 nm Pd particles grown at 90 K). The observed resonances are characteristic of bridge bonded CO only. The terminally bonded CO state can be re-populated at ≥ 1 mbar.

Increasing the pressure to 1, 10 and 200 mbar CO had only a small effect on the peak frequencies and also the ratio of on-top vs. bridge CO was nearly unchanged (Fig. 23a). If the sample temperature is raised to 300 K (Fig. 23b), which is above the desorption temperature of on-top CO on Pd clusters under UHV conditions [3], bridge-bonded CO is the only species that can be observed by SFG at 10^{-7} mbar. However, the on-top adsorption sites can be populated at p ≥ 1 mbar, and at 200 mbar CO a relative on-top/bridge ratio of ~ 0.5 can be obtained, similar to the value at 190 K.

Comparing Figs. 23a and 23b illustrates the pressure- and temperature-dependent adsorption site occupancy of CO on Pd/Al$_2$O$_3$. While at 190 K an extrapolation of the 10^{-7} mbar spectrum to 200 mbar would lead to a satisfactory result, for a temperature of 300 K the prediction would be wrong (absence vs. presence of on-top CO). In addition, the adsorption site occupancy of Pd nanoparticles is influenced by the particle size and surface structure. If the same experiment is carried out on Pd particles grown at 300 K that mainly exhibit well-developed (111) surface facets, the adsorption behavior is again different [10]. These measurements could be repeated several times indicating that significant structural changes of the particles were absent - which seems reasonable at the low temperatures applied. Our adsorption study clearly demonstrates the need to characterize adsorbates under reaction conditions. If this is carried out under several bars of pressure and at higher temperature even more pronounced effects are expected.

6.   PHOTOCHEMISTRY ON METAL AGGREGATES

In addition to thermal studies of reactivity, nano-sized transition metal aggregates lend themselves to photochemical studies, in which the influence of size and morphology on the photochemistry on the surface of the particles is explored [6, 9, 127]. By changing the formation conditions the geometrical and electronic structure of the metal aggregates can be controlled. The cleavage of an intra-molecular bond by laser irradiation is of particular interest for the rather inert molecule methane. Methane will become an increasingly important raw material, when natural petrol sources run out [128]. The primary step of methane conversion to methanol or other hydrocarbons [129] is the cleavage of the C-H bond requiring a dissociation energy of 4.5 eV [130]. Even a higher energy of 8.5 eV is necessary for a photochemical cleavage as the first optical allowed transition leading to bond breaking is situated in the vacuum UV.

On the other hand, it has been recently discovered that methane physisorbed on Pt(111) and Pd(111) single crystal surfaces is readily dissociated into methyl and hydrogen by 193-nm (6.42 eV) ArF excimer laser irradiation, [129, 131, 132]. This strong shift of more than 2.1 eV is surprising in view of the low methane surface interaction of around 230-250 meV. Controlled experiments showed that the electronic states of the metal surfaces play an important role in the photo excitation of methane [131, 133].

By using the same systems as discussed before, the influence of Pd cluster size and morphology on the photochemistry of methane was investigated [134]. The average diameter ranged from 1 to 7 nm as determined by spot profile analysis (SPA)-LEED analysis. We found that the photo reactions strongly depend on the cluster size. In these experiments deuterated methane was photo dissociated by irradiation from excimer laser run at a repetition rate of 4 Hz, a fluence of 2.5 mJ/cm^2 at a photon energy of 6.4 eV, for further experimental details we refer to [135].

As the surface area grows with increasing Pd-deposition a temperature programmed desorption (TPD) spectrum of a saturation coverage of deuterated methane was recorded as a measure of the total surface area available for methane adsorption prior to each photochemistry experiment. For a bare alumina surface no methane desorption peak was observed. The maximum temperature for the TPD spectra did not exceed 300 K. The maximum of the main desorption peak of undissociated methane shifts continuously towards lower temperature with decreasing cluster size. The TPD spectra of the smallest aggregates (1 nm) were shifted by 10 K as compared to the largest ones (7 nm) indicating a decrease of molecule surface interaction with decreasing cluster size. The peak area of the feature of CD_4 (m = 20 amu) was further used as reference for the total changes in methane concentration due to photon induced dissociation and desorption.

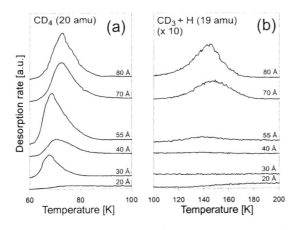

Figure 24. a) Series of (m = 20 amu) TPD spectra after exposure to 1.8×10^{19} photons/cm^2 of 6.4 eV depending on Pd-cluster size. b) number of (m = 19 amu) TPD spectra (recombinative desorption of deuterated methyl plus hydrogen) at otherwise same conditions as in a).

Starting with saturation coverage of deuterated methane the system was exposed to a fixed number of photons impinging on the surface. Figure 24a shows a number of TPD spectra after exposure to 1.8×10^{19} photons/cm^2 as a function of Pd aggregate size and is due to undissociated deuterated methane. The maximum of the TPD spectra for (m = 20 amu) turned out to be very sensitive to the coverage, the average cluster size and also coadsorbates like atomic hydrogen from photo dissociation which explains the different shifts of the main feature in those TPD spectra.

In Fig. 24b the recombinative desorption of CD_3H (m = 19 amu) is displayed. The hydrogen on the surface results from residual gas within the chamber and is due to the high sorptivity of palladium with respect to hydrogen adsorption. As it turned out the background free TPD spectra of CD_3H were valuable for the photo dissociation analysis particularly, in case only small photo dissociation rates were observed. No other reaction products have been observed except for the mentioned recombinative desorption of CD_3+H/D.

The residual amount of methane (Fig. 24a) after laser light exposure if related to the initial coverage is a direct measure for the photon induced depletion rate. This loss is mainly contributed by photo desorption and is plotted in dependence of the cluster size in Fig. 25a. Obviously, for small clusters photo desorption (the negligible contribution of photo dissociated methane will be discussed below) is very pronounced, characterized by a cross section of 1×10^{-19} cm^2. For increasing cluster size the photon induced desorption becomes more and more ineffective and the cross section reaches a value below 1×10^{-20} cm^2, which is also typical from single crystal measurements [136].

The recombinative desorption of CD_3H at around 150 K (Fig 24b) related again to the initial CD_4 coverage before laser light exposure describes the photon reaction rate and is plotted versus the Pd aggregate size in the Fig. 25b. Evidently only above a threshold size of 40 Å photon dissociation starts (cross section of 10^{-22} cm^2) and reaches a value of nearly 10% of the adsorbed methane for the biggest cluster size investigated. This value cannot be exceeded by further photon exposure and is still far below the observed turnover of 80% for Pd(111) [131, 137]. The incomplete turnover of the adsorbed methane by photo dissociation is attributed in case of single crystals to the evolving product concentration on the metal surface [138]. According to IR-measurements the increasing CH_3-coverage reduces the interaction of methane with the surface. As seen above, this self quenching is also observed for photochemistry on the Pd-clusters. However, it occurs at much lower concentrations. For the largest aggregates in our experiments for which the overall photo reaction cross section approaches the value of 7×10^{-21} cm^2 as for the single crystal surface, but *ca.* 85% of the initial coverage remained unreacted.

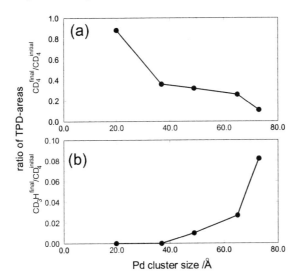

Figure 25. a) Plot of overall reaction loss of undissociated CD_4 normalized to the initial coverage as a function of Pd-cluster size. b) Plot of CD_3H formation ratio in respect to the initial coverage depending on the Pd-cluster size.

Single crystal experiments have shown that the initial excitation step is likely to occur within the adsorbate. Two models are actually discussed in the literature to explain the laser induced dissociation of methane on Pt(111) and Pd(111) [131, 137, 139].

The first model [131] is supported by ab initio cluster calculations for Pt (clusters from 1-10 atoms) [139]. It discusses a mixing of the unoccupied anti-bonding Rydberg state of methane (gas phase value: 8.5 eV) with unoccupied metal states to account for a broadening and a shift of the first electronically excited state so that the photo dissociation at 6.4 eV becomes possible. The excitation energy of this state leading to dissociation to CH_3 + H is strongly depending on the cluster size as the electron redistribution over the surrounding metal plays an important role to stabilize this charge transfer state [139]. Within this model a decrease of the cluster size causes a reduction of the delocalisation resulting in a shift of the excitation energy towards higher energies.

The second model argues that a reduction of the ionization potential of methane due to the interaction with the metal accounts for the possibility to dissociate methane at 6.4 eV [137]. Starting from the vacuum level of Pd the Fermi level of the metal is situated 5.6 eV below the vacuum level according to work function measurements [140]. With reference to the vacuum level the HOMO (highest occupied molecular orbital) state of methane is 12.6 eV below the vacuum level considering the gas phase value [140]. A further attractive force due to the image charge resulting from the ionic excited state further stabilizes this level by a shift towards the vacuum level [137]. Slab LDA calculations on $CH_4/Pd(111)$ revealed that the equilibrium height between the Pd plane to the C nucleus of 3 Å above the surface [137] allows to use the image charge model of Jennison et al. [141] from which an estimate of 1.9 eV can be made for the attractive force. Assuming a low electron tunneling barrier between adsorbate and substrate one obtains an approximate minimum excitation energy of 5.1 eV to produce the dissociative state of cationic methane. Indeed no photo dissociation could be observed on single crystals at excitation energies of 5.0 eV [129, 132]. The image potential will be strongly depending on the electronic structure and thus on the cluster size which explains a decrease in photo dissociation probability. Furthermore, as the image-potential changes scale with $1/r$ (r being the equilibrium height of CH_4 above Pd), the dissociation efficiency might be strongly influenced by the cluster size which governs the actual interaction between methane and the substrate. This is supported by the observation that the peak maxima of TPD spectra of the undissociated methane shift towards lower temperatures with decreasing cluster size indicative for a diminishing interaction between the molecules and the surface thus increasing molecule-surface distance.

Intuitively, one expects an increased lifetime of the excited state concomitant with the quantisation of the electronic states when reducing the cluster size which would increase the photo dissociation probability. However, this effect appears to be less important than the influence of charge delocalisation within the cluster on the photo dissociation threshold.

In summary, photo dissociation, and photo desorption of methane at $\lambda = 193$ nm on the Pd-clusters of various sizes on a thin epitaxial Al_2O_3 support show very pronounced size effects and remarkable differences from the Pd(111) single crystal results. These results strongly suggest that metal clusters of different sizes and shapes provide a unique way for controlling both photochemical and thermal reactions of hydrocarbons at metal surfaces.

7.   SYNOPSIS

We have demonstrated in the present review how surface science tools and ideas can be used to prepare and study model systems for heterogeneous catalysts. In our view it seems possible to bridge the so called materials as well as the pressure gaps between surface science and catalysis. This opens fruitful avenues for surface science. Surface science which was, when Wyn Roberts entered this field, in its infancy has developed into a mature science thanks to scientists of his stature, and there is still room for the development of strategies as well as methods to let our field prosper and keep it very much alive in the future.

8.   ACKNOWLEDGEMENTS

We are grateful to our coworkers, whose names appear in the references, for their contributions to the work summarized in this paper. A number of funding agencies have helped to perform the research described here. Among them are Max-Planck-Gesellschaft, Deutsche Forschungsgemein-schaft, Bundesministerium für Bildung und Forschung, COST, European Union, German Israeli Foundation, as well as the NEDO International Joint Research Grant on Photon and Electron Controlled Surface Processes. We thank them for their support.

9.   REFERENCES

1   12th International Congress on Catalysis, Studies in Surface Science and Catalysis, edited by A. Corma, F.V. Melo, S. Mendioroz and J.L.G. Fierro (Elsevier, Amsterdam, 2000).
2   H.-J. Freund, Faraday Disc., 114 (1999) 1.
3   M. Bäumer and H.-J. Freund, Progr. Surf. Sci., 61 (1999) 127.
4   H.-J. Freund, M. Bäumer and H. Kuhlenbeck, Adv. Catal., 45 (2000) 412.
5   D.W. Goodman, Surf. Rev. Lett., 2 (1995) 9.
6   C.R. Henry, Surf. Sci. Rep., 31 (1998) 231.
7   U. Diebold, J.-M. Pan and T.E. Madey, Surf. Sci., 333 (1995) 845.
8   P.J. Møller and J. Nerlov, Surf. Sci., 309 (1994) 591.
9   H.-J. Freund, Angew. Chem. Int. Ed. Engl., 36 (1997) 452.
10   T. Dellwig, G. Rupprechter, H. Unterhalt and H.-J. Freund, Phys. Rev. Lett., 85 (2000) 776.
11   C.T. Au and M.W. Roberts, Nature, 319 (1986) 206.
12   M.W. Roberts, Appl. Surf. Sci., 52 (1991) 133.
13   R.B. Moyes and M.W. Roberts, J. Catal., 49 (1977) 216.
14   M.W. Roberts, Chem. Soc. Rev., 18 (1989) 451.
15   M.W. Roberts, Chem. Soc. Rev., 25 (1996) 437.
16   R.W. Joyner and M.W. Roberts, Chem. Phys. Lett., 29 (1974) 447.
17   M.W. Roberts and R.S.C. Smart, Surf. Sci., 100 (1980) 590.
18   S.J. Atkinson, C.R. Brundle and M.W. Roberts, Faraday Disc., 58 (1974) 62.
19   M.W. Roberts and R.S.C. Smart, J. Chem. Soc., Faraday Trans. 1, 80 (1984) 2957.
20   M.W. Roberts and R.S.C. Smart, Surf. Sci., 151 (1985) 1.
21   M.W. Roberts and R.S.C. Smart, Chem. Phys. Lett., 69 (1980) 234.
22   H. Schmalzried, Chemical Kinetics of Solids, (Verlag Chemie, Weinheim, 1995).
23   R.K. Grasselli and J.D. Burrington, Adv. Catal., 30 (1981) 133.
24   F. Cavani, G. Centi, F. Trifiro and R.K. Grasselli, Catal. Today, 3 (1988) 185.

25   P.A. Agaskar, L. DeCaul and R.K. Grasselli, Catal. Lett., 23 (1994) 339.

26   B. Tepper, PhD-Thesis, Freie Universität, Belrin, in preparation.

27   A.-C. Dupuis, PhD-Thesis, Humboldt Universität, Berlin, Germany, in preparation.

28   K. Hermann, M. Witko, R. Druzinic, A. Chakrabarti, B. Tepper, M. Elsner, A. Gorschlüter, H. Kuhlenbeck and H.-J. Freund, J. Electron Spectrosc. Relat. Phenom., 99 (1999) 245.

29   M. Figlarz, Progr. Solid State Chem., 19 (1989) 1.

30   G. Renaud, Surf. Sci. Rep., 32 (1998) 1.

31   F. Rohr, M. Bäumer, H.-J. Freund, J.A. Mejias, V. Staemmler, S. Müller, L. Hammer and K. Heinz, Surf. Sci., 372 (1997) L 291.

32   F. Rohr, M. Bäumer, H. J. Freund, J.A. Mejias, V. Staemmler, S. Müller, L. Hammer and K. Heinz, Surf. Sci., 389 (1997) 391.

33   S.K. Shaikhoutdinov and W. Weiss, Surf. Sci., 432 (1999) L627.

34   X.-G. Wang, W. Weiss, S.K. Shaikhoutdinov, M. Ritter, M. Petersen, F. Wagner, R. Schlögl and M. Scheffler, Phys. Rev. Lett., 81 (1998) 1038.

35   N.M. Harrison, X.-G. Wang, J. Muscat and M. Scheffler, Faraday Disc., 114 (1999) 305.

36   K. Wolter, D. Scarano, J. Fritsch, H. Kuhlenbeck, A. Zecchina and H.-J. Freund, Chem. Phys. Lett., 320 (2000) 206.

37   B. Dillmann, F. Rohr, O. Seiferth, G. Klivenyi, M. Bender, K. Homann, I.N. Yakovkin, D. Ehrlich, M. Bäumer, H. Kuhlenbeck and H.-J. Freund, Faraday Disc. 105 (1996) 295.

38   C. Xu, B. Dillmann, H. Kuhlenbeck and H.-J. Freund, Phys. Rev. Lett., 67 (1991) 3551.

39   M. Pykavy, V. Staemmler, O. Seiferth and H.-J. Freund, Surf. Sci., 479 (2001) 11

40   A.F. Carley, D.E. Gallagher and M.W. Roberts, Spectrochim. Acta, 43 (1987) 1447.

41   R.G. Copperthwaite, P.R. Davies, M.A. Morris, M.W. Roberts and R.A. Ryder, Catal. Lett., 1 (1988) 11.

42   H.-J. Freund and M.W. Roberts, Surface Science Reports, 25 (1996) 225.

43   O. Seiferth, K. Wolter, B. Dillmann, G. Klivenyi, H.-J. Freund, D. Scarano and A. Zecchina, Surf. Sci., 421 (1999) 176.

44   M.-C. Wu and P.J. Møller, Surf. Sci., 224 (1989) 250.

45   H. Poppa, Catal. Rev. Sci. Eng., 35 (1993) 359.

46   M. Heemeier, M. Frank, J. Libuda, K. Wolter, H. Kuhlenbeck, M. Bäumer and H.-J. Freund, Catal. Lett., 68 (2000) 19.

47   J. Libuda, F. Winkelmann, M. Bäumer, H.-J. Freund, T. Bertrams, H. Neddermeyer and K. Müller, Surf. Sci., 318 (1994) 61.

48   N. Ernst, B. Duncombe, G. Bozdech, M. Naschitzki and H.-J. Freund, Ultramicroscopy, 79 (1999) 231.

49   K.H. Hansen, T. Worren, S. Stempel, E. Lægsgaard, M. Bäumer, H.-J. Freund, F. Besenbacher and I. Stensgaard, Phys. Rev. Lett., 83 (1999) 4120.

50   N. Nilius, A. Cörper, G. Bozdech, N. Ernst and H.-J. Freund, Progr. Surf. Sci., 67 (2001) 99.

51   G.L. Kellogg, Surf. Sci. Rep., 21 (1994) 1.

52   A. Cörper, G. Bozdech, N. Ernst and H.-J. Freund (unpublished).

53   A. Bogicevic and D.R. Jennison, Phys. Rev. Lett., 82 (1999) 4050.

54   A. Piednoir, E. Perrot, S. Granjeaud, A. Humbert, C. Chapon and C.R. Henry, Surf. Sci., 391 (1997) 19.

55   M. Methfessel, D. Hennig and M. Scheffler, Phys. Rev. B, 46 (1992) 4816.

56   A. Sandell, J. Libuda, P.A. Brühwiler, S. Andersson, M. Bäumer, A.J. Maxwell, N. Mårtensson and H.-J. Freund, Phys. Rev. B, 55 (1997) 7233.
57   A. Bettac, L. Köller, V. Rank and K.H. Meiwes-Broer, Surf. Sci., 402 (1998) 475.
58   M. Valden, X. Lai and D.W. Goodman, Science, 281 (1998) 1647.
59   B.C. Stipe, M.A. Rezaei and W. Ho, Science, 280 (1998) 1732.
60   J.H. Coombs, J.K. Gimzewski, B. Reihl, J.K. Sass and R.R. Schlittler, J. Microscopy, 152 (1988).
61   R. Berndt, in Scanning Probe Microscopy, edited by R. Wiesendanger, Springer Series Nanoscience and Technology (Springer, Berlin, Germany, 1998), page 97.
62   N. Nilius, N. Ernst and H.-J. Freund, Phys. Rev. Lett., 84 (2000) 3994.
63   Optical Properties of Metal Clusters, edited by U. Kreibig and M. Vollmer, Springer Series in Materials Science, Vol. 25 (Springer, Berlin, Germany, 1995).
64   N. Nilius, N. Ernst, H.-J. Freund and P. Johansson, Phys. Rev. B, 61 (2000) 12682.
65   A. Liebsch, Phys. Rev. B, 48 (1993) 11317.
66   H. Schlienz, M. Beckendorf, U.J. Katter, T. Risse and H.-J. Freund, Phys. Rev. Lett., 74 (1995) 761.
67   U.J. Katter, T. Hill, T. Risse, H. Schlienz, M. Beckendorf, T. Klüner, H. Hamann and H.-J. Freund, J. Phys. Chem. B, 101 (1997) 552.
68   T. Risse, T. Hill, J. Schmidt, G. Abend, H. Hamann and H.-J. Freund, J. Chem. Phys., 108 (1998) 8615.
69   T. Hill, M. Mozaffari-Afshar, J. Schmidt, T. Risse, S. Stempel, M. Heemeier and H.-J. Freund, Chem. Phys. Lett., 292 (1998) 524.
70   T. Hill, S. Stempel, T. Risse, M. Baeumer and H.-J. Freund, J. Magn. Mag. Mater., 198-199 (1999) 354.
71   T. Hill, PhD-Thesis, Ruhr-Universität Bochum, 1998.
72   D.R. Rainer and D.W. Goodman, in Chemisorption and Reactivity on Supported Clusters and Thin Films, edited by R. M. Lambert and G. Pacchioni, Series E, Vol. 331 (Kluwer, Dordrecht, 1997), page 27.
73   I. Meusel, J. Hoffmann, J. Hartmann, M. Heemeier, M. Bäumer, J. Libuda and H.-J. Freund, Catal. Lett., (submitted).
74   J. Libuda, I. Meusel, J. Hartmann and H.-J. Freund, Rev. Sci. Instrum., (accepted).
75   E. Gillet, S. Channakhone, V. Matolin and M. Gillet, Surf. Sci., 152/153 (1985) 603.
76   I. Stará, V. Nehasil and V. Matolin, Surf. Sci., 365 (1996) 69.
77   C. Becker and C.R. Henry, Surf. Sci., 352 (1996) 457.
78   C. Becker and C.R. Henry, Catal. Lett., 43 (1997) 55.
79   V.P. Zhdanov and B. Kasemo, J. Catal., 170 (1997) 377.
80   V.P. Zhdanov and B. Kasemo, Surf. Sci., 405 (1998) 27.
81   V.P. Zhdanov and B. Kasemo, Phys. Rev. B, 55 (1997) 4105.
82   C.R. Henry, Surf. Sci., 223 (1991) 519.
83   D.A. King and M.G. Wells, Proc. Roy. Soc. Ser. A, 339 (1974) 245.
84   D.A. King and M.G. Wells, Surf. Sci., 29 (1972) 454.
85   M. Bäumer, J. Libuda and H.-J. Freund, in Chemisorption and Reactivity on Supported Clusters and Thin Films, edited by R. M. Lambert and G. Pacchioni, NATO ASI Series E, Vol. 331 (Kluwer, Dordrecht, 1997), page 61.
86   J. Libuda, PhD-Thesis, Ruhr-Universität Bochum, 1996.
87   T. Engel, J. Chem. Phys., 69 (1978) 373.
88   C.T. Rettner, D.J. Auerbach, J.C. Tully and A.W. Kleyn, J. Phys. Chem., 100 (1996) 13021.
89   J.A. Barker and D.J. Auerbach, Surf. Sci. Rep., 4 (1985) 1.

90   T. Dellwig, J. Hartmann, J. Libuda, I. Meusel, G. Rupprechter, H. Unterhalt and H.-J. Freund, J. Mol. Catal., (accepted).
91   I. Stará, V. Nehasil and V. Matolin, Surf. Sci., 331 (1995) 173.
92   D.L. Weissman, M.I. Shek and W.E. Spicer, Surf. Sci., 92 (1980) L59.
93   P. Legare, L. Hilaire, G. Maire, G. Krill and A. Amamou, Surf. Sci., 107 (1981) 533.
94   D.L. Weissman-Wenicur, M.L. Shek, P.M. Stefan, I. Lindau and W.E. Spicer, Surf. Sci., 127 (1983) 513.
95   L. Surnev, G. Bliznakov and M. Kiskinova, Surf. Sci., 140 (1984) 249.
96   R. Imbihl and J.E. Demuth, Surf. Sci., 173 (1986) 395.
97   B. Oral and R.W. Vook, Appl. Surf. Sci., 29 (1987) 20.
98   M. Milun, P. Pervan, M. Vaajic and K. Wandelt, Surf. Sci., 211-212 (1989) 887.
99   X. Guo, A. Hoffman and J.T. Yates, Jr., J. Chem. Phys., 90 (1989) 5787.
100   B.A. Banse and B.E. Koel, Surf. Sci., 232 (1990) 275.
101   V.A. Bondzie, P. Kleban and D.J. Dwyer, Surf. Sci., 347 (1996) 319.
102   P. Sjövall and P. Uvdal, Chem. Phys. Lett., 282 (1998) 355.
103   P. Sjövall and P. Uvdal, J. Vac. Sci. Technol. A, 16 (1998) 943.
104   E.H. Voogt, A.J.M. Mens, O.L.J. Gijzeman and J.W. Geus, Surf. Sci., 373 (1997) 210.
105   W. Huang, R. Zhai and W. Bao, Surf. Sci., 439 (1999) L803.
106   F.P. Leisenberger, G. Koller, M. Sock, S. Surnev, M.G. Ramsey, F.P. Netzer, B. Klötzer and K. Hayek, Surf. Sci., 445 (2000) 380.
107   M. Eriksson and L.G.M. Pettersson, Surf. Sci., 311 (1994) 139.
108   H. Graoui, S. Giorgio and C.R. Henry, Surf. Sci., 417 (1998).
109   M. Eriksson, L. Olsson, U. Helmersson, R. Erlandsson and L.-G. Ekedahl, Thin Solid Films, 342 (1999) 297.
110   S. Ladas, R. Imbihl and G. Ertl, Surf. Sci., 219 (1989) 88.
111   M.R. Bassett and R. Imbihl, J. Chem. Phys., 93 (1990) 811.
112   T. Engel and G. Ertl, J. Chem. Phys., 69 (1978) 1267.
113   T. Engel and G. Ertl, in The Chemical Physics of Solid Surfaces and Heterogeneous Catalysis, edited by D. A. King and D. P. Woodruff, Vol. 4 (Elsevier, 1982), page 73.
114   I.Z. Jones, R.A. Bennett and M. Bowker, Surf. Sci., 439 (1999) 235.
115   V.P. Zhdanov and B. Kasemo, Surf. Sci. Rep., 39 (2000) 25.
116   J. Libuda, I. Meusel, J. Hoffmann, J. Hartmann, M. Piccolo, C.R. Henry and H.-J. Freund, J. Chem. Phys., 114 (2001) 4669
117   J.A. Jensen, K.B. Rider, M. Salmeron and G.A. Somorjai, Phys. Rev. Lett., 80 (1998) 1228.
118   G.A. Somorjai and G. Rupprechter, J. Phys. Chem. B, 103 (1999) 1623.
119   G. Rupprechter, T. Dellwig, H. Unterhalt and H.-J. Freund, Topics in Catalysis, 15 (2001) 19
120   Y.R. Shen, Surf. Sci., 299/300 (1994) 551.
121   J. Miragliotta, P. Rabinowitz, S.D. Cameron and R.B. Hall, Appl. Phys. A, 51 (1990) 221.
122   M. Bäumer, J. Libuda, A. Sandell, H.-J. Freund, G. Graw, T. Bertrams and H. Neddermeyer, Ber. Bunsenges. Phys. Chem., 99 (1995) 1381.
123   F.M. Hoffmann, Surf. Sci. Rep., 3 (1983) 103.
124   M. Tüshaus, W. Berndt, H. Conrad, A.M. Bradshaw and B. Persson, Appl. Phys. A, 51 (1990) 91.
125   W.K. Kuhn, J. Szanyi and D.W. Goodman, Surf. Sci. Lett., 274 (1992) L611.

126   A.M. Bradshaw and F.M. Hoffmann, Surf. Sci., 52 (1975) 449.

127   C.T. Campbell, Surf. Sci. Rep., 27 (1997) 1.

128   K. Seshan and J.A. Lercher, in Carbon Dioxide Chemistry: Environmental Issues, edited by J. Paul and C.-M. Pradier (The Royal Society of Chemistry, 1994).

129   Y.A. Gruzdkov, K. Watanabe, K. Sawabe and Y. Matsumoto, Chem. Phys. Lett., 227 (1994) 243.

130   K.-P. Huber and G. Herzberg, Molecular Spectra and Molecular Structure, IV. Constants of Diatomic Molecules, (Van Nostrand Reinhold, New York, 1979).

131   Y. Matsumoto, Y.A. Gruzdkov, K. Watanabe and K. Sawabe, J. Chem. Phys., 105 (1996) 4775.

132   K. Watanabe and Y. Matsumoto, Surf. Sci., 390 (1997) 250.

133   K. Watanabe, K. Sawabe and Y. Matsumoto, Phys. Rev. Lett., 76 (1996) 1751.

134   K. Watanabe, Y. Matsumoto, M. Kampling, K. Al-Shamery and H.-J. Freund, Angew. Chem. Int. Ed., 38 (1999) 2192.

135   M. Kampling, PhD-Thesis, Freie Universität, 2000.

136   K. Watanabe, PhD-Thesis, Institute of Molecular Physics, Okasaki, 1997.

137   D.R. Jennison, private communication.

138   J. Yoshinobu, H. Ogasawa and M. Kawai, Phys. Rev. Lett., 75 (1995) 2176.

139   Y. Akinaga, T. Taketsugu and T. Hirao, J. Chem. Phys., 107 (1997) 415.

140   Handbook of Chemistry and Physics, edited by D.R. Lide (CRC Press, Boca Raton, 1997-1998).

141   D.R. Jennison, E.B. Stechel, A.R. Burns and Y.S. Li, Nucl. Instr. Meth. Phys. Res., B101 (1995) 22.

Chapter 7

SURFACE CHEMISTRY OF MODEL OXIDE-SUPPORTED METAL CATALYSTS: AN OVERVIEW OF GOLD ON TITANIA

Douglas C. Meier, Xiaofeng Lai, and D. Wayne Goodman*

Key words: Model catalysts, supported metal clusters, gold clusters, scanning tunneling microscopy, titania, size-dependent electronic structures, nanoclusters, Scanning tunneling spectroscopy, band gaps in metal clusters, quantum size effect, gas-surface reactions, CO oxidation, strong metal-support interaction, Ostwald ripening .

Abstract: Insight into atomic-level surface chemistry of metals on oxide surfaces is vital to understanding heterogeneous catalysis. In this work, scanning tunneling microscopy (STM) in conjunction with traditional surface science techniques is used to study gold metal clusters on a planar titania support. When Au is vapor-deposited onto $TiO_2(110)$ under ultra-high vacuum (UHV) conditions, it grows as three-dimensional (3D) hemispherical clusters on $TiO_2(110)$, indicative of a Volmer-Weber (VW) growth mode. However, at very low coverages (0.01-0.05 ML), quasi-two-dimensional (quasi-2D) Au clusters are observed. Annealing studies reveal that Au clusters form large microcrystals with well-defined hexagonal shapes above 1000 K. Furthermore, an oxygen-induced cluster ripening is observed after $Au/TiO_2(110)$ is exposed to 10.00 Torr O_2 in an elevated pressure reactor. The morphological change of Au clusters induced by O_2 exposure suggests O_2 chemisorption on both the clusters and the TiO_2 substrate at room temperature. Au clusters exhibit a clear bimodal size distribution after O_2 exposure due to Ostwald ripening: some clusters increase in size while others shrink.

Surface Chemistry and Catalysis, Edited by Carley *et al.*
Kluwer Academic/Plenum Publishers, New York, 2002

147

1. INTRODUCTION

The study of model catalysts is a very important research area for surface science. An essential focus of catalytic research is to understand the relationships among the elemental composition, electronic structures, and geometric structures of surfaces, correlating them to specific chemical processes. With the advance in the last thirty years of a variety of ultra-high vacuum (UHV) surface science techniques, powerful methods for examining surfaces at the atomic level have been made available. Accordingly, an integrated approach to catalytic research that combines modern surface techniques with traditional experimental methods can greatly improve our understanding of a wide range of surface catalytic reactions.

In recent years, substantial efforts have been focused on the chemical and physical properties of supported metal clusters and their interactions with oxide supports [1-3]. Metal/oxide systems received considerable attention from both fundamental and applied perspectives due to their importance in catalysis. However, there are still two limitations on the application of UHV surface science techniques to "real world" oxide-supported metal catalysts: the pressure and material gaps [4,5]. The pressure gap problem originates from the difference between the pressures at which almost all surface science methods can work ($P < 10^{-7}$ torr) and those employed in typical industrial processes (atmospheric pressure or higher). Hence, the validity of a correlation between observations under these disparate conditions is in question. The other discrepancy between surface science studies and "real world" studies is the use in surface science of well-ordered single crystals as model samples, while real catalysts represent far more complex surfaces, such as dispersed metal clusters supported on an inert carrier. The differences in the nature of the bulk and highly dispersed metals are the basis for the material gap problem.

1.1 Model Catalysts and Surface Science

Since the mid-1980s, researchers have paid much attention to tackling these problems. Certain features characteristic of real catalysts make the prospect of obtaining a detailed picture of their chemical properties very challenging indeed. First, it is generally impossible to characterize the electronic and structural properties of porous oxide-supported catalysts due to surface charging of the insulating oxide material [5], which leads to poor spectrometer sensitivity and can modify the chemistry of the sample. Second, the structural and chemical properties of many heterogeneous catalysts are non-uniform due to wide particle size ranges and uncertain morphology. Also, problems related to surface contamination and undesirable impurities hinder our efforts to fundamentally describe the nature of catalytic processes.

Considerable progress has recently been made employing planar model catalysts [6,7] composed of well-ordered extended surfaces of catalytic metals prepared under UHV conditions. Surface science studies have been focused on investigating the structural, electronic and chemisorptive properties of low-index single crystal facets, as shown in Fig. 1A. Atomic structure and electronic properties of many metal surfaces are now accessible [8]. An inherent advantage of this approach is the uniform results relating to the physical and chemical properties of surface adsorbate species from some particular binding sites. Additionally, the kinetics of the reaction can be monitored. Some important intermediates, promoters and inhibitors may be identified and correlated to the kinetics results [8,9].

However, there are still certain shortcomings associated with the application of single crystal models. Some important factors special to oxide-supported catalysts, such as metal-support interactions and particle size effects, cannot be clearly settled. It is well documented that both reactivity and selectivity of a catalyst can depend upon the particle size [10].

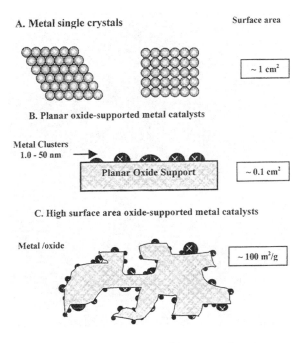

Figure 1. Schematic representation of three types of catalysts.

    The catalyst support is not always a simple mechanical support for the metal particles, since it can modify the electronic properties of the particles and in some cases migrate over the particles as in the case of the strong metal-support interaction (SMSI) [11].

    To bridge the material gap between the single crystal model catalysts and the "real world" oxide-supported catalysts, synthesis and employment of planar oxide supported catalysts can be introduced (Fig. 1B). In this type of model catalyst, a conductive single crystal of oxide material (or oxide thin film) is used as the planar support. Metals of interest are then vapor-deposited onto the support and analyzed using UHV surface science techniques. These models allow for the investigation of metal-support interactions and particle size effects while still featuring many of the advantages of single crystals over powder catalysts [12,13]. Another advantage offered by utilizing model catalysts is that, in principle, control can be exercised over the clusters themselves in terms of size, structure, and coverage. In addition, because the sample is flat and conductive, it is suitable for STM and atomic force microscopy (AFM) studies [14,15].

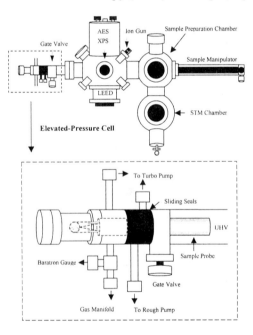

Figure 2. Schematic drawings of UHV surface analysis system and detailed view of the elevated pressure cell/reactor.

Fig. 1 is a schematic representation comparing single crystal, planar oxide supported, and high surface-area oxide-supported metal catalysts. The replacement of a real support by a planar model has many advantages since it is more tractable and easier to study using typical surface science probes. It has been shown by STM that narrow, uniform distributions of cluster sizes can be obtained in the model systems relative to conventional powder catalysts [16,17]. This is clearly desirable when attempting to resolve and characterize particle size effects.

The pressure gap has also been addressed in recent years through the integration of high-pressure cells or micro-reactors into high-vacuum equipment. This employment allows a sample to be transferred between a UHV chamber and a high-pressure cell/reactor *via* a series of differentially pumped sliding seals. By having a smooth outer tube, the outer wall of the sample probe forms a vacuum tight seal with the inner diameter of three spring-loaded teflon seals. The space between successive teflon seals is evacuated such that the pressure in the reactor cell may be increased to atmospheric conditions while UHV is maintained in the main chamber, as shown in Fig. 2. This two-level system is advantageous over traditional vacuum systems because elevated pressure kinetics can be performed, as can sample mounting, without breaking vacuum.

Fig. 3 illustrates our fundamental approach to studying a catalytic system. One can see that the proposed scheme consists of two parts: the field of surface science and the field of heterogeneous catalysis. Starting from the investigation of well-ordered single crystals, one moves from top to bottom (the material gap) through the studies of oxide-supported metal clusters and finally reaches the study of "real world" catalysts. In addition, one has to tackle the pressure gap: from UHV (10^{-8} Torr - 10^{-11} Torr) to high-pressure conditions (1.0 Torr - 1000 Torr). As shown in the sketch, the center circle, which represents planar oxide supported metal clusters, is critical to both the surface science and heterogeneous catalysis.

This review presents our recent model catalyst studies of gold clusters on planar titania supports, featuring the extensive use of scanning tunneling microscopy (STM) in characterizing surface morphologies and monitoring surface reactions. We describe the nucleation and growth of Au clusters on the $TiO_2(110)$-(1×1) surface in UHV. The thermal stability and electronic properties of the clusters are subsequently discussed. Finally, morphological changes of this model catalyst induced by surface reactions are examined under elevated pressure conditions, and an Ostwald ripening mechanism is proposed. This review emphasizes that limited dimensionality of metal clusters and chemical properties of oxide supports play significant roles in altering heterogeneous catalyst reactivity.

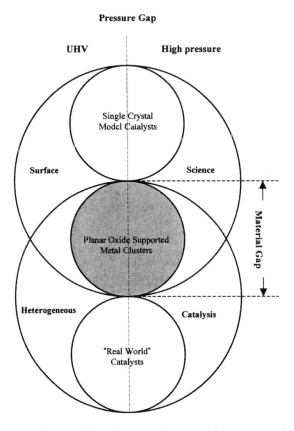

Figure 3. Schematic drawing of the relationship between surface science and heterogeneous catalysis.

2.   EXPERIMENTAL

All experiments are performed in a combined elevated-pressure cell/multi-technique UHV analysis chamber (base pressure, $\sim 1.0 \times 10^{-10}$ Torr). This system is equipped with a double-pass cylindrical mirror analyzer for Auger electron spectroscopy (AES) and X-ray photoelectron spectroscopy (XPS), low energy electron diffraction (LEED) optics (Perkin-Elmer), and a quadrupole mass spectrometer (UTI Instruments) for residual gas analysis. STM images are acquired using an Omicron UHV STM-1 instrument with an inherent magnetic damping system. The system consists of four main parts: an elevated-pressure cell, surface analysis chamber, sample preparation chamber, and STM chamber. A detailed diagram of this system is shown in Fig. 2.

The sample preparation chamber contains multiple metal dosers (Pd, Au, Ag, Al, Ti, etc.) which are used for vapor-depositing metal clusters onto the planar supports using resistive heating. There is also a leak valve connected to the gas manifold, which is used to introduce O_2, CO, or other gases or gas mixtures. Two tungsten filaments are used for electron beam heating. By using a wobble stick, the sample can be transferred to the STM chamber where the general surface morphology is studied. The surface analysis chamber contains the facilities for AES, XPS, LEED, and ion sputtering. The elevated pressure cell is located at the end of the system and is isolated from the main chamber by a gate valve. This design allows the sample to be linearly transferred into the cell for high-pressure exposures or catalytic studies.

In this study, Au clusters with diameters between 1.0 – 7.0 nm are grown by physical vapor deposition of high purity metal from a resistively heated source onto the oxide surface. The average metal cluster size depends upon the flux, deposition time, total dose, and substrate temperature [16,17]. The doser was constructed by wrapping high purity Au wire around a W wire (0.010" diameter, H. Cross Co.). This doser functions by resistive heating and was thoroughly outgassed prior to use; the system pressure during evaporation never exceeds 2×10^{-10} Torr. The flux of the Au doser is calibrated by measuring relative AES peak intensities for Au adsorbed on either Re(0001) or Mo(110) surfaces and further verified using STM. Metal coverages are reported in monolayers (ML), with one ML corresponding to ~ 1.4×10^{15} atoms/cm^2.

A W-5%Re / W-26%Re (Type C, H. Cross Co.) thermocouple is glued to the edge of the TiO$_2$ crystal using high temperature ceramic adhesive (AREMCO 571). The thermocouple is used to measure the surface temperature and to calibrate a pyrometer (OMEGA OS3700) prior to disconnection of the thermocouple for STM measurements. The pyrometer is then used to measure the temperature in subsequent annealing experiments. The emissivity for TiO$_2$ is approximately 0.50 [18].

3.   RESULTS AND DISCUSSION

3.1   Nucleation and Growth of Au on TiO$_2$(110)

Metal/oxide systems play important roles in a wide range of applications and greatly influence the behaviour of catalysts and semiconductor devices. An understanding of the interfacial properties between metal clusters and oxide substrates is of both fundamental interest and practical importance. Initial stages of the interaction between metal and oxide involve adsorption of metal atoms, surface atom diffusion, and the nucleation and growth of islands. In a recent review, Campbell [1] showed that at low temperature, the growth of metal on oxides is controlled by kinetics, leading to flat quasi-2D islands. At high temperature, it is controlled by thermodynamics, leading to 3D shapes. Under equilibrium conditions, the growth of metal layers on oxides as well as other substrates is controlled by energetic and thermodynamic parameters. Significant progress has been made in recent years with respect to understanding the metal/oxide interface. Numerous metals (including Ag [17,19,24], Al [25-28], Au [17,19,20,29-37], Ca [38], Cr [39,40], Cu [40-46], Fe [40,46-49], Hf [50], Ir [51,52], K [53-57], Na [54,58-64], Ni [65,66], Pd [16,17,67-71], Pt [72-76], Rh [77-81], Ti [82,83] and V [84]) have been deposited on TiO$_2$(110) and studied by various surface science techniques including X-ray and ultraviolet photoelectron spectroscopies (XPS and UPS), AES, LEED, LEIS, and STM.

Nanometer-sized Au clusters supported on oxides have recently been found to be active for a number of catalytic reactions such as low temperature CO oxidation, NO reduction, and hydrogenation and partial oxidation of hydrocarbons [29,37,85-106]. These observations have promoted extensive studies of the catalytic properties of supported Au. Haruta's group demonstrated that the catalytic properties of Au depend on the choice of oxide supports, preparation methods, and in particular, the size of the Au clusters [86,89-94,100-103,105]. For example, Au clusters with either spherical or hemispherical shapes can be prepared by deposition-precipitation or impregnation methods. The hemispherical Au clusters show much higher activity for CO oxidation than those clusters with a spherical shape. It was proposed that the more extensive perimeter interface of the hemispherical Au clusters (as compared to spherical Au clusters) provides more active sites for CO and O$_2$ adsorption [94].

The growth of Au on TiO$_2$(110) has also been studied by Madey et al. [33,34] using LEIS, XPS and LEED. It was shown that Au grows initially as two-dimensional islands (2D) or as flat three-dimensional islands (so-called quasi-2D islands) and forms three-dimensional (3D) islands at a Au coverage of one monolayer (ML) or higher. XPS results indicate no evidence of an interface reaction at any Au coverage or substrate temperature. The Au/TiO$_2$ system shows very good thermal stability upon annealing to 800 K; no encapsulation of Au clusters by TiO$_x$ (0<x<2) was observed under these conditions.

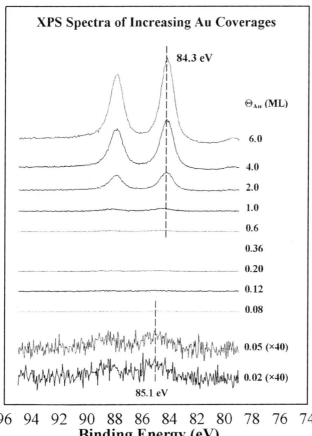

Figure 4. XPS spectra of Au 4f peaks on TiO$_2$(110) surface at various coverages, 300 K, Al Kα anode (hv = 1486.6 eV).

In this study, the nucleation and growth of Au on is thoroughly examined using STM, AES, XPS, and other UHV surface techniques. The TiO$_2$(110) substrate is an ideal support due to the use of its high surface area analogue in real-world catalyst systems; furthermore, a series of morphological studies of the substrate itself have been accomplished, characterizing (1×1) and (1×2) phases as well as oxygen induced titania cluster formation [16,17,29]. Consequently, not only can the adsorbed gold be characterized, but its affect upon the support can be characterized as a function of treatment and conditions as well.

Fig. 4 shows XPS spectra of the Au 4f at 300 K at increasing coverages. A linear increase of Au XPS intensity with respect to Au coverage is apparent, suggesting a 3D growth mode for Au on TiO$_2$(110). More interestingly, the Au 4f core levels shift toward lower binding energy from 85.1 to 84.3 eV when Au coverages are increased from 0.02 to 2.0 ML and remain the same (84.3 eV) from 2.0 to 6.0 ML. The binding energy shift can be due to emission from larger Au clusters. The higher binding energy of the Au 4f feature for lower coverages can be either attributed to finite cluster size effects, or a preference of small Au cluster growth on surface defect sites/step edges rather than flat terraces.

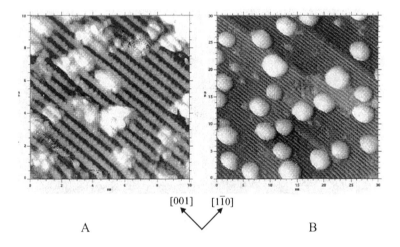

[001]   [1$\bar{1}$0]

A   B

Figure 5. STM images (2.0 V, 2.0 nA) of Au clusters on the TiO$_2$(110)-(1×1) surface. A: both quasi-2D Au clusters (lengths of 1-2 nm and heights of 1-2 atomic layers) and 3D Au clusters (lengths of 2-3 nm and heights ≥ 3 atomic layers) are observed for 0.1 ML Au coverage. B: only hemispherical 3D clusters (diameters of 3-5 nm and heightsof 1.0-2.5 nm) are observed for a 1.0 ML Au coverage.

At present, direct imaging of Au clusters on TiO$_2$(110) is feasible; therefore, STM provides considerable detail regarding the nucleation and growth mechanism. Fig. 5 shows two constant current topographic (CCT) STM images (+ 2.0 V bias voltage, 2.0 nA feedback current) of Au clusters (5A, 0.1 ML; 5B, 1.0 ML) on TiO$_2$(110) at room temperature. These images represent two different stages of Au clusters grown on the substrate: a quasi-2D mode, as described by Madey et al. [29] and a hemispherical 3D mode. Generally, the quasi-2D mode yields clusters with characteristic heights of 1-2 atomic layers. Low Au coverage leads to the formation of small, quasi-2D clusters with varying shapes, as is apparent in Fig. 5A. These quasi-2D clusters typically have diameters from 0.6-2.0 nm and heights of 1 to 2 atomic layers.

Most of the Au clusters are elongated along the [001] direction and are located on top of the bright rows of the substrate. Some clusters have a linear shape and are located above one [001] row while others are rectangular and are located above two [001] rows. The Au-Au distance in the rectangular cluster along the [1$\bar{1}$0] direction is 0.65 nm, corresponding to the lattice constant of the TiO_2(110) substrate in the same direction. These dimensions are indicative of a pseudomorphic growth of the Au clusters.

Assuming the cluster density is the same as that of bulk Au metal (19.3 g/cm^{-3}), the number of atoms contained in a single cluster can be calculated using the cluster size measured by STM. This calculation suggests that the number of Au atoms in the small quasi-2D clusters does not exceed 40 atoms. However, in the center of the image, some larger clusters have heights corresponding to three atomic layers or more and exhibit irregular shapes, which indicate the beginning of a phase transition from 2D to flat 3D clusters.

It is noteworthy that the (1×1) surface structure around the small clusters is still observed after Au cluster deposition, and that much of the exposed substrate appears to be free of defects. This observation implies that Au atoms nucleate on surface defect sites first. Many small Au clusters in Fig. 5A are located on top of the bright substrate rows that arise from five-fold-coordinated Ti atoms; thus, it appears that once defect sites are saturated, Au atoms nucleate on top of Ti cations between two bridging oxygen rows.

Fig. 5B was acquired after a 1.0 ML Au deposition. The formation of 3D clusters is seen, having diameters of 3.0 - 5.0 nm and heights of 1.0 - 2.5 nm. A volume calculation yields ~ 250 atoms for the 3.0×1.0 nm^2 cluster and ~ 2,000 atoms for the 5.0 × 2.5 nm^2 cluster. Even though atomic resolution is achieved on the TiO_2(110) substrate with a constant spacing of 0.65 nm between the individual rows, close inspection of the individual 3D clusters did not reveal any ordered atomic structure.

The two forms of Au clusters in Fig. 5 represent two different stages: a cluster-nucleation stage and a cluster-growth stage. The only difference between these two systems is the Au coverage, which suggests competition between nucleation of sites and cluster growth. At low coverages, Au atoms adsorb mainly on surface defect sites and form 2D clusters that then become nuclei for further growth. Increased deposition of Au atoms results in the formation of 3D clusters.

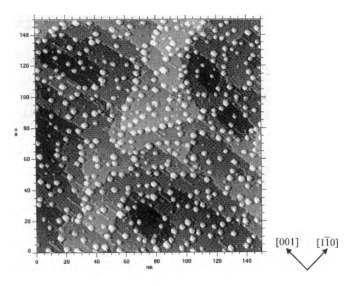

Figure 6. General morphology of Au clusters on the TiO₂(110)-(1×1) surface (2.0 V, 2.0 nA). The dosing flux was 0.083 ML min⁻¹ and the Au coverage was approximately 0.50 ML. Hemispherical Au clusters are observed to preferentially nucleate and grow along the substrate step edges. The raised rows in the [001] direction are from a small amount of reconstructed TiO₂(110)-(1×2).

For a better view of the general morphology of Au clusters on the $TiO_2(110)$ surface, an expanded STM image is shown in Fig. 6. In this image, hemispherical clusters with a very narrow size distribution are observed to preferentially grow along the step edges. However, clusters on the flat terraces are also observed. At 0.50 ML Au coverage, more than 60% of the substrate area is still unoccupied and separated by mono-atomic steps. This observation coupled with the observation of 3D clusters indicates an island, or Volmer-Weber (VW), growth mode for Au on $TiO_2(110)$. These findings are consistent with our previous conclusions from ISS measurements [107].

One advantage of STM over other surface techniques is that it provides us very accurate measurements of individual cluster size. Cluster growth of Au on $TiO_2(110)$ was also measured with STM. Fig. 7 shows six different STM images with Au coverages of 0.10, 0.25, 0.50, 1.0, 2.0, and 4.0 ML. The 3D growth mode of Au clusters on $TiO_2(110)$ is apparent. With increasing coverage, the Au clusters grow larger and gradually cover most of the surface. However, even at 4.0 ML, some portions of the TiO_2 substrate are still visible.

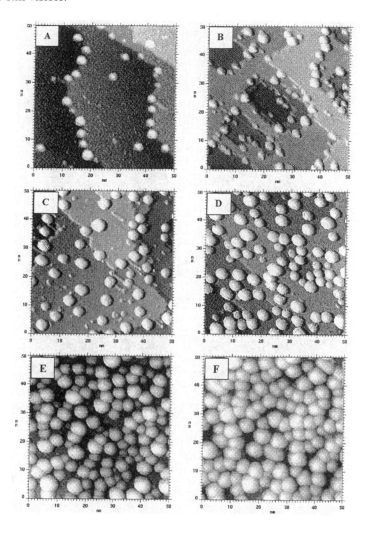

Figure 7. A set of 50×50 nm^2 STM images (2.0 V, 1.0 nA) of TiO$_2$(110)-(1×1) with different Au coverages: (A) 0.10 ML, (B) 0.25 ML, (C) 0.50 ML, (D) 1.0 ML, (E) 2.0 ML, (F) 4.0 ML. With increasing coverage, Au clusters grew larger and gradually covered the majority of the surface.

At a relatively low coverage of 0.10 ML, hemispherical 3D clusters are observed with diameters of 2.0 - 3.0 nm and heights of 1.0 - 1.5 nm;

however, the clusters mainly grow along the step edges. Well-dispersed quasi-2D clusters, having a height of 0.3 - 0.6 nm and a diameter of 0.5 - 1.5 nm, can also be seen on the terraces. These results indicate a bimodal distribution of cluster sizes, correlating with a 2D to 3D cluster growth transition. Furthermore, the 3D Au clusters gradually transform from a flat shape (height ≈ 1/3 diameter) to a hemispherical shape (height ≈ 1/2 diameter) as cluster size increases from 2.5 nm to 5.0 nm.

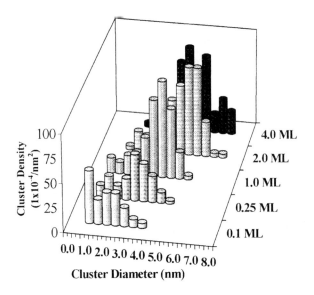

0.10 ML, d_{ave} = 2.0 nm   2.00 ML, d_{ave} = 4.5 nm
0.25 ML, d_{ave} = 2.9 nm   4.00 ML, d_{ave} = 5.4 nm
1.00 ML, d_{ave} = 3.7 nm

Figure 8. Size distributions for Au clusters at different coverages from 0.10 ML to 4.0 ML on the TiO₂(110)-(1×1) surface.

The Au cluster size and density can be carefully measured and averaged over different scanning regions. The overall size distributions of Au/TiO₂(110) for different coverages can be obtained, as shown in Fig. 8. The average cluster sizes (in diameter) increase from 2.0 nm for 0.10 ML Au to 5.4 nm for 4.0 ML Au.

With increasing Au coverage, the cluster size increases steadily while the cluster density remains essentially constant (Fig. 9). A rapid increase in cluster density is observed upon deposition of 0.10 ML Au. With a further increase in the Au coverage from 0.10 ML to 0.25 ML, the cluster density increases by ~ 30%, and then increases again by ~ 50% from 0.25 ML to 1.0 ML Au. The cluster density reaches a maximum at ~ 2.0 ML, at which point ~ 70% of the substrate surface is covered by Au clusters. At higher Au coverages, the cluster density begins to decline due to cluster coalescence and agglomeration. These results indicate that ~ 60% of the nucleating sites are formed at a coverage of 0.25 ML.

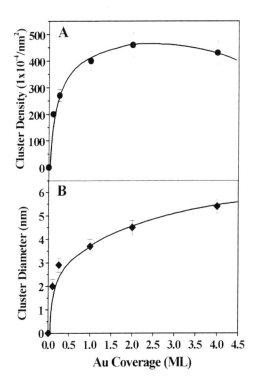

Figure 9. Plots of the cluster density and cluster diameter versus Au coverage on TiO₂(110).

The cluster size (measured as a function of diameter) increases continuously with Au coverage. Increasing the amount of metal deposited from 0.10 ML to 2.0 ML increases the cluster size from 2.0 nm to 4.5 nm. A rapid increase in Au cluster size is observed for very low coverages (below 0.10 ML), corresponding to the nucleation and growth of quasi-2D clusters, which typically have a cluster size below 2.0 nm. Higher Au coverages result in a rapid decline in the rate of cluster size growth, corresponding to a quasi-2D to 3D growth transition. It should also be mentioned that the increase in average cluster size is not proportional to the increase in cluster

volume. For example, a 22% increase in average cluster size from 3.7 nm
(1.0 ML) to 4.5 nm (2.0 ML) corresponds to an 80% increase in the average
cluster volume. The correlation between Au coverage and Au cluster size
indicates that the vacuum deposition method can be tailored to produce a
specific size range of Au clusters, which is very useful for experimental
catalysis.

Madey et al. [29] reported a study of Au on $TiO_2(110)$ and found
quasi-2D cluster growth for fractional monolayer Au coverage. It was found
that the intensity of the substrate Ti and O signals drops by ~ 15% at the
break point of the LEIS curve and hence suggests that the transition from
quasi-2D to 3D clusters occurs at ~ 0.15 ML Au. A clear 3D island growth
mode is observed at a 1.0 ML Au coverage. No chemical interaction
between the Au clusters and $TiO_2(110)$ substrate is detected by either LEIS
or XPS measurements. The STM results presented here for Au/$TiO_2(110)$ are
qualitatively similar: (i) quasi-2D clusters at coverages less than 0.10 ML
Au; (ii) a bimodal distribution and transition of 2D to 3D growth mode at
0.10 ML coverage; and (iii) the full growth of 3D clusters at coverages
above 0.25 ML. However, STM indicates a clear beginning of 3D cluster
growth at ~ 0.10 ML, which is a lower Au coverage than reported previously
[29].

3.2   Thermal Stability of Au Clusters on $TiO_2(110)$

Thermal stability of metal clusters on oxide supports during
annealing is of primary concern for structure-sensitive surface reactions. For
transition metals on TiO_2, there is no significant change in interfacial
reactivity between 300 and ~ 575 K. Au and Pd have been reported to form
large microcrystals with specific orientation upon annealing to over 800 K
[16,17,34]. Jupille and colleagues have observed the growth of Ag clusters
from 2D shapes to 3D islands when the direct deposition temperature is
increased from 300 K to 620 K [22-24]. No sign of strong interaction
between the metal and TiO_2 is observed for Ag, Au, and Pd. However, for
group VIII metals (Fe, Ni, Rh, Ir), encapsulation of metal clusters by a
reduced TiO_{2-x} layer is observed by LEIS and XPS upon annealing to
temperatures of several hundred °C [48,73]. For supported catalysts (metals
on TiO_2), encapsulation is often observed upon heating in a reducing
atmosphere (H_2). It has been suggested that diffusion of vacancies from the
bulk of (partially reduced) TiO_2 may play an important role in the many
observations of encapsulation under UHV conditions [108].

The thermal stability of Au clusters on $TiO_2(110)$ is studied by e-
beam annealing the Au/$TiO_2(110)$ sample to elevated temperatures, then
imaging with STM. Since oxygen exposures at elevated temperatures

significantly change TiO_2 surface morphology, similar experiments have also been performed on the $Au/TiO_2(110)$ system. Figs. 10 and 11 show the effects of O_2 exposure on the morphology and size of 0.10 ML Au on $TiO_2(110)$ at different temperatures. The Au clusters (bright spots) accommodate the step edges of the $TiO_2(110)$ substrate. Following deposition of Au at 300 K, O_2 was leaked into the UHV chamber during STM imaging. Fig. 10A was acquired at 1×10^{-6} Torr O_2 after several hours of exposure at room temperature. Compared to the freshly prepared surface as shown in Fig. 7A, the surface textures obtained immediately after Au deposition and O_2 exposure show no change of the Au cluster size distribution and density.

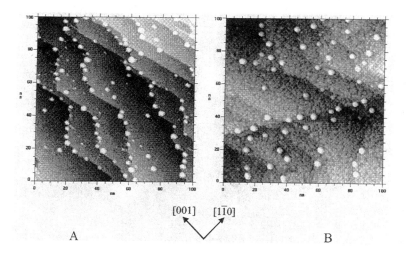

[001]   [1$\overline{1}$0]

A   B

Figure 10. Topographic STM images (1000×1000 Å, 2.0 V, 2.0 nA) of 0.10 ML Au clusters on the $TiO_2(110)$-(1×1) surface. A: 0.10 ML Au exposed to 1×10^{-6} Torr O_2 for several hours, room T. B: 0.10 ML Au annealed to 650 K in 2×10^{-8} Torr O_2 for 10 min.

The search for the influence of O_2 on small Au clusters has also been performed at elevated temperatures. Fig. 10B was acquired after 0.10 ML Au was annealed in 2×10^{-8} Torr O_2 at 650 K for 10 min.

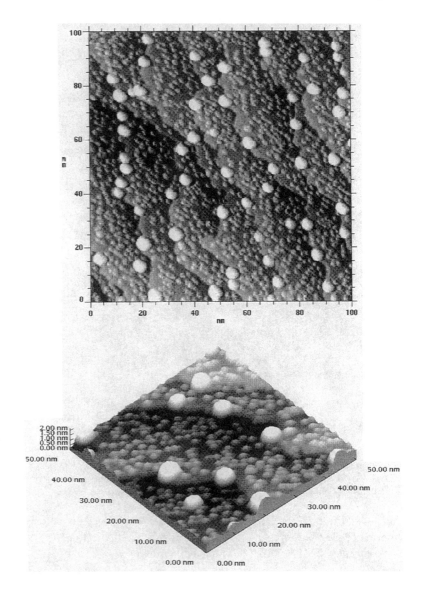

Figure 11. STM images (2.0 V, 2.0 nA) of 0.10 ML Au clusters on the TiO$_2$(110)-(1×1) surface after annealing to 800 K in 2×10^{-8} Torr O$_2$ for 30 seconds. Topographic view (top) and 3D view (bottom).

There is only a slight increase in the cluster size and a very small degree of reduction in the cluster density. However, STM reveals a significant modification of the substrate morphology. The whole surface is mildly disordered by vacancies and defects, although the terraces can still be seen. This roughening effect is similar to the pattern prepared by O_2 treatment (2×10^{-8} Torr) of the clean substrate under the same annealing conditions (see Fig. 12A).

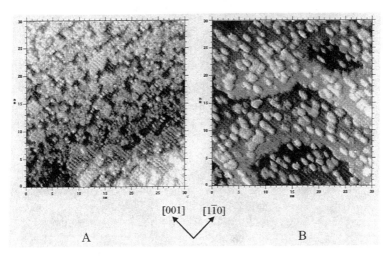

$[001]$ $[1\bar{1}0]$

A   B

Figure 12. CCT STM images (2.0 V, 2.0 nA) of the clean $TiO_2(110)$ surface after 12L exposure of O_2 at (image A) 650 K and (image B) 800 K.

After O_2 exposure at 650 K, the 0.10 ML $Au/TiO_2(110)$ sample was further annealed to 800 K for 30 seconds. Morphological changes are shown in Fig. 11. Small TiO_2 clusters appear with a very homogeneous size distribution of 1-2 nm oriented along the [001] direction. The islands are well distributed over the terraces and cover ~ 70% of the whole surface. This feature is also similar to that prepared using only O_2 treatment, as shown in Fig. 12B. The Au clusters grown are distinctly larger than the TiO_2 islands. This suggests that the Au clusters consist of atoms that may migrate on the $TiO_2(110)$ surface. The migration and coalescence were activated by 10 minutes annealing at 650 K and proceeded further when the sample was annealed to 800 K for 30 seconds. Although tremendous roughening of the TiO_2 surface is observed in STM images (Figs. 10, 11, and 12), XPS measurements reveal no detectable change in the chemical composition of the substrate; it remains only slightly oxygen deficient. This confirms that the oxygen roughening effect only influences the outermost surface; the bulk chemical structure remains unchanged.

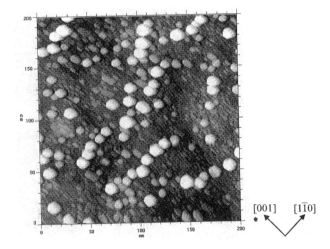

Figure 13. A topographic STM image (2.0 V, 1.0 nA, 2000×2000Å) acquired after briefly annealing 4.0 ML Au on TiO₂(110) to 1100 K.

It is also notable that Au clusters are thermally stable after momentarily annealing at 800 K; only a slight cluster size increase is observed. However, annealing to 800 K for longer periods or higher temperatures leads to agglomeration and continued growth of Au clusters. Fig. 13 shows a 200 nm × 200 nm STM image of 4.0 ML Au deposited on the TiO_2 surface annealed to 1100 K for 30 seconds [17]. The surface morphology changes substantially, as the originally flat $TiO_2(110)$ terraces cannot be clearly distinguished in the image of the annealed surface. Rectangular Au crystallites exhibit a dramatic elongation in the [001] direction parallel to the row structure of the substrate and cover much of the surface. This feature can be explained by an anisotropic diffusion coefficient of Au crystallites: Au atoms are more mobile in the close-packed [001] direction [77]. Moreover, additional Au atoms agglomerate on top of the Au crystallites and form large microcrystals with sizes of 10 - 15 nm and heights of 4.0 - 7.0 nm. These are characterized by a well-defined hexagonal shape and are aligned with two of the six edges parallel to the [001] direction. This result suggests that the microcrystals consist of close-packed (111) oriented planes in accordance with the results of Cosandey, et al. [34].

A line profile along the [001] direction is shown in Fig. 14 for one of the large Au microcrystals and shows a 14 nm diameter and a 6.0 nm height, corresponding to 34,000-38,000 Au atoms. The top surface is atomically flat over 5 nm. Cosandey, et al. [34], used electron backscattered diffraction (EBSD) in combination with high resolution scanning electron microscopy

(HRSEM) to determine the orientation of $Au/TiO_2(110)$ and suggested that Au microcrystals grow with a <111> orientation. The (1×1) substrate structure cannot be clearly distinguished, which suggests a strong interaction at higher temperatures between the small clusters and the surface. This interaction could be in the form of encapsulation of the clusters by the substrate, but no experimental evidence was found to support this. No appreciable change in the size or shape of the Au clusters is found below 800 K, but the Auger Au/Ti ratio decreases significantly in this temperature range, suggesting the loss of Au by evaporation and the formation of large Au microcrystals. Neither ISS studies of Au clusters on a TiO_2 thin film [107] nor LEIS studies of $Au/TiO_2(110)$ [33] show any significant interaction of the Au clusters with the TiO_2 substrate.

3.3   Electronic Properties of Au Clusters on $TiO_2(110)$

Size-dependent electronic structures of supported nanoclusters have attracted both theoretical and experimental attention in recent years. Photoelectron spectroscopies (UPS and XPS) have been common tools in this field [109]. However, the relatively low spatial resolution (*i.e.* on a μm scale) of spectroscopic information makes analysis thereby of individual nanoclusters impossible.

Scanning tunneling spectroscopy (STS) combined with the simultaneous CCT STM imaging is a superb method for the determination of local electronic structure of surfaces on a nanometer scale [110]. The tunneling dependence can be exploited to produce a map of the local density of states (DOS) under the tip by varying the applied voltage and measuring the tunneling current. Both occupied and unoccupied electronic states can be probed using this method. By holding the tip position and tunneling gap constant and measuring the tunneling current as a function of the bias voltage, *I-V* spectra providing information on the chemical environment of a single atom can be obtained.

In this technique, the feedback loop is locked at an equilibrium gap spacing determined by the applied bias and tunneling current for a short period of time (in the order of milliseconds, too short of a time period for the tip-sample distance to drift significantly). Under different sample biases, tunneling current may flow either from the surface to the tip (measuring the density of occupied states of the surface) or from the tip to the surface (measuring the density of unoccupied states of the surface). Thus, for biasing conditions in which the tip injects electrons into unoccupied sample states (positive sample bias), the density of states of the tip can be taken to be a constant with respect to the voltage.

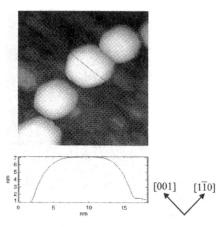

Figure 14. An expanded view of Au hexagonal microcrystals with a line profile along the [001] direction. The image scale is 400×400 Å, the bias voltage is 2.0 V, and tunneling current is 2.0 nA.

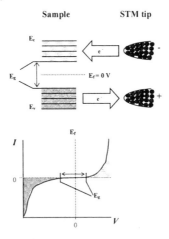

Figure 15. Simplified scanning tunneling spectroscopy energy diagram. In this example the sample is a semiconductor with a band gap (E_g) between the valence (E_v) and conduction (E_c) band edges. Tunneling occurs between the Fermi levels of the tip and the sample, with electrons tunneling out of the one that is negative. The amount of current reflects the density of states (DOS).

Under these biasing conditions, the structure in the spectrum corresponds largely to the spectrum of unoccupied sample states. The idealized band structures in Fig. 15 schematically illustrate current flow in this situation. For semiconductors or insulators, the band gap (E_g) is the difference in energy between the conduction band edge (lowest point of the conduction band) and the valence band edge (highest point of the valence band). Since STS tunnels both into the valence band and out of the conduction band, the band gap can be measured as a zero-current period in the *I-V* curve. For metals, the valence band and conduction band overlap; therefore, the measured band gap is 0 eV. These *I-V* curves can then be correlated with the corresponding real space features on the surface. The topographic scan can be used to correlate the STS curve with a particular geometric feature and also provide a check for tip stability during the measurements.

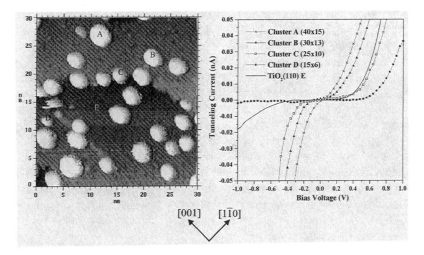

Figure 16. A CCT STM image (2.0 V, 2.0 nA) and the corresponding STS data acquired for Au clusters of varying sizes on the $TiO_2(110)-(1 \times 1)$ surface. For reference, the STS curve of the $TiO_2(110)$ substrate is also shown.

The ability to perform high quality STS measurements and simultaneously obtain topographic images allows for determination of the electronic structure of individual metal clusters at the microscopic scale. Band gaps in metal clusters have been previously observed by First et al. [111,112] for Fe clusters on the GaAs(110) surface and by Suzuki et al. [113] for Al clusters on the GaAs(110) surface. For Fe on GaAs(110), Fe clusters consisting of about 13 atoms show non-metallic character while clusters consisting of more than 35 atoms have metallic properties. Larger Fe clusters with volumes of about 1.0 nm^3, corresponding to about 85 atoms, exhibit fully metallic characteristics. Similar observations are made for Cs on GaAs(110) and Cs on InSb(110), where one- and 2D Cs overlayers exhibit a band gap at the Fermi level whereas a 3D Cs overlayer shows metallic behavior [111,114,115]. Interestingly, the band gap of about 0.6 eV for the 2D Cs overlayer on InSb(110) is nearly identical to that observed for a 2D Cs overlayer on GaAs(110). The measured energy gap may be characteristic of the 2D Cs overlayer. Since the bulk band gap of InSb is only 0.16 eV, the Cs on InSb(110) overlayer system provides an outstanding example where a single layer of adsorbed metal atoms opens up a band gap larger than that of the underlying semiconductor substrate. However, the onset of metallic character may depend on the particular overlayer system under examination. In the case of Al on GaAs(110), clusters consisting of as many as 200 atoms still show non-metallic character [113].

The electronic structures of Au clusters on TiO$_2$(110) are studied using STS. Fig. 16 shows a CCT STM image (top) and four STS curves (bottom) acquired for Au clusters of varying sizes on the TiO$_2$(110) surface. The *I-V* curves are obtained from four Au clusters ranging in size from 15 × 6 Å to 40 × 15 Å. An additional STS curve from the TiO$_2$(110) substrate is shown for reference. The curves exhibit distinctly different behaviours, which can be attributed to cluster size dependent variations of the local density of states near the Fermi level. Curve E, acquired for the TiO$_2$ substrate, shows a band gap of ~ 1.0 eV. However, the band gap for the substrate as measured by STS is directly related to the number and conductivity of oxygen vacancies and defect sites on the surface and thus varies with different preparation methods. Depending on the surface and subsurface defect densities created by extensively heating and sputtering the TiO$_2$ sample, surface band gaps of 0.8 to 3.0 eV are observed for TiO$_2$(110) under various preparation conditions.

As discussed above, Au clusters exhibit a transition from quasi-2D to 3D growth at increasing coverages. STS curves are shown for both quasi-2D and 3D clusters. Cluster D, a cluster of ~ 35 atoms having a 15 Å diameter and 6 Å height (~ 2 atomic layers), has a > 1.5 eV band gap, which is larger than that of the TiO$_2$(110) substrate. Similar *I-V* curves are observed for other quasi-2D clusters with diameters smaller than 20 Å and heights of 1-2 atomic layers. These results suggest a weak interaction between adjacent Au atoms in the quasi-2D clusters since the clusters still exhibit nonmetallic properties and indicate that the Au-substrate interaction dominates the geometric and electronic structure of the quasi-2D clusters.

A slight increase in the cluster volume and size significantly changes the localized electronic structure of Au clusters. For example, cluster C, (25 Å in diameter and 10 Å in height, ~ 3 atom layers, ~ 170 atoms), shows a different *I-V* curve with a 0.6 eV band gap. An even larger cluster size of 30

× 13 Å (~ 4 atom layers, ~ 340 atoms/cluster) leads to further band gap reduction to 0.3 eV. For an Au cluster size of 40 × 15 Å (~ 5 atom layers, ~ 660 atoms/cluster), the *I-V* curve shows a fully metallic electronic structure. Increasing the cluster diameter and/or cluster height further causes no apparent change in the resulting *I-V* curve. It should be noted that the largest change in the *I-V* curves occurs between clusters C and D, where the nonmetal to metal transition begins. These results indicate that the nonmetal to metal transition occurs over a cluster diameter range of 20 - 40 Å and a height of ~ 2 atom layers, and that the bulk-like electronic structure is well-developed at a cluster diameter over 40 Å. This behavior was previously observed in our laboratory for different metal clusters (Ni [116], Au [107], Pd [16]) supported on various oxides (TiO$_2$, MgO, Al$_2$O$_3$). Therefore, it has been established that STS measurements can yield local electronic state information on specific individual nano-clusters with high spatial resolution.

3.4   Quantum Size Effect

Several mechanisms can give rise to an apparent band gap for small metal clusters in the *I-V* curve. In order to rule out the possibility of electronic charging, *I-V* curves were measured for identical Au clusters at the same sample bias of 2.0 V and six different tunneling currents (0.2 - 4 nA). No obvious change is observed for the measured band gap. These results indicate that electronic charge dissipation is a relatively fast process under our experimental conditions and that the observed band gap is not due to the oxide substrate. A second possible explanation is that the apparent band gap is due to a Coulomb Blockade [117]. However, a Coulomb Blockade should either give rise to a staircase behavior in the *I-V* curve for single electron charging, $e^2/2C$, that exceeds kT or the *I-V* curve will be smeared out when $kT > e^2/2C$ [118,119]. In the present experiments, only a single step near the Fermi level is observed at either negative or positive bias voltages up to 4 eV; therefore, in our present studies, the observed band gap is dominated by an intrinsic depletion of the density of the states near the Fermi level within the small metal clusters. This depletion of the density of states can be due either to quantum size effects [111,113] or to spillover of the oxide (or oxygen) from the support to the metal clusters. However, the magnitude of the band gap measured here is most consistent with that expected for a quantum size effect.

A simplified illustration for the quantum size effect is presented in Fig. 17. For a molecular solid, the intermolecular interactions are much weaker than the intramolecular bonding energies. Therefore, the bulk properties of a molecular solid can be regarded as the sum of individual molecular contributions. However, inorganic semiconductors and metals are different. For a metal crystal, electronic excitation is usually delocalized over lengths longer than the lattice constant. As the size of the metal crystallite decreases and approaches the de Broglie wavelength of an electron (~ 5 Å), its electronic properties start to change. The electronic wave functions normal to the plane become exponentially damped, and the energy level spectrum narrows and splits into sub-bands. The splitting of the orbital bands induce the appearance and enlargement of the band gap between the valence and conduction bands. As a result, the band gap can be detected and measured in the STS *I-V* curve, signifying non-metallic properties. The de Broglie wavelength only matches 1-2 atoms in height in a metal cluster; therefore, this phenomenon typically occurs at the transition from a 3D cluster to a one or two-dimensional cluster. These and other physical phenomena arising from the lowering of dimensionality are known as the quantum size effect [111,120-125].

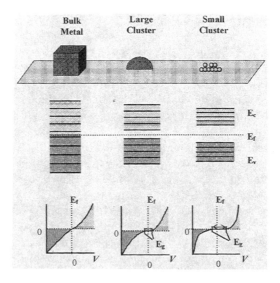

Figure 17. Model illustrations of the quantum size effect for the appearance of a band gap for small metal clusters.

However, for clusters in such a small size regime, a large percentage of the atoms are on or near the surface. The existence of this vast interface between the cluster and the surrounding medium can also have a profound effect on cluster properties. For example, some reactive metals might have stronger bonding to the substrate atoms or are more readily oxidized than others. Therefore, caution should be taken since the size regimes in which the nonmetal-to-metal transition occurs may vary for different metals.

In Fig. 18, the *I-V* characteristics of various Au clusters are shown; their band gaps are plotted as a function of cluster size. These data show $TiO_2(110)$ supports with coverages ranging from 0.10 ML to 4.0 ML. While there is scatter in the data, an obvious trend toward a non-zero band gap at lower cluster sizes is apparent. The Au clusters > 4.0 nm have a fully metallic electronic structure. The onset of the band gap occurs as the cluster size decreases to 3.5 nm. The band gap continues to increase steadily to > 1.0 eV as the cluster sizes decrease to 1.5 - 2.0 nm. On the other hand, the small quasi-2D clusters (< 2.0 nm) exhibit a nonmetallic electronic structure. It is obvious that a metal-to-nonmetal transition occurs as the cluster size is decreased from 4.0 nm to 2.0 nm.

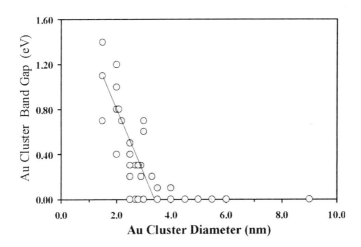

Figure 18. Size dependent nonmetal-to-metal transition for Au clusters. The transition size regime is ~ 2.0 - 4.0 nm.

These results demonstrate the viability of STS as a means of investigating electronic properties in supported metal systems, a powerful complement to traditional photoelectronic and photoemission spectroscopies.

3.5   Gas-Surface Reactions on $Au/TiO_2(110)$

Recently, metal clusters on $TiO_2(110)$ have received considerable attention as model catalysts due to novel interactions between gas molecules and these systems. There are numerous chemisorption studies on metal/TiO_2

with an emphasis on simple adsorbates such as CO, O_2, H_2, H_2O, etc. Much of the work has focused on the possible effect of TiO_2 surface defects on adsorption and dissociation of adsorbed molecules [126-142].

The adsorption and interaction of H_2O, O_2 and H_2 with $TiO_2(110)$ surfaces have been examined by XPS, LEIS and synchrotron radiation photoemission [133,127]. It has been found that the dissociative adsorption of H_2O and H_2 increases the number of oxygen-defects on the nearly perfect surfaces. Using LEED, XPS, EELS, and TPD, Rocker, et. al. [134] has shown that surface defects are thermodynamically stable at high temperatures and act as donors and specific adsorption sites for H_2 and CO. The observation that defective $TiO_2(110)$ does not enhance CO chemisorptive capacity suggests that CO adsorbs more strongly on lattice Ti sites rather than the anionic O sites. Yates postulated that enhanced CO bonding occurs *via* the interaction of the O moiety of CO with the anion vacancy site while primary adsorbate bonding occurs *via* the C moiety on Ti cation sites [137].

In the case of "real world" catalysts, which often consist of small metal particles supported on oxide substrates, one is not only concerned with changes associated with the substrate surface, but also with changes in the supported metal clusters. It is now technologically feasible to control the density and the size of metal clusters by depositing atoms on clean oxide single crystals or thin films. However, the morphology of the metal clusters depends not only on the preparation conditions of the model catalyst but also on the subsequent treatments experienced by the catalyst. Therefore, it is necessary to characterize the morphology of the clusters before, during and after the catalytic reactions.

The physical and chemical properties of small supported metal clusters (< 100 nm) are known to be markedly size dependent. These size dependencies can be attributed to intrinsic electronic cluster size effects resulting from changes in the average atom coordination number and enhancements in the metal-support interaction. Accordingly, many catalytic reactions exhibit cluster size dependencies with respect to activity and selectivity.

Chemisorption of an atom or a molecule often promotes reorganization of the substrate atoms near the adsorption site. Depending on the strength of the surface-adsorbate bond, the surface may experience varying degrees of restructuring originating from either weak local relaxation or massive transportation of surface atoms. Since the driving force is the minimization of the surface free energy, making strong adsorbate-substrate bonds will consequently weaken the bonding of the substrate atoms. Metal clusters contain a relatively small number of atoms; thus, surface atoms can be easily influenced because of their low coordination number. As a result, supported nanoclusters are extremely vulnerable to adsorbate-induced morphological changes.

Previous studies have used surface spectroscopies to interpret the chemisorption and catalytic reaction mechanism; however, STM has rarely been used for direct imaging of structural changes of the metal-TiO_2 surface or the chemisorptive sites for CO and O_2. Quite recently, chemisorption experiments conducted by Solymosi and coworkers [77,79] show that STM is an excellent tool for monitoring the structural changes of the metal-TiO_2 surface. It was reported that exposures of Rh/TiO_2(110)-(1×2) to CO between 10^{-3} and 10^{-1} mbar resulted in a significant agglomeration of Rh clusters [79]. This phenomenon was attributed to the formation of Rh-CO bonds (185 kJ mol^{-1}) that promote disruption of the weaker Rh-Rh bonds (44.5 kJ mol^{-1}). Similar effects have also been observed for Ir/TiO_2(110)-(1×2) [77].

The ultimate goal of studying thermodynamic and chemisorptive properties of metal clusters on TiO_2(110) using STM and related surface spectroscopies is to correlate the structural and chemical properties to catalytic activity in metal-oxide support systems. Understanding these relationships will allow for tailoring of supported catalysts to meet specific reaction requirements.

Quite recently gold has become the subject of intensive research because of its unusual catalytic properties. It was discovered that gold, deposited as nanometer-sized clusters on reducible metal oxides such as TiO_2, exhibits an extraordinarily high activity for various surface reactions: low temperature oxidation of CO, partial oxidation of hydrocarbons, hydrogenation of unsaturated hydrocarbons, and reduction of nitrogen oxides [86-106]. For most of these reactions, the activity and selectivity of supported Au clusters is observed to be markedly size dependent. Despite considerable effort, there is no consensus as to the specific mechanism responsible for this pronounced structure sensitivity.

CO and O_2 adsorption on model Au/TiO_2 systems have been investigated by several groups [29,89-106]. O_2 is assumed to adsorb on the TiO_2 support, migrate across the surface, and react with CO molecules at the edges of nano-sized Au islands. The diameter and density of the Au islands and the TiO_2 support play a crucial role in catalytic activity. In order to address basic issues of structure sensitivity of CO oxidation over supported Au catalysts, the reaction of CO and O_2 on Au clusters of varying size supported on TiO_2(110)-(1×1) was investigated at reaction conditions similar to those employed in actual technological applications.

Fig. 19 shows the influence of 10.00 Torr O_2 exposure on a 0.25 ML Au/TiO_2(110) surface [19]. Following deposition of Au at 300 K (Fig. 19A), Au clusters are imaged as bright protrusions with a relatively narrow size distribution. The clusters, with an average size of ~ 2.6 nm in diameter and ~ 0.7 nm in height, are observed to adhere to the step edges of the $TiO_2(110)$ substrate. After surface characterization, the $Au/TiO_2(110)$ sample was then transferred into the elevated pressure cell. A $CO:O_2$ (2:1) gas mixture was loaded into the cell at a total pressure of 10.00 Torr for 120 min at 300 K. Fig. 19B shows the effect of a 10.00 Torr $CO:O_2$ exposure to Au clusters: the number and density of the Au clusters is greatly reduced and the average size of the clusters increases to ~ 3.6 nm in diameter and ~ 1.4 nm in height. In addition to the significant agglomeration of the Au clusters, extremely small clusters (~ 1.5 nm in diameter) also form.

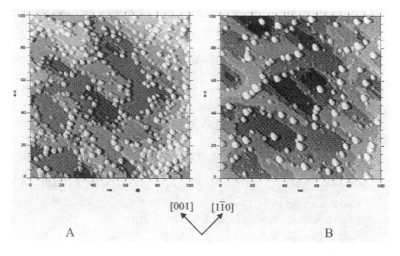

Figure 19. Topographic STM images (100 × 100 nm², 2.0 V, 2.0 nA) of 0.25 ML Au clusters on the $TiO_2(110)$-(1×1) surface. A: Fresh 0.25 ML Au deposited at room T. B: 0.25 ML Au exposed to 10.00 Torr $CO:O_2$ (2:1) for 120 min at room T.

The size distribution of 0.25 ML $Au/TiO_2(110)$ before and after the $CO:O_2$ exposure are presented in Fig. 20, clearly showing significant morphological changes. Before the $CO:O_2$ exposure, the Au clusters have a homogeneous size distribution with a maximum of 2.5 - 3.0 nm in diameter, which is in the metal-to-nonmetal transition range for Au clusters *via* STS. A bimodal distribution is apparent after the $CO:O_2$ exposure. The large feature at ~ 4.0 nm corresponds to fully metallic Au clusters according to the STS measurements, while the small peak at ~ 1.5 nm suggests fully nonmetallic Au clusters. Bear in mind that a portion of the small clusters may arise from the TiO_2 substrate, (as was observed in the oxygen-induced surface growth of small titania clusters), which may explain why there is a much higher density of these small clusters.

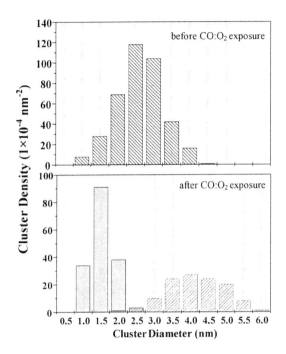

Figure 20. Size distributions of 0.25 ML Au/TiO₂(110)-(1×1) before and after 10.00 Torr CO:O₂ (2:1) exposure for 120 min at room T.

XPS measurements before and after CO exposure show no change in the chemical composition of Au clusters on the $TiO_2(110)$-(1×1) surface; however, the $TiO_2(110)$ surface is oxidized after the $CO:O_2$ exposure (Fig. 21). The small shoulder at the low-binding energy side of the XPS Ti 2p peak due to the presence of Ti^{3+} species is completely absent after the 120 minute $CO:O_2$ exposure at 300 K.

The influence of a separate exposure of CO, O_2 and $CO:O_2$ at a total pressure of 10.00 Torr on the morphology and the size distribution of Au at 300 K is illustrated in Fig. 22. The adsorption of CO has no effect on the surface structure of the Au/$TiO_2(110)$-(1×1) (Fig. 22B). In addition, XPS measurements performed before and after the CO exposure confirm that the chemical composition of Au/$TiO_2(110)$-(1×1) remains unchanged. STM images measured following the O_2 and $CO:O_2$ exposures reveal a similar modification of the morphology of Au/$TiO_2(110)$-(1×1); the number and density of the Au clusters is greatly reduced and the average cluster size is increased. Furthermore, oxidation of the $TiO_2(110)$-(1×1) surface is observed by XPS after the O_2 and $CO:O_2$ exposures (Fig. 21).

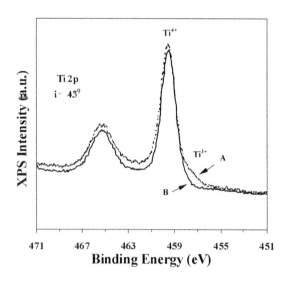

Figure 21. XPS spectra for TiO_2 substrate before and after $CO:O_2$ exposure. A: Slightly oxygen deficient TiO_2 surface prepared by cycles of sputtering and annealing to 1100 K. B: Fully oxidized TiO_2 surface after 10.00 Torr $CO:O_2$

Because all of the structural and surface chemical changes upon exposure to O_2 and $CO:O_2$ are identical and since there are no detectable changes after exposure to CO, we conclude that the $Au/TiO_2(110)$ surface exhibits an exceptionally high reactivity toward O_2 at 300 K that promotes the sintering of Au clusters. Although O_2 adsorption on atomically flat metal single crystals of Au is a highly activated process with an extremely low sticking probability at 300 K [143], Au nanoclusters apparently can activate O_2 and produce atomically adsorbed oxygen atoms [103].

These STM results provide compelling evidence that $TiO_2(110)$-(1×1) supported Au nanoclusters are exceptionally reactive toward O_2. The effect of this high reactivity is sintering of the clusters at 300 K. Our results of O_2 adsorption on $Au/TiO_2(110)$-(1×1) are attributed to the unique properties of nanosized Au clusters. It should be pointed out that the O_2-induced morphological changes were size dependent as these are not observed for clusters larger than ~ 4.0 nm. Fig. 23 shows the surface morphologies of 1.6 ML $Au/TiO_2(110)$ before and after the O_2 exposure. The average Au cluster size is ~ 4.2 nm, large enough to be fully metallic. In this case, Au clusters are less reactive and sensitive to O_2 adsorption. Comparing these two images, Au cluster sizes and density are remarkably identical. No obvious morphological changes are observed.

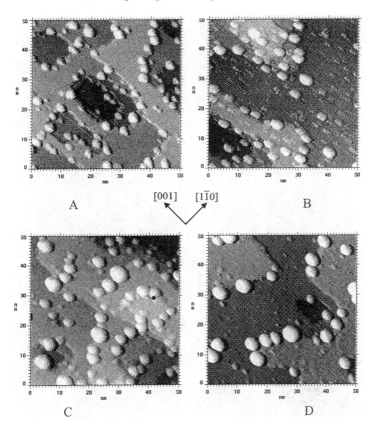

Figure 22. A series of CCT STM images (2.0 V, 2.0 nA, 50 × 50 nm²) of 0.25 ML Au/TiO₂(110)-(1×1) surface. A: Fresh 0.25 ML Au/TiO₂(110). B: After 120 min exposure to 10.00 Torr CO at room T. C: After 120 min exposure to 10.00 Torr O₂ at room T. D: After 120 min exposure to 10.00 Torr CO:O₂ (2:1) at room T.

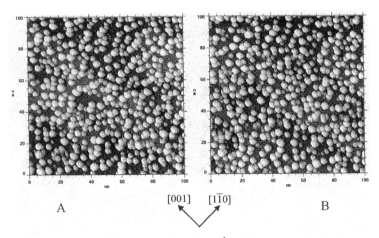

[001] [1Ī0]

A   B

Figure 23. Topographic STM images (100 × 100 nm², 2.5 V, 0.3 nA) of 1.60 ML Au clusters on the TiO₂(110)-(1×1) surface. A: Fresh 1.60 ML Au deposited at room T. B: 1.60 ML Au exposed to 10.00 Torr O₂ for 120 min at room T.

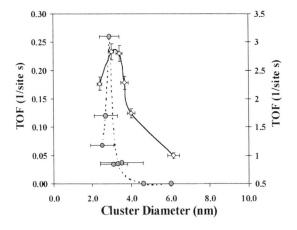

Figure 24. CO oxidation turnover frequencies (TOFs) as a function of the Au cluster size supported on TiO₂. A: The Au/TiO₂ catalysts were prepared by deposition-precipitation method, and the average cluster sizes were measured by TEM, 300K, M. Haruta, et al. (178). B: The CO:O₂ mixture was 1:5 at a total pressure of 40 Torr, 350 K.

These results reflect similarities in structure sensitivity of CO oxidation over Au/TiO₂. Maximum activity is exhibited at an average ~ 3.0 nm Au cluster size on these totally different TiO₂ supports and under different preparation and treatment conditions. In fact, this behavior has been observed for other structure sensitive reactions as well. For example, in ethane hydrogenolysis over supported Ni catalysts, activity increases with increasing cluster size and reaches a maximum at an average ~ 2.5 nm cluster size. However, the activity decreases as larger clusters are used for the reaction [144]. The surprising result is that the maximum activity coincides with clusters in the metal-to-nonmetal transition range as measured by STS. Clusters that are either big enough to be fully metallic or small enough to be fully nonmetallic are not catalytically active for CO oxidation.

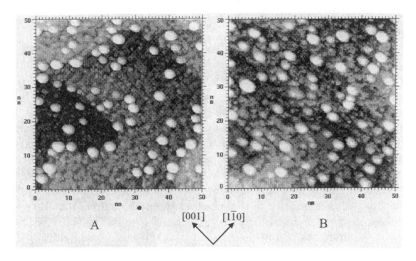

Figure 25. Topographic STM images (50 × 50 nm^2, 2.0 V, 2.0 nA) of 0.25 ML Au clusters on the O$_2$-roughened TiO$_2$(110) surface. A: Fresh 0.25 ML Au on the O$_2$-roughened TiO$_2$(110). B: Au/TiO$_2$(110) after exposing to 10.00 Torr O$_2$ for 120 min.

A strong metal-support interaction (SMSI) is a possible factor that could influence the catalytic activity of the metal clusters supported on reducible metal oxides such as Au/TiO$_2$. In particular, an increasing contribution from the SMSI effect to catalytic activity may be involved at small cluster sizes. However, this contribution would depend on the support material and therefore result in a different catalytic activity if the support is replaced by another with different surface chemical properties. Recently, Okumura et al. [93] compared the specific activities of Au/SiO$_2$, Au/Al$_2$O$_3$, and Au/TiO$_2$ and found that the catalytic activity is essentially independent of the support material provided that the Au cluster size was sufficiently small. Since SiO$_2$ and Al$_2$O$_3$ are covalently bonded metal oxides and have surface chemical properties different from TiO$_2$, it appears that the activity of the low-temperature CO oxidation on supported Au nanoclusters is not determined by the SMSI effect.

As discussed above, the activity of the reaction between CO and O_2 over supported Au nanoclusters strongly depends on cluster size. It is therefore of crucial importance to technological applications to explore the factors governing the cluster size, such as deposition methods, pretreatments, and wetting of Au on the surface of the support. Even though the TiO_2 supported Au catalysts exhibit a high activity for low temperature CO oxidation, the catalysts are often very rapidly deactivated [37]. Usually the model Au/TiO_2 catalyst is deactivated after running the CO:O_2 (1:5) reaction for ~ 120 min at 40.00 Torr.

The activity of CO oxidation on Au/TiO_2 is size dependent (decreasing as the cluster size increases). Dramatic morphological changes in the Au clusters are observed for CO oxidation over 0.25 ML Au/TiO_2(110) at 300 K (as seen in Figs. 20 and 22). Hence, it is likely that the deactivation is due to agglomeration of the Au clusters induced by their strong interaction with O_2. The oxidation of the slightly oxygen deficient TiO_2 surface after the 120 minute CO:O_2 exposure (see Fig. 21) likely lowered the activity of Au/TiO_2 even further because the fully oxidized stoichiometric TiO_2 surface can no longer adsorb O_2 at 300 K [107]. Oxidation of the TiO_2 surface during CO oxidation also provides direct evidence that the deactivation is not caused by encapsulation of Au clusters by reduced Ti suboxides as, for example, in the case of Pt/TiO_2(110) [73].

As discussed before, the relatively mild O_2 treatment (2×10^{-8} Torr, 650 K for 10 min) affects only the top surface of the TiO_2(110) substrate. The TiO_2 surface experiences a roughening transition while the chemical state remains unchanged. A difference in general disorder is obscured when comparing a CCT STM image of 0.25 ML Au on TiO_2(110)-(1×1) after deposition of Au at 300 K and that of a subsequent O_2 exposure of 2×10^{-8} Torr at 650 K for 10 min (Fig. 25A), with one for which no O_2 treatment is made (Fig. 7B). The cluster density and size distribution of the Au clusters, however, are identical for both surfaces. More interestingly, while this O_2 treated Au/TiO_2(110) surface is exposed to CO:O_2 (2:1) for 120 min at a total pressure of 10.00 Torr at 300 K, the TiO_2 surface becomes oxidized after the reaction. No change in morphology of the Au clusters is seen (Fig. 25B). Apparently, the roughness of the surface hinders any migration of Au atoms on the terraces due to a high diffusion barrier for Au to move over the steps. Hence, it can be concluded that the O_2 treated TiO_2 surface stabilizes small Au nanoclusters and prevents the sintering at elevated O_2 pressure. It is quite possible that a high-temperature reduction, calcination and low-temperature reduction procedure creates a similar kind of a rough TiO_2 phase on a high-surface area Au/TiO_2 catalyst. This surface would exhibit a higher degree of resistance toward sintering of the Au clusters during CO oxidation at low temperatures [95,98,99].

These results indicate that the pronounced structural sensitivity of CO oxidation on Au/TiO$_2$ originates from quantum size effects associated with the supported Au clusters. Furthermore, these results suggest that the catalytic properties of supported clusters may deviate significantly from the bulk metal as one dimension of the clusters becomes smaller than three atomic layers. The tailoring of the properties of small metal clusters by altering the cluster size and support material could prove to be universal for a variety of metals and will likely be useful in the design of nanostructured materials for catalytic applications.

3.5.1 Ostwald Ripening Process

Since oxide-supported metal catalysts are typically used in elevated O$_2$ pressure conditions, an analysis of these induced morphological changes is of general interest to the catalysis community. In general, cluster growth of supported metal catalysts can proceed by two processes. First, clusters can migrate along the surface until they collide with other clusters, resulting in coalescence. Second, cluster growth can occur by intercluster transport, or Ostwald ripening, which is driven by capillary action. In this case, the reduction of the total surface free energy by intercluster transport occurs such that certain clusters grow larger at the expense of other clusters [145]. Thus, in light of the bimodal size distribution observed following O$_2$ exposure, Ostwald ripening is likely the cause of cluster growth. Regardless of the cause, cluster growth results in catalysts with decreased active surface areas, leading to a decline in catalytic activity.

The Ostwald ripening process is illustrated in Fig. 26. Intercluster transport of atomic (or molecular) species can occur by either surface diffusion along the substrate or *via* vapor phase transport. Under vacuum or reducing conditions, the transport between Au clusters can only occur in the form of free metallic Au atoms, and the driving force is related to the Au vapor pressure. However, the Au vapor pressure depends exponentially on the energy required to break Au-Au metal bonds and subsequently transfer a Au atom to the vapor phase (*i.e.* the sublimation energy ΔH_{subl}(Au) $\approx$ 370 kJ mol^{-1}). Obviously, such a high-energy barrier suggests that intercluster transport by free Au atoms will be very slow at room temperature. This result is in accordance with our STM observations that Au clusters are generally stable in UHV conditions.

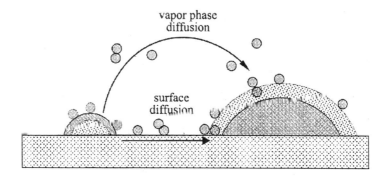

Figure 26. Schematic illustration of Ostwald ripening process: atomic intercluster transport leads to a bimodal size distribution.

In an oxidizing environment, the situation is quite different. For example, Wynblatt [146] showed that growth of Pt particles in O_2 environments occurs *via* formation of volatile PtO_2. Platinum oxide has a lower sublimation energy than platinum metal and therefore serves as a mechanism by which intercluster transport can occur. Unfortunately, to the best of our knowledge, no vapor pressure or sublimation energy data are available for gold oxide, rendering it difficult to directly compare such values with Au. We have shown, however, similar bimodal distributions in oxygen-exposed silver clusters on titania, and the thermodynamics of that system suggests Ostwald ripening is feasible if conducted *via* a silver oxide (Ag_2O) intermediate. Furthermore, intrinsic electronic considerations governing the size-dependent reactivity of silver particles with oxygen molecules provide a complete rationalization for the observed bimodal distribution in that system [19]. Analogous arguments are expected to apply to the Au/TiO_2 systems.

In summary, we observe O_2-induced morphological changes for Au under elevated-pressure O_2 exposure. The size-dependent reactivity to O_2 can be attributed to intrinsic electronic properties of Au clusters, which show the maximum reactivity to be within the metal-to-nonmetal transition region. The morphological changes undergo an Ostwald ripening process in which certain clusters grow at the expense of other clusters, leading to a bimodal size distribution.

4. CONCLUSIONS

Deposition of gold onto titania provides a convenient way to model important aspects of supported metal catalysts and represents a promising method of simulating the "real world" catalysts in an experimentally tractable manner. Of concern are issues that are keys to understanding the relationship between surface structure and catalytic activity/selectivity, such as intrinsic cluster size effects and metal-support interactions. Characterization studies carried out on this model focusing on structural, electronic, and chemical properties as a function of cluster size have been related to parallel studies of surface reactions. These unique systems have provided a direct connection of well-defined model system studies to "real world" processes, allowing the "pressure and material gaps" to be bridged simultaneously.

5. ACKNOWLEDGEMENTS

We acknowledge with pleasure the supports of this work by the Department of Energy, Office of Basic Energy Sciences, Division of Chemical Sciences and the Robert A. Welch Foundation. We thank Dr. Charles Chusuei for helpful discussions.

6. REFERENCES

1. Campbell, C. T., *Surf. Sci. Rep.* 27, 1 (1997).
2. Rainer, D. R., Xu, C., and Goodman, D. W., *J. Mol. Catal. A: Chemical* 119, 307 (1997).
3. Henry, C. R., *Surf. Sci. Rep.* 31, 231 (1998).
4. Rodriguez, J. A. and Goodman, D. W., *Surf. Sci. Rep.* 14, 1 (1991).
5. Goodman, D. W., *Surf. Rev. Lett.* 2, 9 (1995).
6. Xu, X. and Goodman, D. W., *J. Phys. Chem.* 97, 683 (1993).
7. Somorjai, G. A., *Surf. Sci.* 299/300, 849 (1994).
8. Somorjai, G. A., "Introduction to Surface Chemistry and Catalysis", Wiley, New York, 1994.
9. Sault, A. G. and Goodman, D. W., *Adv. Chem. Phys.* 76, 153 (1989).
10. Che, M. and Bennett, C. O., *Adv. Catal.* 36, 55 (1989).
11. Persaud, R. and Madey, T. E., *in* "The Chemical Physics of Solid Surfaces and Heterogeneous Catalysis", (King, D. A., and Woodruff, D. P., Eds.), Vol. 8, pp 407, Elsevier, Amsterdam, 1997.
12. Gunter, P. L. J., Niemantsverdriet, J. W. H., Ribeiro, F. H., and Somorjai, G. A., *Catal. Rev. Sci. Eng.* 39, 77 (1997).
13. Goodman, D. W., *Chem. Rev.* 95, 523 (1995).
14. Crew, W. W. and Madix, R. J., *Surf. Sci.* 319, L34 (1994).

15. Schröder, U., Mc Intyre, B. J., Salmeron, M., and Somorjai, G. A., *Surf. Sci.* 333, 337 (1995).
16. Xu, C., Lai, X., Zajac, G. W., and Goodman, D. W., *Phys. Rev. B* 56, 13464 (1997).
17. Lai, X., St.Clair, T. P., Valden, M., and Goodman, D. W., *Prog. Surf. Sci,* 59, 25 (1998).
18. "Handbook of Chemistry and Physics", (Weast, R. C., Astle, M. J., and Beyer, W. H., Eds.), 68th edition, p. E-395, CRC Press, Boca Raton, 1988.
19. Lai, X., St.Clair, T. P., and Goodman, D. W., *Faraday Discuss.* 114, 279 (1999).
20. Chusuei, C. C., Lai, X., Luo, K., and Goodman, D. W., *Topics in Catal.* 14, 71 (2001)
21. Luo, K., St.Clair, T. P., Lai, X., and Goodman, D. W., *J. Phys. Chem. B* 104 3050 2000.
22. Martin, D., Creuzet, F., Jupille, J., Borensztein, Y., and Gadonne, P., *Surf. Sci.* 377, 958 (1997),
23. Abriou, D., Gagnot, D., Jupille, J., and Creuzet, F., *Surf. Rev. Lett.* 5, 387 (1998).
24. Martin, D., Jupille, J., and Borensztein, Y., *Surf. Sci.* 404, 433 (1998).
25. Dake, L. S. and Lad, R. J., *Surf. Sci.* 289, 297 (1993).
26. Dake, L. S. and Lad, R. J., *J. Vac. Sci. Technol. A* 13(1), 122 (1995).
27. Carroll, D. L., Liang, Y., and Bonnell, D., *J. Vac. Sci. Technol. A* 12, 2298 (1994).
28. Lai, X., Xu, C., and Goodman, D. W., *J. Vac. Sci. Technol. A* 16(4), 2562 (1998).
29. Valden, M., Lai, X., and Goodman, D. W., *Science* 281, 1647 (1998).
30. Zhang, L., Persaud, R., and Madey, T. E., *Phys. Rev. B* 56(16), 10549 (1997).
31. Xu, C. and Goodman, D. W., *Chem. Phys. Lett.* 263, 13 (1996).
32. Yakshinskiy, B., Akbulut, M., and Madey, T. E., *Surf. Sci.* 390, 132 (1997).
33. Zhang, L., Cosandey, F., Persaud, R., and Madey, T. E., *Surf. Sci.* 439, 73 (1999).
34. Cosandey, F., Persaud, R., Zhang, L., and Madey, T. E., *Mat. Res. Soc. Symp. Proc.* 440, 383 (1997).
35. Bondzie, V. A., Parker, S. C., and Campbell, C. T., *J. Vac. Sci. Technol. A* 17, 1717 (1999).
36. Valden, M. and Goodman, D. W., *Israel J. Chem.* 38, 285 (1998).
37. Valden, M., Pak, S., Lai, X., and Goodman, D. W., *Catal. Lett.* 56, 7 (1998).
38. Nörenberg, H. and Harding, J. H., *Phys. Rev. B* 59(15), 9842 (1999).
39. Pan, J. M., Diebold, U., Zhang, L., and Madey, T. E., *Surf. Sci.* 295, 411 (1993).
40. Pan, J. M., Maschhoff, B. L., Diebold, U., and Madey, T. E., *Surf. Sci.* 291, 381 (1993).
41. Møller, P. J. and Wu, M. C., *Surf. Sci.* 224, 265 (1989).
42. Wu, M. C. and Møller, P. J., *Surf. Sci.* 235, 228 (1990).
43. Diebold, U., Pan, J., and Madey, T. E., *Phys. Rev. B* 47(7), 3868 (1993).
44. Wagner, M., Kienzle, O., Bonnell, D. A., and Ruhle, M., *J. Vac. Sci. Technol. A* 16, 1078 (1998).
45. See, A. K., Thayer, M., and Bartynski, R. A., *Phys. Rev. B* 47, 13722 (1993).
46. See, A. K. and Bartynski, R. A., *Phys. Rev. B* 50, 12064 (1994).
47. Pan, J. M. and Madey, T. E., *Catal. Lett.* 20, 269 (1993).
48. Pan, J. M. and Madey, T. E., *J. Vac. Sci. Technol. A* 11, 1667 (1993).
49. Deng, J., Wang, D., Wei, X., Zhai, R., and Wang, H., *Surf. Sci.* 249, 213 (1991).
50. Diebold, U., Pan, J. M., and Madey, T. E., *Surf. Sci.* 333, 845 (1995).
51. Berkó, A. and Solymosi, F., *Surf. Sci.* 411, L900 (1998).
52. Berkó, A., Klivenyi, and G., Solymosi, F., *J. Catal.* 182, 511 (1999).
53. Hayden, B. E. and Nicholson, G. P., *Surf. Sci.* 274, 277 (1992).
54. Souda, R., Hayami, W., Aizawa, T., and Ishizawa, Y., *Surf. Sci.* 285, 265 (1993).
55. Lad, R. J. and Dake, L. S., *Mater. Res. Soc. Symp. Proc.* 238, 823 (1992).
56. Calzado, C. J., San Miguel, M. A., and Sanz, J. F., *J. Phys. Chem. B* 103, 480 (1999).
57. Bredow, T., Apra, E., Catti, M., and Pacchioni, G., *Surf. Sci.* 418, 150 (1998).

58. Onishi, H., Aruga, T., Egawa, C., and Iwasawa, Y., *Surf. Sci.* 199, 54 (1988).
59. Murray, P. W., Condon, N. G., and Thornton, G., *Surf. Sci.* 323, L281 (1995).
60. Gutierrez-Sosa, A., Walsh, J. F., Lindsay, R., Wincott, P. L., and Thornton, G., *Surf. Sci.* 435, 538 (1999).
61. Hird, B. and Armstrong, R. A., *Surf. Sci.* 431, L570 (1999).
62. San Miguel, M. A., Calzado, C. J., and Sanz, J. F., *Surf. Sci.* 409, 92 (1998).
63. Nerlov, J., Christensen, S. V., Weichel, S., Pedersen, E. H., and Møller, P. J., *Surf. Sci.* 371, 321 (1997).
64. Onishi, H. and Iwasawa, Y., *Catal. Lett.* 38, 89 (1996).
65. Wu, M. C. and Møller, P. J., *Surf. Sci.* 279, 23 (1992).
66. Onishi, H., Aruga, T., Egawa, C., and Iwasawa, Y., *Surf. Sci.* 233, 261 (1990).
67. Stone, P., Bennett, R. A., Poulston, S., and Bowker, M., *Surf. Sci.* 435, 501 (1999).
68. Bennett, R. A., Stone, P., and Bowker, M., *Catal. Lett.* 59, 99 (1999).
69. Bredow, T. and Pacchioni, G., *Surf. Sci.* 426, 106 (1999).
70. Stone, P., Poulston, S., Bennett, R. A., and Bowker, M., *Chem. Commun.* 13, 1369 (1998).
71. Evans, J., Hayden, B. E., and Lu, G., *Surf. Sci.* 360, 61 (1996).
72. Steinrück, H. P., Pesty, F., Zhang, L., and Madey, T. E., *Phys. Rev. B* 51, 2427 (1995).
73. Pesty, F., Steinrück, H. P., and Madey, T. E., *Surf. Sci.* 339, 83 (1995).
74. Schierbaum, K. D., Fischer, S., Wincott, P., Hardman, P., Dhanak, V., Jones, G., and Thornton, G., *Surf. Sci.* 391, 196 (1997).
75. Fischer, S., Schierbaum, K. D., and Göpel, W., *Vacuum*, 48, 601 (1997).
76. Schierbaum, K. D., Fischer, S., Torquemada, M. C., de Segovia, J. L., Román, E., and Martín-Gago, J. A., *Surf. Sci.* 345, 261 (1996).
77. Berkó, A., Ménesi, G., and Solymosi, F., *Surf. Sci.* 372, 202 (1997).
78. Berkó, A. and Solymosi, F., *Surf. Sci.* 400, 281 (1998).
79. Berkó, A., Ménesi, G., and Solymosi, F., *J. Phys. Chem.* 100, 17732 (1996).
80. Berkó, A., and Solymosi, F., *J. Catal.* 183, 91 (1999).
81. Poirier, G. E., Hance, B. K., and White, J. M., *J. Phys. Chem.* 97, 6500 (1993).
82. Mayer, J. T., Diebold, U., Madey, T. E., and Garfunkel, E., *J. Elec. Spec. & Rel. Phenom.* 73, 1 (1995).
83. Rocker, G. and Göpel, W., *Surf. Sci.* 181, 530 (1987).
84. Zhang, Z. and Henrich, V. E., *Surf. Sci.* 277, 263 (1992).
85. Park, E. D. and Lee, J. S., *J. Catal.* 186, 1 (1999).
86. Boccuzzi, F., Chiorino, A., Tsubota, S., and Haruta, M., *J. Phys. Chem.* 100, 3625 (1996).
87. Minicò, S., Scirè, S., Crisafulli, C., Visco, A. M., and Galvagno, S., *Catal. Lett.* 47, 273 (1997).
88. Dekkers, M. A. P., Lippits, M. J., and Nieuwenhuys, B. E., *Catal. Lett.* 56, 195 (1998).
89. Hayashi, T., Tanaka, K., and Haruta, M., *J. Catal.* 178, 566 (1998).
90. Haruta, M., *Catal. Surveys of Jpn.* 1, 61 (1997).
91. Tsubota, S., Nakamura, T., Tanaka, K., and Haruta, M., *Catal. Lett.* 56, 131 (1998).
92. Bamwenda, G. R., Tsubota, S., Nakamura, T., and Haruta, M., *Catal. Lett.* 44, 83 (1997).
93. Okumura, M., Nakamura, S., Tsubota, S., Nakamura, T., Azuma, M., and Haruta, M., *Catal. Lett.* 51, 53 (1998).
94. Haruta, M., *Catal. Today*, 36, 153 (1997).
95. Lin, S. D., Bollinger, M., and Vannice, M. A., *Catal. Lett.* 17, 245 (1993).
96. Fukushima, K., Takaoka, G. H., Matsuo, J., and Yamada, I., *Jpn. J. Appl. Phys.* 36, 813 (1997).

97. Cant, N. W., and Ossipoff, N. J., *Catal. Today*, 36, 125 (1997).
98. Liu, Z. M., and Vannice, M. A., *Catal. Lett.* 43, 51 (1997).
99. Bollinger, M. A. and Vannice, M. A., *Appl. Catal. B* 8, 417 (1996).
100. Sakurai, H., and Haruta, M., *Appl. Catal. A* 127, 93 (1995).
101. Haruta, M., Yamada, N., Kobayashi, T., and Iijima, S., *J. Catal.* 115, 301 (1989).
102. Haruta, M., Tsubota, S., Kobayashi, T., Kageyama, H., Genet, M. J., and Delmon, B., *J. Catal.* 144, 175 (1993).
103. Iizuka, Y., Fujiki, H., Yamauchi, N., Chijiiwa, T., Arai, S., Tsubota, S., and Haruta, M., *Catal. Today,* 36, 115 (1997).
104. Yuan, Y., Asakura, K., Wan, H., Tsai, K., and Iwasawa, Y., *Catal. Lett.* 42, 15 (1996).
105. Tsubota, S., Cunningham, D. A. H., Bando, Y., and Haruta, M., in "Preparation of Catalysts VI", (Poncelet, G., et al., Eds.), pp. 227-235, Elsevier, Amesterdam, 1995.
106. Grunwaldt, J. D. and Baiker, A., *J. Phys. Chem. B* 103, 1002 (1999).
107. Xu, C., Oh, W. S., Liu, G., Kim, D. Y., and Goodman, D. W., *J. Vac. Sci. Technol. A* 15, 1261 (1997).
108. Gao, Y., Liang, Y., and Chambers, S. A., *Surf. Sci.* 365, 638 (1996).
109. Goodman, D. W., Kelley, R. D., Madey, T. E., and Yates, Jr. J. T., *J. Catal.* 63, 226 (1980).
110. "Scanning Tunneling Microscopy and Spectroscopy", (Bonnell, D. A., ed.), VCH, Weinheim, 1993.
111. First, P. N., Stroscio, J. A., Dragoset, R. A., Pierce, D. T., and Celotta, R. J., *Phys. Rev. Lett.* 63, 1416 (1989).
112. First, P. N., Dragoset, R. A., Stroscio, J. A., Celotta, R. J., and Feenstra, R. M., *J. Vac. Sci. Technol. A* 7, 2868 (1989).
113. Suzuki, M. and Fukuda, T., *Phys. Rev. B* 44, 3187 (1991).
114. Whitman, L. J., Stroscio, J. A., Dragoset, R. A., and Celotta, R. J., *Phys. Rev. Lett.* 66, 1338 (1991).
115. Whitman, L. J., Stroscio, J. A., Dragoset, R. A., and Celotta, R. J., *Phys. Rev. B* 44, 5951 (1991).
116. Xu, C., Lai, X., and Goodman, D. W., *Faraday Discuss.* 105, 247 (1996).
117. Wiesendanger, R., "Scanning Probe Microscopy and Spectroscopy", Cambridge University Press, Cambridge, 1994.
118. Andres, R. P., Bein, T., Dorogi, M., Feng, S., Henderson, J. I., Kubiak, C. P., Mahoney, W., Osifchin, R. G., and Reifenberger, R., *Science*, 272, 1323 (1996).
119. Schönenberger, C., Van Houten, H., and Donkersloot, H. C., *Europhys. Lett.* 20, 249 (1992).
120. Ciraci, S. and Batra, I. P., *Phys. Rev. B* 33, 4294 (1986).
121. Wang, Y. and Herron, N., *J. Phys. Chem.* 95, 525 (1991).
122. Bolotov, L., Tsuchiya, T., Nakamura, A., Ito, T., Fujiwara, Y., and Takeda, Y., *Phys. Rev. B* 59(19), 12236 (1999).
123. Beckmann, A., *Surf. Sci.* 349, L95 (1996).
124. Wojciechowski, K. F. and Bogdanów, H., *Surf. Sci.* 397, 53 (1998).
125. Huang, W. C. and Lue, J. T., *Phys. Rev. B* 49(24), 17279 (1994).
126. Henrich, V. E. and Cox, P. A., "The Surface Science of Metal Oxides", Cambridge University Press, Cambridge, 1994.
127. Pan, J. M., Maschhoff, B. L., Diebold, U., and Madey, T. E., *J. Vac. Sci. Technol. A* 10, 2470 (1992).
128. Ferris, K. F. and Wang, L. Q., *J. Vac. Sci. Technol. A* 16, 956 (1998).
129. Wang, L. Q. and Ferris, K. F., *J. Vac. Sci. Technol. A* 16, 3034 (1998).

130. Wang, L. Q., Ferris, K. F., Shultz, A. N., Baer, D. R., and Engelhard, M. H., *Surf. Sci.* 380, 352 (1997).
131. Barteau, M. A., *Chem. Rev.* 96, 1413 (1996).
132. Cocks, I. D., Guo, Q., Patel, R., Williams, E. M., Roman, E., and de Segovia, J. L., *Surf. Sci.* 377, 135 (1997).
133. Kurtz, R. L., Stockbauer, R., and Madey, T. E., *Surf. Sci.* 218, 178 (1989).
134. Göpel, W., Anderson, J. A., Frankel, D., Jaehnig, M., Phillips, K., Schäfer, J. A., and Rocker, G., *Surf. Sci.* 139, 333 (1984).
135. Yanagisawa, Y. and Ota, Y., *Surf. Sci. Lett.* 254, L433 (1991).
136. Göpel, W., Rocker, G., and Feierabend, R., *Phys. Rev. B* 28, 3427 (1983).
137. Lu, G., Linsebigler, A., and Yates, Jr., J. T., *J. Phys. Chem.* 98, 11733 (1994).
138. Linsebigler, A., Lu, G., and Yates, Jr. J. T., *J. Chem. Phys.* 103, 9438 (1995).
139. Hadjiivanov, K., Lamotte, J., and Lavalley, J. C., *Langmuir*, 13, 3374 (1997).
140. Beck, D. D., White, J. M., and Ratcliffe, C. T., *J. Phys. Chem.* 90, 3123 (1986).
141. Beck, D. D., White, J. M., and Ratcliffe, C. T., *J. Phys. Chem.* 90, 3132 (1986).
142. Beck, D. D., White, J. M., and Ratcliffe, C. T., *J. Phys. Chem.* 90, 3137 (1986).
143. Canning, N. D. S., Outka, D., and Madix, R. J., *Surfactant Sci.* 141, 240 (1984).
144. Coulter, K., Xu, X., and Goodman, D. W., *J. Phys. Chem.* 98, 1245 (1994).
145. Wynblatt, P., Dalla Betta, R. A., and Gjostein, N. A., in "The Physical Basis for Heterogeneous Catalysis", (Drauglis, E., and Jaffee, R. I., Eds.), pp501, Plenum Press, New York, 1975.
146. Wynblatt, P., *Acta Metall.* 24, 1175 (1976).

Chapter 8

NANOSCALE CATALYSIS BY GOLD

G. U. Kulkarni, C. P. Vinod and C. N. R. Rao

Key words: Gold catalysis, CO oxidation, metal clusters, metal to non-metal transition, Mössbauer and photoelectron spectroscopy, Scanning tunneling microscopy and spectroscopy, temperature programmed desorption.

Abstract: Recent work related to gold catalysis has been discussed addressing the important issue of how the nobleness of gold breaks down at nanometric sizes when in contact with oxidic supports. The high reactivity of gold catalysts in comparison to other metal catalysts is illustrated by reactions such as oxidation of CO and reduction of NO under ambient conditions, as well as epoxidation and hydrochlorination of unsaturated hydrocarbons. Investigations carried out on gold catalysts using a variety of spectroscopy and microscopy techniques are discussed along with the general mechanism of the catalytic process. The observation of maximum reactivity at a cluster size of 2-3 nm, coincident with the size-induced metal to non-metal transition in gold forms the central theme of the article.

1. INTRODUCTION

Gold is one of the truly noble metals. From times immemorial, it has been used in ornamental and coinage industry. The unique nature of gold can be rationalised in terms of the relativistic orbital contraction[1]. Electrons in atoms with high atomic numbers such as gold are influenced by the increased nuclear point charge and reach speeds approaching that of light. This relativistic effect results mainly in outer s-orbital contraction. Figure 1 demonstrates the relativistic contraction for the 6s orbital of the heavy element as a function of the atomic number Z. The element gold (Z=79) is at a pronounced local minimum. The unique yellow color of gold derives from the relativistic effects. Hammer and Norskov[2] have shown that the low activity exhibited by a gold surface in adsorbing molecules is related to the degree of orbital overlap and the degree of filling of the antibonding states. These factors determine the strength of the adsorbate-metal interaction and the energy for dissociation. Using density functional theory, they have pointed out that the nobleness of gold is due to the weak coupling of the adsorbant electronic states with the d band of gold. This coupling is not strong enough to push the antibonding state above the Fermi level of the metal upon adsorption resulting in a repulsive interaction. The activity of gold in surface reactions is so poor that even reactive molecules like O_2 and H_2 do not adsorb on its surfaces[3,4]. Indeed for many years, gold catalysts were investigated more in the hope of boosting or modifying the known catalytic activity and selectivity of other metals, for example in dehydrocyclization and hydrogenolysis reactions[5].

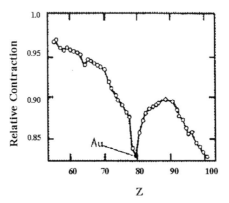

Figure 1. Relativistic contraction (r_{rel}/r_{nonrel}) for the 6s orbitals of the heavy elements as a function of the atomic number Z. Gold (Z=79) represents a pronounced local minimum (from ref. 1).

The nobleness of gold breaks down when it is dispersed to nanometric dimension especially on an oxidic support. In their seminal work, Haruta *et al*[6,7] showed that gold in the form of fine particles can be a potential catalyst for oxidation of CO at ambient temperature. High activity of gold has been registered in hydrochlorination of ethyne as well[8]. Other possibility that has been indicated is the production of hydrogen peroxide[9] where work to till date has been predominantly with Pd based catalysts. In recent years, gold catalysis has attracted growing interest from scientists and engineers. In 1994, a special interdisciplinary basic research institute was founded in Osaka, devoted solely to the research of nanoscale gold catalysts.

This article gathers some of the current research activity in gold catalysis with special focus on particle-size dependence and support interaction. A few reactions that have been widely studied are also listed.

2. GOLD CATALYSIS

2.1 Low temperature oxidation of carbon monoxide

$$2CO + O_2 \rightarrow 2CO_2$$

The oxidation of CO to CO_2 is one of the most extensively studied reactions. Any catalyst which favors the reaction at ambient temperatures would find immense applications for air purification in automobile industry and in breathing apparatus. The industrial catalyst used at present is hopcalite, a mixed oxide of copper and manganese. It deactivates quite rapidly and is unsuitable for long term use.

As mentioned before, Haruta *et al*[6,7] showed that supported gold catalysts could be effective for the carbon monoxide oxidation at ambient temperature. They noted that the gold catalysts prepared by the conventional impregnation mechanism were less active than the platinum group metal catalysts prepared in a similar way, while those prepared by the co-precipitation method were entirely different in nature. Calcination of the coprecipitates in air at 400°C produced ultrafine gold particles smaller than 10 nm which were uniformly dispersed on the transition metal oxides. Among them, $Au/\alpha\text{-}Fe_2O_3$, Au/Co_3O_4 and Au/NiO catalysts with gold loading of ~ 5% were highly active for CO oxidation[10] (and also for H_2 oxidation). The activity was appreciable even at a temperature as low as ~ 70°C. These catalysts were able to oxidize CO completely at 30 °C even under a relative humidity of 76%. Lin *et al*[11,12] found that Au/TiO_2 catalysts are an order of magnitude more active than Au/SiO_2 catalysts.

Studies by Gardner *et al*[13] confirmed the high activity of gold catalysts in comparison with other supported transition metal catalysts. They have shown that Au supported on MnO_x and CeO_x are exceptionally active compared with their Ag, Ru or Pd counterparts. Figure 2 summarizes their results obtained for different reaction temperatures At 75°C, Au/MnO_x sustains nearly 100% CO_2 yield over several hours. Excellent activity is also observed at 50° and 30°C. Moreover, the activity profiles exhibit negligible decay over the entire test period. Au/MnO_x is clearly the most active catalyst examined in this study. Other catalysts based on Ag, Pd and Ru showed much less activity and deactivited faster. Compared to other metal catalysts,

Au catalysts are more structure sensitive, the activity being critically dependent on the deposition method employed, structure and chemical composition of the support as well as pretreatment of the catalyst[14,15]. Hutchings *et al*[16] adopted a prolonged aging process during catalyst synthesis, which further improved deactivation.

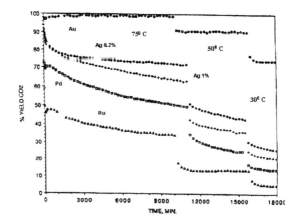

Figure 2. CO oxidation activity of 5 atom% Au/MnOx, 0.2wt% Pd/MnOx, 2wt% Ru/MnOx, 0.2wt% Ag/MnOx, and 1.0wt% Ag/MnOx at 75°, 50°, and 30°C as a function of time. The data were collected using 0.15 gm of catalyst and a reaction mixture of 1% CO, 0.5% O_2 and 2% Ne in helium atmosphere at 1atm pressure (from ref. 13).

2.2 Catalytic reduction of NO

The catalytic reduction of nitrogen oxides from automobile exhaust has been one of the most important scientific challenges during the last few decades. The most practical and the convenient method for removing nitric oxide (NO) being the catalytic reduction using the unburnt compounds such as carbon monoxide and hydrocarbons from the exhaust. Untill now only noble metals such as Rh, Pd and Pt supported on oxides like CeO_2 and Al_2O_3 were used as catalytic components due to their durability under oxidizing and reducing conditions. Again, Haruta and coworkers[17] found supported gold particles to be superior for the reduction of nitric oxide by hydrocarbons such as propane, propene, ethane and ethene. The addition of carbon monoxide and hydrogen to the reactant gas was found to increase the conversion of NO to nitrogen. The reduction of nitric oxide to nitrogen is assumed to take place *via* the formation of nitrogen dioxide through the oxidation of NO which is then reduced by the hydrocarbons to nitrogen. A mixture of Mn_2O_3 and Al_2O_3 as support material seems to offer the best catalytic performance[18].

Besides catalytic materials, sensors for nitrogen oxides based on gold have also been developed[19].

2.3   Hydrochlorination of ethyne

Heterogeneously catalyzed addition of hydrogen chloride to ethyne is used for the industrial manufacture of vinyl chloride[16]

$$HC\equiv CH + HCl \rightarrow CH_2=CHCl$$

Mercuric chloride supported on carbon or silica is usually the industrial catalyst. It is known to deactivate quite rapidly under reaction conditions and is also toxic.

The rate-determining step involves the addition of HCl to a surface ethyne complex,

$$MCl_n + HC\equiv CH + HCl \rightarrow MCl_n.HC\equiv CH.HCl$$

Active catalyst should therefore be able to form surface metal-ethyne and metal-HCl complexes. A range of metal chlorides have been tried out. There have been attempts to correlate the activity of a metal with its electron affinity and oxidation state of the cation. However, Hutchings[20] more recently, proposed that the standard electrode potential should be a more suitable parameter for the correlation of catalytic activity. The importance of this finding is that such a correlation can be used predictively since any metal cation with a higher electrode potential than Hg^{2+} would be expected to give enhanced catalytic activity. On this basis, gold cation was predicted to be the most active catalyst. Research in this direction has confirmed this prediction. Gold catalysts are about three times more active than mercuric chloride catalyst.

2.4   Epoxidation of propylene

$$CH_3CH=CH_2 + O_2 + H_2 \rightarrow CH_3CH-CH_2 + H_2O$$
$$\diagdown \diagup$$
$$O$$

Propylene oxide, which is one of the most important chemical feedstocks for producing resins such as polyurethane, is produced industrially in two-stage processes[21]. The direct synthesis of PO by the use of molecular oxygen has long been desired. Despite considerable effort, no economically viable route has yet been found. A few years ago, Hayashi *et al*[22] provided the first evidence for the vapor phase oxidation of propylene to propylene oxide in presence of O_2 and H_2 using a catalyst comprising of gold deposited on TiO_2.

Combination of Au and TiO_2 seems indispensable for the selective oxidation. Over Au/TiO_2, the presence of H_2 not only enhances the oxidation of propylene but also leads to the epoxidation producing propylene oxide with selectivity above 90%. It appears that the reaction pathway can be tuned, by controlling the size of the Au particles and switches from oxidation to hydrogenation at a critical size of around 2 nm. Although no direct experimental evidence was obtained, it is suggested that the active oxygen species are produced by the reductive activation of molecular oxygen with molecular hydrogen at the Au particles- TiO_2 interface.

Besides the above mentioned reactions, gold catalysts are shown to be good in the selective hydrogenation of acetylene[23] and in the low-temperature water-gas shift reaction[24].

3.   MORPHOLOGY AND ELECTRONIC STRUCTURE

It is well known fact that electronic structure plays an important role in deciding the reactivity of metal clusters. The underlying mechanism which makes gold catalytically active in reduced dimensions has been a subject matter of several spectroscopy and microscopy investigations in the last few years[25]. Few studies focus on the electronic nature of nanometric gold particles, while other studies deal with the particle-support interaction in connection with the catalytic activity.

On a small particle, the surface atoms have incomplete coordination leading to dangling bonds which play crucial role in catalytic activity. As the particle size decreases, the ratio of surface atoms to those in the interior increases rapidly. This is nicely evidenced by Stievano *et al*[26] by carrying out a detailed Mössbauer study of supported gold particles. [197]Au Mössbauer spectra from gold clusters of different sizes are shown in Figure 3. A sample with gold layer of thickness of 65 nm showed only of a single line with isomer shift and linewidth matching closely that of bulk metallic gold. With decreasing gold coverage, the peak becomes asymmetric, showing the presence of additional components in the spectra. The bottom spectrum in the figure compares well with that of Au_{55} cluster[27]. The increase in the intensity of the quadrupole splitting with decreasing particle size is attributed to the atoms on the surface of the particle, as their fraction increases with decreasing particle size. The presence of two quadrupole components in the Mössbauer spectra has been designated by them as due to different positions of gold atoms on the surface. This may arise from the different number of

gold neighbours to which the different type of surface atoms are coordinated. Thus, the authors were able to find a clear distinction between
surface and core atoms for gold particles less than 6 nm size using
Mössbauer spectroscopy.

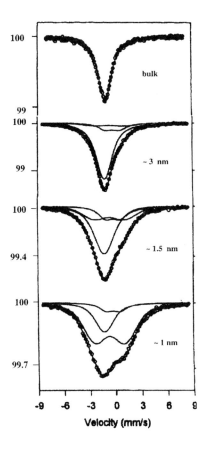

Figure 3. [197]Au Mössbauer spectra of gold nanoparticles of different sizes deposited on a
Mylar film by resistive evaporation of the metal. The mean sizes were determined by electron
microscopy (from ref. 26).

Electron spectroscopy studies have been carried out to look at the
electronic and other properties while going from the bulk metal to small
clusters. Several workers have tried to understand the nature of the metal
clusters by studying the variation in the core-level binding energy of the
metal deposited on solid substrates with the coverage[28-31]. In Figure 4 is

shown a typical x-ray photoelectron spectrum of gold cluster deposited on a graphite substrate. It is clear from the figure that the core-level in case of small clusters has shifted towards a higher binding energy value by as much as 1 eV. In general, an increase in core-level binding energy accompanying a decrease in metal coverage is taken to reflect a decrease in core-hole screening. This is considered as a clear manifestation of the occurrence of a metal-to-nonmetal transition in these tiny clusters. Accompanying diagram shows the variation of the shift in the Au $4f_{7/2}$ binding energy, ΔE, of Au clusters (relative to the bulk metal value) with cluster size. It is clear from the figure that clusters higher than 6 nm show no shift in core-level binding energy and are bound to exhibit bulk metallic properties.

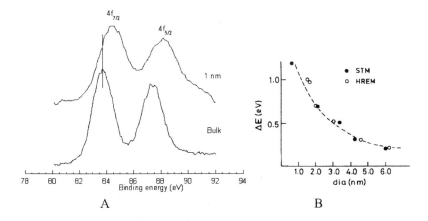

Figure 4. A. Au 4f region for clusters deposited on a graphite substrate. The spectrum for bulk Au is also given for comparison. The mean diameter is estimated to be 1 nm based on electron and scanning tunneling microscopy. The 1 nm cluster exhibits a positive shift of ~ 0.8 eV with respect to the gold foil. B. shows the variation in the shift in binding energy relative to the bulk metal value with the average diameter of the clusters. (from ref. 28).

A few years ago, Roulet *et al*[32] investigated the valence band of gold clusters grown on various alkali halides. The x-ray photoelectron valence band spectra of Au clusters of various sizes are given in Fig.5 along with the bulk Au spectrum. The VB spectrum remains similar to the bulk value down to 4 nm cluster size. Below this value, the 5d band moves to a higher binding energy value along with a band narrowing. These results along with the results from core-level spectroscopy summarize that cluster size has a pronounced effect on the electronic structure of Au clusters, smaller clusters having discretized bands.

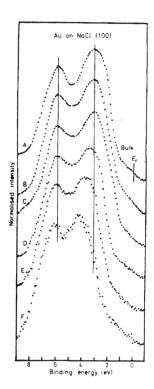

Figure 5. X-ray valence band spectra of Au clusters of varying diameters deposited on a NaCl(100) surface using resistive evaporation of the metal. The sizes were determined using electron microscopy. Normalized XP spectra of the Au VB for (A) bulk metal and (B-F) clusters on NaCl(100) surfaces. Cluster diameters are B, 60; C, 40; D, 30; E, 25 and F, 15 Å (from ref. 32).

Direct information on the nonmetallic gap states in metal clusters can be obtained from by scanning tunneling spectroscopy (STS). This technique provide desired structural sensitivity and spatial resolution making it possible to carry out tunneling spectroscopic measurements on individual clusters. In a careful experiment, STS measurements of Au clusters deposited on a graphite substrate were carried out by Vinod *et al*[33] in ultra high vacuum condition. Figure 6 shows tunneling images of the gold clusters. The background graphite lattice is clearly seen in the high magnification scan. The adjoining diagram shows the I-V curves for cluster of two different sizes along with that of the substrate. For a given bias, the larger cluster seems to produce more tunnel current. The derivative curve obtained from the I-V showed features ascribable to the nonmetallic gap states in these tiny clusters. A 0.5 nm cluster showed a gap of 50 meV, while 0.8 nm cluster gave a smaller gap, ~30 meV. Above 2 nm the derivative curves were featureless indicating that the non-conducting gap had vanished and the clusters were becoming metallic.

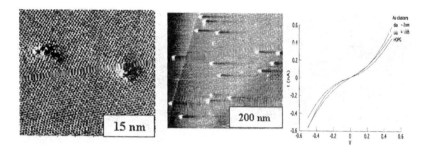

Figure 6. STM image of vacuum deposited Au clusters on a graphite substrate. The large clusters (~ 10 nm) are visible in the large area scan. Small clusters become visible at lower scan area, where substrate lattice is also clearly seen. The I-V curves for the corresponding clusters are shown along with that of the substrate (from ref. 33).

Goodman and coworkers have recently carried out a detailed investigation of Au clusters deposited on TiO_2 using tunneling microscopy (STM) and also measured activity of these clusters for CO oxidation[34,35]. A possible correlation between the electronic properties and the specific activity of Au clusters was probed by tunneling spectroscopy. In Figure 7, the apparent activation energy (E_a) for CO oxidation is shown as a function of the cluster diameter. They observed Au clusters of diameter ~ 3 nm to be catalytically most active for CO oxidation. Interestingly, their STS measurements also predict a metal-nonmetal transition in this cluster size regime. In Figure 7 are also shown the tunneling measurements in terms of band gap of the Au clusters as a function of cluster diameter. A metal to non-metal transition is apparent as the cluster size approaches approximately ~ 4.0 nm in diameter (~ 400 atoms), the approximate size at which the onset of catalytic activity is observed for CO oxidation. These data strongly suggest that the specific activity of Au/TiO_2 interface is determined by the unique electronic structure of Au clusters which exhibit a band gap of ~0.4 eV on TiO_2 as measured by tunneling spectroscopy. An analysis of the cluster morphology revealed that clusters with optimum activity were two atomic layers thick.

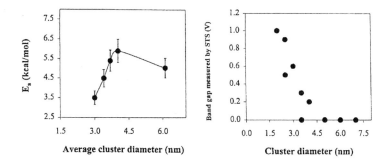

Figure 7. Apparent activation energy (E_a) as a function of average Au cluster diameter. The Au/TiO$_2$ surfaces were prepared by vapor deposition of Au on to a thin film of TiO$_2$ supported on a Mo(100) surface. A gas mixture containing CO and O$_2$ in 1:5 proportion at a pressure of 40 torr was used. The products were analyzed by gas chromatography. Adjoining figure shows gold cluster band gap measured as a function of the cluster size on Au/TiO$_2$(001)/Mo(100). The band gaps were measured for various Au coverages ranging from 0.5 to 3.0 ML corresponding to cluster diameters ranging from 1.5 to 7.0 nm (from ref. 35).

Mavrkakis *et al*[36] in their paper titled "Making Gold less noble" report that the catalytic activity of highly dispersed Au particles may in part be due to high step densities on the small particles and/or strain effects due to the mismatch at the Au-support interface. The structure and morphology of tiny gold particles on MgO substrate have been elucidated by Giorgio *et al*[37], in their high resolution microscopy work.

Bondzie *et al*[38] obtained a coverage of 0.35 L of adsorbed oxygen on Au islands deposited on a TiO$_2$(110) surface, exposing the surface to gaseous oxygen near a hot filament. Their desorption experiments showed a higher O$_2$ desorption temperature (741 K) from ultrathin Au islands than from thicker particles (545 K) clearly implying that O$_a$ bonds more strongly to small particles. Accordingly CO oxidation was found to be faster as the adsorption strength increased. Rao and coworkers[39] have carried out a XPS study on the interaction of CO with Au clusters of ZnO and TiO$_2$ deposited in-situ in vacuum. Figure 8 depicts the C1s region for Au/ZnO. It is quite evident that Au clusters of diameter 2.5 nm clearly show a feature at 285 eV in the C1s spectrum indicative of molecular CO on the cluster surface. The CO coverage is at best ~0.05 ML[40]. As the size of the cluster reaches 5 nm, only a weak adsorption is seen, whereas a 10 nm cluster behaves more like bulk gold showing no tendency to adsorb CO. Such cluster size dependent adsorption is known in the case of other metals as well[41-44]. Au/ZrO$_2$ system has been investigated by Boccuzzi *et al*[45] using infrared spectroscopy, uv-visible and electron microscopy. The authors report that CO is adsorbed on metallic sites on a fresh catalyst samples while that subjected to repetitive reduction-oxidation cycles, showed less activity. The stoichiometry of the substrate surface also may play an important role[46,47].

Recently, Heiz *et al*[48,49] studied the reactivity of gold clusters deposited on MgO substrate employing temperature programmed reaction (TPR) and Fourier transform infrared (FTIR) spectroscopy. The CO oxidation reaction was followed using the technique of isotope labeling. The TPR spectra of $^{13}C^{16}O^{18}O$ during the oxidation of $^{13}C^{16}O$ with $^{18}O_2$ are shown in Figure 9, Carbon monoxide was seen to get oxidized on small gold clusters at ~ 140 K as well as above 200 K, which the authors ascribe to two different mechanisms. The temperature ranges vary somewhat for different nuclearity clusters.

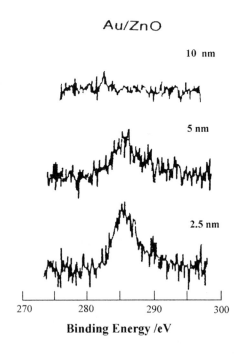

Au/ZnO

10 nm

5 nm

2.5 nm

270   280   290   300

Binding Energy /eV

Figure 8. C1s spectra recorded after exposing Au particles supported on a ZnO substrate to 2000L of CO at room temperature. The feature at ~285 eV corresponds to molecularly adsorbed CO. The diameters have been obtained from tunneling measurements (from ref. 39).

What is noteworthy is that clusters up to a nuclearity of 7 are not at all active and the CO production depends sensitively on the cluster size for Au_8 to Au_{20}, while a thick Au film does not catalyze the reaction. These results clearly show that gold becomes reactive when going to small clusters, in contrast to the distinct nobleness of gold in the bulk limit. Thus, using gas-phase clusters and depositing them with low energies on well characterized oxide surfaces, model catalysts can be obtained, which are ideal for studying the size effects in catalytic reactions[50-52].

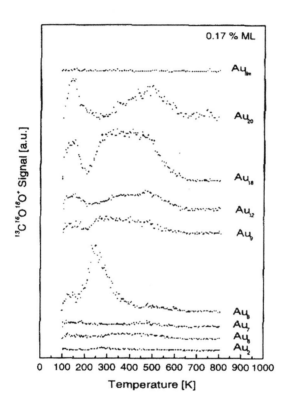

Figure 9. CO$_2$ production ($^{13}C^{16}O^{18}O$) on gold clusters of varied nuclearity as a function of temperature measured by temperature-programmed desorption, after exposing the clusters to $^{13}C^{16}O$ and $^{18}O_2$ mixture at 90 K (from ref. 48).

Generally accepted mechanism[15] to this date involves first, the activation of oxygen at the vacancies on the oxide surface either during pretreatment of the catalyst or from the (CO+O$_2$) gas feedstock.

$$\square_o + O_2 + e^- \rightarrow \square_o{-}O_2{}^-{}_{ads}$$

This process strongly depends on the nature of the oxide support. For example, a TiO_x surface activates oxygen much more easily than a TiO_2 surface,[46] needless to mention differences between the other metal oxides. The adsorbed oxygen then spills over on the gold cluster surface.

$$2Au + \square_o - O_2^-{}_{ads} \rightleftarrows 2Au - O_{ads} + e^- + \square_o$$

This step critically depends on the particle size. All literature reports agree that the active size regime is 2-4 nm,[36,53] which coincides with metal-nonmetal transition in gold[33-35]. It appears that the disappearance of the metallic electronic state with the emergence of bound states is very important for the catalytic activity. On the other hand, very small clusters (< Au_8) provide very little surface to interact[52]. The activated oxygen reacts with reversibly adsorbed CO, which is presumably a very fast process and it efficiency depends on the reaction temperature.

$$Au + CO \rightleftarrows Au - C \equiv O$$

$$Au - C \equiv O + Au - O_{ads} \rightarrow 2Au + CO_2$$

4. SUMMARY

In nanoscale catalysis, gold does not remain noble. Gold particles measuring 2-4 nm dispersed on oxide supports such as TiO_2, MgO, Al_2O_3 and ZrO_2 exhibit high activity at ambient temperatures for CO oxidation and NO reduction, reactions which are of utmost importance in automobile industry and in issues related to earth's environment. Gold catalysts are also potential candidates for other industrial reactions such as the hydrogenation of alkenes, oxidation of propylene and hydrochlorination reactions. Gold is a more viable replacement for platinum group metals, because of the relative abundance and ease of storage. Moreover, gold catalysts are not easily deactivated compared to other catalysts.

5. REFERENCES

1. Schmidbaur H., Gold Bull. 23 (1990) 11.
2. Hammer B. and Norskov J. K., Nature 376 (1995) 238.
3. Wang J., Voss M. R., Busse H. and Koel B. E., J. Phys. Chem B 102 (1998) 4693.
4. Pireaux J. J., Chtaib M., Delure J. P., Thiry P. A., Liehr M. and Caudano R., Surf. Sci. 141 (1984) 211.
5. Schaik J. R. H., Dressing R. P. and Ponec V., J. Catal. 38 (1975) 273.
6. Haruta M., Kobayashi T., Sano H. and Yamada N., Chem. Lett (1987) 405.

7. Haruta M., Tsubota S., Kobayashi T., Kageyama H., Genet M. J. and Delmon B., J.Catal. 144 (1993) 175.
8. Nkosi B. Coville N. J., Hutchings G. J., Adams M. D., Friedl J. and Wagner F., J.Catal. 128 (1991) 366.
9. Oliviera P. P., Patritio E. M. and Sellars H., Surf. Sci. 313 (1994) 25.
10. Haruta M., Yamada N., Kobayashi T. and Ijima S., J. Catal 115 (1989) 301.
11. Lin S. and Vannice M. A., Catal. Lett. 10 (1991) 47.
12. Lin S. D., Bollinger M. and Vannice M. A., Catal. Lett. 17 (1993) 245.
13. Gardner S. D., Hoflund S. B., Schryer D. R., Schryer J., Upchurch B. T. and Kielin E. J., Langmuir 7 (1991) 2135.
14. Bamwenda G. R., Tsubota S., Nakamura T. and Haruta M., Catal. Lett. 44 (1997) 83
15. Grunwaldt, J. D. and Baiker, A., J. Phys. Chem. 103 (1999) 1002.
16. Hutchings, G. J., Gold. Bull. 108 (1996) 2973.
17. Ueda, A., Ohshima, T., and Haruta, M. Appl. Catal B 12 (1997) 81.
18. Ueda, A and Haruta, M. Gold. Bull. 32 (1999) 1.
19. Baratto, C., Sberveglieri, G., Comini, E., Faglia, V., Ferrara, V. L., Lancellotti, L., Francia, G. D., Quercia, L., Guidi, V., Boscarino D., and Rigato, V., in (Bio) Chemical Sensors and Systems 105.
20. Hutchings, G. J., J. Catal 96 (1985) 292.
21. Ainsworth, S. J., C & EN (1992) March 2, 9.
22. Hayashi, T., Tauaka, K. and Haruta, M. J. Catal 178 (1998) 566.
23. Jia, J., Haraki, K., Kondo, J. N., Domen, K. and Tamaru, K., J. Phys. Chem B 104 (2000) 11153.
24. Sakurai, H., Ueda, A., Kobayashi, T. and Haruta, M. Chem. Comm (1997) 271.
25. Rao, C. N. R., Kulkarni, G. U., Thomas, P. J. and Edwards P. P., Chem. Soc. Rev. 29 (2000) 27.
26. Stievano, L., Santucci, S., Lozzi, L., Calozero, S. and Wagner F. E., J. Non Crystalline Solids 234 (1998) 644.
27. Schmid, G., 'Clusters and Colloids, from theory to applications', VCH, Weinheim, 1994.
28. Aiyer, H. N., Vijayakrishnan, V., Subbanna, G. N. and Rao C. N. R., Surf. Sci. 313 (1994) 392.
29. Rao, C. N. R., Vijayakrishnan, V., Aiyer, H. N., Kulkarni, G. U. and Subbanna, G. N., J.Phys. Chem. 97 (1993) 11157.
30. Wertheim, G. K. and Dicenzo, S. B., Phys. Rev. B. 37 (1988) 844.
31. Mason M. G., in: Clusters Model for Surface and Bulk Phenomena, eds Pacchioni *et al* (Plenum, New York, 1992).
32. Roulet, H., Mariot J. M., Dufour, G. and Hague, C. F., J. Phys. F. Metal Phys. 10 (1980) 1025.
33. Vinod, C. P., Kulkarni, G. U. and Rao C. N. R., Chem Phys. Lett. 289 (1998) 329.
34. Valden, M., Lai, X. and Goodman, D. W., Science 281 (1998) 1647.
35. Valden, M., Pak, S., Lai, X. and Goodman, D. W., Catal. Lett. 56 (1998) 7.
36. Mavrikakis, M., Stoltze, P. and Norskov, J. K., Catal. Lett. 64 (2000) 101.
37. Giorgio, S., Chapon, C., Henry, C. R., Nihoul, G. and Penisson, J. M., Phil. Mag. 64 (1991) 87.
38. Bondzie, V. A., Parker, S. C. and Campbell, C. T., Catal. Lett. 63 (1999) 143.
39. Vinod, C. P., Kulkarni, G. U. and Rao, C. N. R. (to be published).
40. Zhang, L., Persaud, R. and Madey, T. E; Phys. Rev B 56 (1997) 10549.
41. Bukhtiyarov, V. I., Carley, A. F., Dollard, L. A. and Roberts, M. W; Surf. Sci. 381 (1997) L605.

42. Carley, A. F., Dollard, L. A., Norman, P. R., Pottage, C. and Roberts, M. W., J. Elect. Spec. Rel. Phen. 99 (1999) 223.
43. Rao, C. N. R., Vijayakrishnan, V., Santra, A. K. and Prins, M.W. J., Angew. Chem. Int. Ed. Engl. 31 (1992) 1062.
44. Vinod, C. P., Harikumar, K. R., Kulkarni, G. U and Rao, C. N. R., Topics in Catalysis 11/12 (2000) 293.
45. Boccuzzi, F., Cerrato, G., Pinna, F. and Strukul, G., J. Phys. Chem. 102 (1998) 5733.
46. Magkoev, T. T., Rosenthal, D., Schroder, S. L. M. and Christmann, K., Tech. Phys. Lett. 26 (2000) 894.
47. Gardner, S. D., Hoflund, S. B., Davison, M. R., Laitinen, H. A., Schuyer, D. R. and Upchurch, B. T; Langmuir 7 (1991) 2141.
48. Heiz, U., Abbet, S., Sanchez, A., Schneider, W. D., Hakkinen, H. and Landmann, U., Proceedings of the Nobel Symposium 2000 World Scientific 2000.
49. Heiz, U. and Schneider, W. D., J. Phys. D. Appl.Phys. 33 (2000) R 85.
50. Heiz, U., Sanchez, A., Abbet, S., and Schneider, W. D., Chem. Phys. 262 (2000) 189.
51. Heiz, U., Sanchez, A., Abbet, S., and Schneider, W. D; Eur. Phys. J. D. 9 (1999) 1.
52. Sanchez, A., Abbet, S., Heiz, U., Schneider, W.D., Hakkinen, H., Barnett, R. N. and Landman, U., J. Phys. Chem. 103 (1999) 9573.
53. Jia, J., Haraki, K., Kondo, J. N., Domen, K. and Tamaru, K., J. Phys. Chem. 104 (2000) 11153.

Chapter 9

CATALYSIS FROM ART TO SCIENCE
Reflections on Half of a Century of Research on Heterogeneous Catalysis

Wolfgang M. H. Sachtler

Key words: Acid Catalysts, Carbenium Ions, Ligand Effect, Hydrogen Spillover, Hydrogen Transfer - intermolecular, Catalysis by Alloys, Adsorption on Metal Crystal Faces, Zeolite Encaged Clusters, Catalytic Reforming, Catalytic Alkylation, Catalytic Hydrogenation, Fischer-Tropsch Catalysis, Catalytic Oscillations, Substituted Cyclopropane Intermediates

Abstract: The edifice called "Science of Heterogeneous Catalysis" rests on two cornerstones: (1) the chemical interaction between surface atoms and reactants, and (2) the atomic geometry of the adsorbing sites and their environment. Fundamental research on surface science and catalysis of transition metals, their alloys, strong acids, bifunctional catalysts, and zeolite-based systems has revealed additional concepts of catalytic activity and selectivity. Substituted cyclopropane intermediates are key to both acid catalyzed and metal catalyzed isomerizations. New results on the hydrogenation of nitriles to amines illustrate the importance of intermolecular H-transfer on metal surfaces, analogous to the known hydride transfer in acid catalyzed alkane conversions. Limitations of the hydrogen spillover concept have been identified. Promising new fields seem to open up in the areas of zeolite encaged clusters and transition metal oxo-ions, including binuclear ions with oxo and peroxo bridges that are similar to active groups in enzyme catalysts. While catalytic oscillation of N_2O decomposition over zeolite encaged iron and cobalt complexes is still incompletely understood, this phenomenon shows clearly that events occurring in different zeolite cages can be perfectly synchronized.

Surface Chemistry and Catalysis, Edited by Carley *et al.*
Kluwer Academic/Plenum Publishers, New York, 2002

1.   INTRODUCTION

The flame of catalysis was ignited, literally, on July 27 of the year 1823, when Johann Wolfgang Döbereiner, working in his lab in Jena, Germany, directed a flow of hydrogen from a Kipp apparatus onto some platinum black on a filter paper. Within seconds the platinum grew hot and a flame appeared. He repeated this experiment many times and noted that the propensity of platinum to ignite a hydrogen/air mixture did not decrease even after thousand runs. To appreciate his excitement we should remember that matches did not yet exist; making fire was a rather laborious job. If yours went out you might take a candle and see a friendly neighbor to borrow a flame.

The miracle of obtaining a flame by simple contact of a cold gas with a cold solid drew much attention. Jöns Jacob Berzelius admitted that he could not explain this new phenomenon, so he did what people often do in such a situation: he coined a new, impressive-sounding word [1]. He chose the term *CATALYSIS*. He speculated that some mysterious force, the *"vis catalytica"*, was responsible. Justus von Liebig was vehemently opposed to the idea of postulating a new *"vis occulta"* in a time when science tried to rid itself of mystical beliefs, but he did not come up with a better explanation. Science in those days just did not provide an appropriate concept.

Though the mechanism of catalytic reactions long remained obscure, pioneers realized its importance for industry and were inspired to invent new processes. Table 1 shows a brief list of some of the better known industrial processes invented during the twentieth century and entirely based on catalysis. Clearly, our modern world can hardly be imagined without them. The ammonia synthesis process, developed by F. Haber and K. Bosch, paved the way for the production of nitrate fertilizers, without which the present world population would suffer severe famine. How would the "Battle of Britain" in 1940 have ended, without the high-octane fuel being available to the RAF, and not their adversary? This fuel was manufactured by the then new *"catalytic reforming"* process invented by Eugène Houdry; its octane number was boosted further with *iso*-octane, produced in another brand new process, *"catalytic alkylation"*, invented by Herman Pines and Vladimir N. Ipatieff [2]. We better leave it to a science fiction writer to describe how the present world would look without industrial catalysis.

Most of the inventions were made with very limited knowledge of the atomic processes at the surface of a catalyst, and each one triggered research to shine light into that darkness. The results of such investigation led to tentative models that helped to improve the catalysts and gave some guidance to other inventors working on similar processes. This mutual fertilization of research for better understanding and innovative inventions has led us a long way from the *"vis occulta"* of Berzelius *via* Sir Eric Rideal's book *"Concepts in Catalysis"* [3], to the more modern view of heterogeneous catalysis as a branch of the physical sciences.

Table 1. 1900-2000 Inventions based on Heterogeneous Catalysis

Date	Process	Inventor
1900	Hydrogenation of unsaturated acids over Raney Nickel	Sabatier, Senderens
1913	Ammonia Synthesis	Haber Bosch
1925	Methanol Synthesis	BASF
1931	Ethylene Epoxidation	Lefort
1930s	Conversion of CO + H_2 to Hydrocarbons	Fischer, Tropsch
1930s	Hydroformylation Synthesis of oxygenates from CO + H_2	Roelen
1930s	Hydrocarbon Reforming over Solid Acids	Pines, Ipatieff, Houdry, Oblad
1941	Fluidized Bed Catalytic Cracking (FCC)	Murphree
1949	Platforming	Haensel
1952	Polymerization with Alkylated $TiCl_3$	Ziegler
1955	Stereoscopic Polymerization	Natta
1960	Selective Oxidation and Ammoxidation of Olefins	Callahan, Grasselli
1964	Hydrocarbon Conversion over Acidic Zeolites	Rabo, Plank, Rosinsky
1970s	"Three-Way" Auto-Emission Catalyst	Lester, Shelef
1980	Methanol to Gasoline over H-ZSM-5	Chang, Silvestri
1989	Synthesis of Fluorohydrocarbons	Manzer
1990	Epoxidation of Butadiene	Monnier

An early breakthrough was Irving Langmuir's finding that short range chemical forces are responsible for the attachment of molecules to the surface of a metal. S. Brunauer, P.H. Emmett and E. Teller developed a useful method to measure the surface area of solids by counting the number of molecules of known size which form an adsorbed "monolayer" [4]. Sir Hugh Taylor [5] showed that not all surface atoms are equally active, but that practical catalysts often function by virtue of some "active sites". Gwathmey and Leidheiser [7] showed that different crystal faces of a given metal can act very differently in catalysis. The consequence that adsorption should also be crystal-face specific was confirmed when the *field emission microscope* was invented by Erwin Müller in 1937 [8]. For the first time, scientists could visually follow the movement of molecules over the surface of a metal tip.

Much progress in understanding heterogeneous catalysis was made after the Second World War, though key observations had already been reported in the 1930s. The new concepts were triggered by four, initially unrelated findings. One was the big breakthrough in the catalysis of petroleum refining. The textbook dogma of unreactive paraffins evaporated when Herman Pines, Vladimir Ipatieff and Eugène Houdry found that strong acids are able to catalyze novel hydrocarbon reactions, including alkylation, isomerization and cracking. Acid catalysis is now the most important application of catalysis, outweighing all others by the tons of products, their commercial value and the amounts of the catalyst used. A second revolution took place in the laboratories where research on metal surfaces triggered an avalanche of new spectroscopic and physical methods. The atomic details of such surfaces and their interaction with adsorbing molecules no longer had to be derived from indirect data; direct, reproducible results could be measured by new highly sophisticated techniques. A third hotspot for improved understanding of catalysis was discovered on transition metal oxides, such as vanadates or molybdates, which catalyze selective oxidations. This is the field where Piet Mars and Hendrik van Krevelen showed that catalysis often consists of a series of "quite ordinary" chemical steps. A fourth source of inspiration were the new findings in metal-organic chemistry, such as the discoveries of Karl Ziegler and Giulio Natta who demonstrated not only that polymerization was possible at atmospheric pressure, but also that the atomic geometry of the catalyst sites preempts the mechanical qualities of the product. A new logical link emerged between optical activity, asymmetric catalysis and mechanical hardness of synthetic polymers.

Recently, M. Wyn Roberts in a remarkable paper gave his view how catalysis developed [9]. He wisely added the subtitle "some personal reflections", knowing that every scientist who is actively involved in the generation of new insights will see the scenery of his science from his perch inside this field, not as a traveler in outer space. In the present paper, dedicated to Professor Wyn Roberts, with whom I share many fond memories, I will give another personal view. I will address the question: *what did we learn* from the eruption of experimental results? How has *understanding* of catalysis changed?

To illustrate this evolution of concepts I will limit this paper to three areas of catalysts: (1) strong acids, (2) transition metals and their alloys, and (3) zeolites. This limitation will leave other important fields unmentioned, such as selective oxidation of hydrocarbons and other organic molecules over transition metal oxides. The paper will not discuss methods to characterize catalysts or adsorbates in detail. It should however be mentioned, that in retrospect most conclusions on molecular mechanisms of catalytic reactions in this paper were obtained by means of two most powerful methods: (1) infrared spectroscopy and (2) isotopic labeling.

Future generations might take the technical progress of the 20th century for granted and merely ask us: what did you do to save this planet from the disastrous consequences of air pollution, water pollution and global warming? We will then point to the Three-Way catalyst in our cars, the deep desulfurization of our fossil fuels and, perhaps, the NO_x reduction over zeolite based catalysts. Only the last mentioned subject will be briefly mentioned in this paper as a small example of the present world-wide effort to use catalysis for saving the environment of our planet.

2.   CATALYSIS OVER STRONG ACIDS

Early in the 1930s, Herman Pines observed that shaking a petroleum fraction with strong sulfuric acid not only removed the water-soluble components but also triggered a new class of chemical reactions. Together with Vladimir N. Ipatieff he systematically studied reactions of hydrocarbons over strong liquid and solid acids. Their work thoroughly revolutionized the field of hydrocarbon conversion and resulted in a number of important new industrial processes [10,11,12,13,14]. In the same period, Eugène Houdry found that the quality of gasoline could be markedly improved by blowing the vapor over certain heated solids. This discovery led to the *"catalytic reforming"* process. Its industrial production was started on June 6, 1936. In 1939 the world production was 10,000 bbl/day [15]; in 1978 the capacity for catalytic cracking in the US alone had reached 5,000,000 bbl/day. This process now outperforms all other industrial catalytic processes. Approximately 35% of all crude petroleum feeds are processed through a "cat cracker". The first *Fluid bed Catalytic Cracking,* or FCC, unit was introduced in 1941. The early catalysts were clay minerals; later, halided alumina or silica/alumina was used. After 1961 zeolite catalysts were introduced that displayed an even higher acid strength and a higher selectivity for the desired gasoline fractions. P.B. Weisz and his associates at Mobil played a dominant role in this development [16]. At present most refineries use these strong and robust acidic catalysts. The technical development and the growing understanding of catalytic cracking has been described by major actors in the field, including V. Haensel [17] R.A. Shankland [18] and H. Heinemann [19,20]. *Liquid* acids, in particular concentrated sulfuric acid and liquid hydrogen fluoride, were mainly used in the early work of Pines and Ipatieff on catalytic alkylation, but *solids* of the Friedel Crafts type were soon included in their work. Solid acids are the backbone of present day cracking and isomerization processes and their numerous variants, including *"Shape Selective Cracking"* [21] but they also catalyze the *"Methanol to Gasoline"*, or *MTG,* process [22]. As a consequence of the large scale of these industrial processes, even minor byproducts dominate their specific markets. An example is hydrogen, which is formed when alkanes are dehydrocyclized to aromatics or coke deposits. Process variables such as the pressure (33-40 atm), the hydrocarbon/steam ratio and alkaline catalyst promoters have been optimized to obtain high hydrogen production [23].

Crucial for the interaction between acid catalysts and hydrocarbons is the basicity of olefins and aromatics, which enables them to react with strong Lewis and Brønsted acids. For instance, an olefin reacts with a Brønsted acid site to form a *"carbenium ion"*

$$R_2C=CR'_2 + H^+ = R_2C\text{-}CR'_2 \qquad\qquad (1)$$
$$\underset{H}{\overset{|\ (+)}{}}$$

Already in his 1948 paper, Ipatieff, quoting Whitmore [24] assumes that these ions are the key players in this chemistry. The mechanism of their isomerization has been studied by numerous authors, including D.M. Brouwer and H. Hogeveen [25,26] who showed that carbenium ions with five or more carbon atoms easily isomerize *via* a substituted cyclo-propyl ion intermediate: The three- membered ring subsequently opens to form a *secondary* carbenium ion that easily isomerizes, *via* a hydride shift, to a *tertiary* carbenium ion.

$$
\begin{array}{ccccccc}
CH_3 & & CH_2 & & CH_3 & & CH_3 \\
| & \xLongrightarrow{} & /\;H+\;\backslash & \xLongrightarrow{} & | & \Rightarrow & | \\
R\text{-}C\text{-}CH_2\text{-}R & & R\text{-}CH\text{-}CH\text{-}R' & & R\text{-}CH\text{-}\text{-}CH\text{-}R' & & R\text{-}CH_2\text{-}\text{-}C\text{-}R' \quad (2)\\
(+) & & & & (+) & & (+)
\end{array}
$$

Unlike the mechanistic models proposed before, Brouwer's mechanism avoids the energetically very unfavorable formation of a *primary* carbenium ion. The fact that *n*-butane is <u>not</u> isomerized to *iso-butane* under conditions where *n*-pentane and all higher *n*-alkanes isomerize easily, is a strong argument in favor of this mechanism. As the model shows, a *n*-C_4 carbocation, though also able to form a three-membered ring, lacks the ability to open it to a *sec.* cation of *iso*- C_4 structure. Only a *primary* cation would lead to an *iso*-C_4 structure. As *prim.* cations have a much higher energy than *sec.* cations, this route is not open at the temperature of these experiments. The cyclic C_4 ion can, however, open to another *sec.* ion of *n*-

C_4 structure. Indeed, isotopic labeling confirmed that isomerization of C^*H_3-CH_2-CH_2-CH_3 to CH_3-C^*H_2-CH_2-CH_3 is quite rapid.

Isomerization of *n*-butane to *iso*-butane becomes possible, however, if a path is opened to a mechanistic detour, as proposed by Guisnet [27,28]. In this case the C_4 carbocation has to first react with an olefin, such as a butene molecule. The resulting C_8 carbocation can easily isomerize without passing through the stage of a *prim.* carbenium ion. If the isomerized product decomposes into two C_4 fragments, one of which (or both) having an *iso* C_4 structure, a skeletal isomerization is achieved. This chemistry is indeed taking place over sulfated zirconia, as was shown by Adeeva *et al.* [29,30] using the double labeled molecule $^{13}CH_3$-CH_2-CH_2-$^{13}CH_3$. The sulfated zirconia is a strong Brønsted acid; its acid strength was shown to be of the same order as that of conventional acid zeolites [31].

While carbenium ions are easily formed from olefins and strong acids, an additional step is necessary to make this reaction path available to alkanes. Ipatieff and Schmerling [11] assumed that a H atom is transferred with its electron pair from an alkane, such as isobutane, to a carbenium ion. This was confirmed by isotopic labeling in elegant work of Otvos *et* al. who studied the H/D exchange of *iso*-butane in D_2SO_4 [32]. They found that only the H atoms in the three methyl groups were exchanged against D, but the H atom at the *tertiary* C atom, though known to be most reactive, was apparently not exchanged. They knew that a trace amount of olefin eagerly reacts with the acid D^+ sites

$$(CH_3)_2C{=}CH_2 \ + D^+ \ \rightleftharpoons \ [(CH_3)_2C\text{-}CH_2D]^+ \tag{3}$$

but that process would ultimately only lead to complete H/D exchange of the <u>eight</u> H atoms in the olefin or the <u>nine</u> H atoms in the carbenium ion. A different process is required to convert either of these species into an *iso*-butane molecule. In conformity with Ipatieff's view, they assumed that a H⁻ ion was transferred from another alkane molecule to the carbenium ion

$$[(CD_3)_3C]^+ + \ H\text{-}C(CH_3)_3 \ \rightleftharpoons \ [C(CH_3)_3]^+ + (CD_3)_3C\text{-}H \tag{4}$$

The authors confirmed this mechanism by adding D-isopentane, $(CH_3)_2CDC_2H_5$, as another donor of D atoms. This led indeed to H/D exchange of all 10 H atoms in *iso*-butane. It follows that the catalytic cycle of alkane isomerization over strong acids is opened and closed by hydride shift from an alkane to a carbenium ion. For converting an infinite number of alkane molecules it suffices that each skeleton resides for a short while in the state of the carbenium ion. In this state it can isomerize or form a ring of five or six carbon atoms or react with another alkene. Alkane conversion over acid catalysts thus always includes three crucial steps:

1. Proton addition to an olefin
2. Skeletal isomerization of carbocations *via* cyclic intermediates,
3. H transfer from an alkane to a carbocation.

While these three steps describe the essence of acid catalyzed isomerizations, additional phenomena are of importance in acid catalyzed reactions. In brief they are summarized as follows:

- Carbenium ions can react with olefins to form larger carbenium ions, while the reverse process leads to cracking of larger entities. Cationic polymerization and catalytic cracking are based on this general concept.
- Many solid acids expose not only Brønsted acid sites, but also Lewis acid sites at their surface. Actually, aluminum chloride, a typical Friedel-Crafts catalyst, was the first commercial catalyst used for the conversion of heavier petroleum fractions to lighter hydrocarbons. Since a solid exposing acidic O-H groups and incompletely coordinated Al^{3+} ions at its surface will chemisorb ammonia on both types of acid sites, the two adsorbates are characterized by IR spectroscopy. On Brønsted sites ammonia shows the spectroscopic signature of ammonium ions, NH_4^+, but on Lewis acid sites a typical Lewis acid-Lewis base complex $Al-NH_3$ will be formed. As chemisorption of water transforms Lewis into Brønsted sites, the relative concentration of either site depends on the extent of calcination of the solid. It has been proposed that alkanes are dehydrogenated over these sites possibly *via* intermediate radicals [33].
- Acid strengths of solid catalysts are measured in various ways. Adsorption of a base, followed by temperature programmed desorption, provides data on the activation energy for desorption, which is often assumed equal to the heat of adsorption. More reliable is the spectroscopic identification of the interaction between the acid site and a weak base. For instance, in the adsorption of a CO molecule on a Lewis acid site, electrons in the antibonding level of the $C\equiv O$ bond become stabilized, *i.e.*, the $C\equiv O$ bond becomes stronger. Knözinger *et al.* use the resulting shift of the IR band to higher wave numbers as a measure of the acid strength [34,35]. With Brønsted acid sites the perturbation by acetonitrile of the bands characteristic for surface OH groups reveals the acid strength, as shown by Neyman *et al.* [36].
- G. Olah studied hydrocarbons in "superacids" such as $H[SbF_6]$[37]. Although the basicity of alkanes is much lower than that of alkenes, they can form adducts with protons of superacids. The result is a **carbonium** ion. Their chemistry is distinctly different from that of carbenium ions. For instance isobutane can exchange all its ten H atoms against D.

Sommer, therefore proposes to use this as a criterion to distinguish *superacids* and normal strong acids as present, for instance, in zeolites, where the carbenium ion chemistry of normal strong acids is operating. In normal strong acids, *iso-* butane in contact with the deuterated acid will exchange the nine H atoms in its three methyl groups, but superacids enticing carbonium ion chemistry will exchange all ten H atoms against D [38]. This fact can be used as an operational definition of superacids [39]. Unfortunately, there is no consensus on this definition. Some authors used the isomerization of isobutane as an operational definition. This must be strongly discouraged, as it is well known that a normal acid can achieve this isomerization by using the detour *via* dimerization to an octane intermediate, as shown above.

- Release of a H_2 or a CH_4 molecule converts a carbonium ion to a carbenium ion.

$$[(C_nH_{2n+1})_3CH-CH_3]^+ => [(C_nH_{2n+1})_3C]^+ + CH_4 \qquad (5)$$

- As neither hydrogen nor methane formation are expected on the basis of classical carbenium ion chemistry, Haag and Dessau concluded from the formation of these molecules under industrial conditions that alkanes form carbonium ions also on the surfaces of industrial solid acid catalysts [40]. At high temperature the carbonium ions will be converted to carbenium ions. Theoretical work by Kramer *et al.* shows that in the H/D isotope exchange between a hydrocarbon and a Brønsted proton a symmetrical intermediate state is passed in which an H and a D atom are both bonded to the same C atom and simultaneously to two neighboring oxygen ions of the catalyst surface [41]. With this chemistry, methane can exchange its H atoms against the D atoms of a deuterated acid catalyst [42].

- The stable chemisorption complex of an olefin on a solid acid such as an H- zeolite is not a classical carbenium ion. Kazansky [43,44,45] concludes from quantum mechanical calculations that a $C_nH_{2n+1}^+$ ion held by mere Coulombic energy to the surface of a zeolite is less stable than a structure which includes the strong interaction of one C atom with an O^- ion of the zeolite, thus resembling an alkoxy group. Likewise, J. Haw *et al* [46] conclude from solid state NMR results that the model of a free carbenium ion held to the surface by electrostatic forces is a rather poor representation of the actual complex. Korel *et al* [47]. prefer the term "complexed olefin". Still, most structural predictions for the isomerization of aliphatic molecules that are based on the carbenium ion model also permit easy rationalization of the experimental results on solid acids.

3. **METAL ASSISTED ACID CATALYSIS; MECHANISTIC COMMONALITIES OF ACID AND METAL CATALYZED REACTIONS**

An important breakthrough in industrial hydrocarbon conversion catalysis was achieved by Vladimir Haensel at Universal Oil Products (UOP). He found that deposition of a small amount of platinum on an acidic alumina catalyst strongly improves the quality of the product and the life of the catalysts [48] The platinum helps to establish thermodynamic equilibrium between hydrogen alkenes, dienes, alkynes, aromatics and the corresponding saturated molecules. It also helps hydrogenating coke precursors and thus increases catalyst life [17]. The need for continuously hydrogenating deposits explains the apparent paradox that a process that is designed to *de*-hydrocyclicize alkanes to aromatic compounds is carried out under a substantial pressure of hydrogen. A brief discussion of bifunctional catalysts, exposing both acid and metal sites, will therefore be in order in this chapter. Moreover, we shall briefly mention that some mechanistic principles of acid catalysis have been found to be applicable to metal catalyzed reactions.

Transition metals have the unique ability to dissociatively chemisorb H_2 molecules. Within this group of metals, platinum is unique to combine high activity for hydrogenation with very low activity to hydrogenolysis of hydrocarbons. In an atmosphere rich in H_2 a small amount of platinum thus suffices to keep the steady state concentration of dienes low and thus control formation of carbonaceous deposits. The mechanistic question is this: do the unsaturated molecules have to make chemical contact with the H-covered platinum, or can they be saturated by reacting with H atoms "spilling over" from the Pt surface onto the solid acid?

While this question may still be considered controversial by some, the present author is offering his conclusions based on thermodynamic and experimental facts. There is no doubt that <u>hydrogen spillover</u> is thermodynamically favorable and experimentally proven for systems consisting of Pt particles in contact with a <u>reducible oxide</u> such as WO_3, MoO_3 or Fe_2O_3. [49,50,51]. Solid acids, such as zeolites, or silica/alumina are, however, NOT easily reducible. The jump of an H atom from a Pt surface to a non-reducible oxide, where it is held only by van der Waals-London forces, would be an extremely endothermic step. As Figure 1 shows, a ΔH value of + 470 kJ/mol of H_2 is estimated for this step. This is considered prohibitive at the temperatures of interest for hydrocarbon conversion catalysis. While protons easily migrate over such surfaces, electrons do not migrate in insulators. Partially reduced oxides of W, Mo or Fe are, of course, semiconductors, whereas SiO_2, Al_2O_3 or zeolites are such poor conductors of electricity that they may be called insulators. It thus follows that H spillover from Pt is thermodynamically permitted onto a semiconducting (=reducible) oxide, but for all practical purposes forbidden onto an insulating (=non-reducible) oxide. This concept was probed in recent work with zeolite or silica supported transition metals in intimate physical contact with Fe_2O_3. The results of this research showed unambiguously that whenever the presence of Pt (or another transition metal) enhanced the reducibility of the Fe_2O_3, it was because the platinum migrated out of the zeolite onto the reducible Fe_2O_3. If the transition metal could not migrate, the reduction characteristics were identical with that of pure Fe_2O_3. We therefore conclude that no H spillover of catalytic significance takes place on typical platforming catalysts.

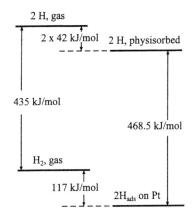

Figure 1. Approximate energy levels of H atoms in gas phase, physisorbed on insulating surface, and chemisorbed on transition metal. All numbers in kJ/mole H_2

Hydrocarbons are isomerized both on acid and on transition metal catalysts. In both cases, cyclic intermediates are formed. The group of F. Gault studied isomerizations of alkanes on platinum and other metals by using ^{13}C labeled molecules to identify the reaction mechanism [55]. They distinguish two main mechanisms which they call the *"bond shift"* and the *"cyclic"* mechanism. In fact, both mechanisms make use of a carbon-metalla-cycle including at least one metal atom of the catalyst surface. The former mechanism uses an intermediate with three C atoms and an unspecified number of metal atoms. In the latter mechanism, the assumed intermediate is a cycle of five C atoms and presumably one metal atom. We prefer the nomenclature used by Ponec in discussing the mechanistic consequences of Gault's results, whereas Gault himself refers to the 3C and the 5C complexes respectively [56]. Brouwer's three- membered ring carbocation is a substituted cyclopropane molecule with one extra proton interacting with the π-electrons of the C_3 ring. Similarly, one can write Gault's 3C intermediate as a substituted cyclopropane molecule with one metal atom interacting with the C_3 ring. Likewise, there is at least a formal analogy between the 1,5 ring closure and opening on an acid site and on a metal *via* the 5C intermediate.

Also the hydride transfer step of alkane conversions *via* carbenium ions appears to have a close analogue in the catalysis by metals. When ethene reacts with an excess of D_2 at a nickel surface, the prevailing initial product is C_2H_6, NOT $C_2H_4D_2$. With the H/D exchange of ethylamine or diethyl amine over Pt, Rh or Ru, it has been found that all C-bonded H atoms are easily exchanged against D, but the N-bonded H atoms, normally the most reactive ones, are seemingly not exchanged . The reason is, again, that these reactive atoms leave the molecule during the formation of the adsorption complex, for instance for diethylamine on ruthenium:

$$(C_2H_5)_2 \text{ N-H} + 2 \text{ Ru} \implies (C_2H_5)_2 \text{ N-Ru} + \text{Ru-H.} \tag{6}$$

After H/D exchange in the ethyl groups, the N-Ru bond is broken by H transfer from another molecule, for instance:

$$(C_2D_5)_2 \text{ N-Ru} + (C_2H_5)_2 \text{ N-H} \implies (C_2D_5)_2 \text{ N-H} + (C_2H_5)_2 \text{ N-Ru.} \tag{7}$$

The analogy with the observations by Otvos *et al.* on the H/D exchange of *iso*-butane will be obvious. Also in the formation of amines by "adding" D_2 to a nitrile, the product contains mainly H atoms bonded to the N atom of the amine, while D atoms are bonded to C-atoms [57]. Clearly, the intermolecular H transfer from the methyl group of an acetonitrile molecule to the N atom of a nitrene-like chemisorption complex is more efficient than simple addition of adsorbed D atoms in hydrogenolizing the bond between the N atom and the metal surface [58].

A common element of equation (4), when applied to solid acids, and equation (7) is that the adsorption step for a molecule is also the desorption step for an adsorbed moiety. This chemistry can thus also be considered as a rather extreme example of the phenomenon which Yamada and Tamaru called "adsorption assisted desorption" [59].

4. ADSORPTION AND CATALYSIS ON WELL DEFINED METAL SURFACES

Although acid catalysts form the lion's share of the world's industrial use of heterogeneous catalysis, most fundamental research on catalysis has been focused not on acids but transition metals. Even earlier, Horiuti and Polyani [60] studied H/D exchange between D_2 and ethanol over platinum black. The formation of HD from H_2 and D_2, called isotopic equilibration, was widely used as a probe reaction for metal catalysts. This research led to the conclusion that in contact with a nickel film, hydrogen molecules dissociate into two adsorbed H atoms even at liquid nitrogen temperature.

Early research on chemisorption and catalysis of clean metal surfaces focused either on the electronic interactions between a transition metal and an adsorbate or on the mechanism of model reactions by making use of isotopic labelling. More recently the dazzling opportunities presented by modern surface science methods are exploited and much research is done on single crystals.

Otto Beeck's often quoted paper *"Catalysis - a challenge to the physicist"* and his presentation at the Discussions of the Faraday Society in 1950 [62] illustrate the conceptual ideas that fascinated surface scientists in the 1950s. In physics, the electron band model had been launched to explain phenomena such as electric conductivity or magnetism of transition metals. As the same transition metals were also the best known hydrogenation catalysts, including those for the hydrogenation of N_2 to ammonia and the hydrogenation of carbon monoxide to Fischer-Tropsch hydrocarbons, it was tempting to expect a simple causal relationship between catalysis and this electron band model. Georg-Maria Schwab applied these ideas to catalysis by alloys; Dowden and Eley applied them to single metals. Suhrmann's group studied photoelectric emission, first from a platinum foil cleaned in UHV, and later from metal films [63]. The objective of that work (of which the present author's thesis was a part) was to quantitatively measure the change in work function caused by chemisorption of certain molecules. This work was extended to the electric conductivity of metal films, a parameter which was also found to change upon chemisorption [64]. In the same period, Mignolet in Belgium observed that chemisorption, but also physical adsorption of xenon, causes pronounced changes in the work function of the metal [65]. These changes were measured by means of a vibrating condenser which registered the contact potential between the metal face of interest and an inert metal.

In the same period, other authors exploited the opportunities to unravel reaction mechanisms by isotopic labeling. The groups of Charles Kemball [66] Robert Burwell, Jr. [67] Geoffrey Bond [68,69] John Rooney [70] and the present author [71,72,73] used H/D exchange of alkanes and other molecules, while isomerization of ^{13}C labeled hydrocarbons was studied by Otto Beeck [74] and later most extensively by François Gault and his group [55]. Very important general conclusions were drawn from this work concerning the fission and formation of C-H and C-C bonds.

From this early work the catalysis community learned a number of simple lessons:

- Chemisorption by transition metals of H_2 or alkanes is dissociative at all temperatures of interest in catalysis.
- With these adsorbates, the chemical bond between the metal surface atoms and the adsorbed moieties is mainly covalent, i.e, the dipole moment of the adsorption bond is very small.
- Metal-metal bonds are weakened by chemisorption [75]. The contribution of surface atoms to the "collective properties" of the metal, such as electric conductivity, is strongly reduced (often to near zero) by formation of a chemisorption bond [76,77].
- Crystal faces of a given metal strongly differ in their propensity to chemisorb a given molecule. Models suggest that the sites where strong bonds are broken consist of a rather large *ensemble* of typically five or more atoms [78,79].
- Molecules hitting a metal surface often migrate over considerable distances before they (or their fragments) find a site of high heat of adsorption. "Landing sites" and chemisorption sites are not necessarily identical [80].
- None of the "collective" parameters of transition metals which describe their electric conductivity and the phenomenon of ferromagnetism correlates with their catalytic activity or selectivity.

The last statement shows a certain resignation. It concedes that no correlation of predictive value links catalysis to the electronic band model for the collective phenomena of conductivity and magnetism of transition metals. Catalysis, described earlier as a challenge to the physicist, came home to roost on the roof of the chemist. As predicted by Sabatier and A.A. Balandin [81] the catalytic activity of metals for a given reaction could better be correlated with the heats of adsorption of reactants and products. If activity is plotted against an appropriate enthalpy term, a "volcano-shaped curve" will result. This was indeed found to be the case for the decomposition of formic acid, a reaction which is catalyzed by a large number of metals [82,83,84]. The resulting volcano-shaped curve is shown in Figure 2.

The study of adsorption on well-defined single crystal faces was initially confined to metal tips in the field emission microscope (FEM). The early results clearly demonstrated the strong crystal face specificity of chemisorption and even physical adsorption. The groups of Gomer [85] Ehrlich [86] and the present author [87] used FEM for research of adsorption on metals including platinum. The obvious commonality between chemisorption of atoms such as oxygen or nitrogen and the "chemisorption" of metal atoms, *i.e.*, crystal growth from the vapor, became a valuable guide for understanding crystal face specificity of chemisorption, and more generally, *structure sensitivity* of heterogeneous catalysis. I. Stranski had given an atomic interpretation of the Gibbs Wulff principle, which relates the equilibrium shapes of crystals to the surface energies of their crystal faces [88,89]. To some extent, the same principles appear applicable to chemisorption on crystal faces.

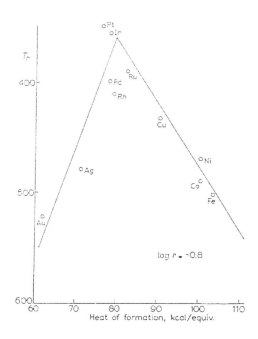

Figure 2. Activity of various transition metals for formic acid vapor decomposition, plotted versus the heat of formation of their formates. T_r = temperature at which log rate = -0.8 (from "The Mechanism of Heterogeneous Catalysis", J.H. de Boer *et al. Eds.*, Elsevier 1960, p. 42).

Individual surface atoms became visible when the field ion microscope was invented by E.W. Müller [90]. This instruments revealed that chemisorption on a well-ordered surface can induce thorough surface reconstruction, leading to disorder, in other words chemisorption is *"corrosive"* [91]. This reconstruction process may lead to a new ordering of atoms, detectable by LEED.

The modern epoch of single crystal research started in the 1960s when a plethora of new instruments and methods became available, starting with LEED, AES, and a group of electron spectroscopies. An avalanche of fascinating results descended upon the community of scientists interested in catalysis. More atomic details of the chemisorption process were revealed than anybody had ever imagined. No attempt will be made in the present paper to give even a nearly complete overview. We merely mention the trend setting research by Gabor Somorjai [92] Wayne Goodman [93] M. Wyn Roberts [94,95] David A. King [96] Gerhard Ertl [97,98] H. Ibach [99] and Bruce Koel [100]. As mentioned in the Introduction, this paper will focus instead on the conceptual progress in the areas of chemisorption and catalysis. Here are some answers:

- Chemical work with isotopic labeling and kinetic analysis proves that dissociation of N_2 is the first and crucial step in ammonia synthesis over Fe catalysts. Brill showed that ammonia synthesis over an iron catalyst was overwhelmingly determined by the (111) face of this bbc metal [101]. At this rather open face are "hollow" sites, called C_7 *sites,* which expose coordinatively unsaturated iron atoms of three different atomic layers. Research by Spencer *et al.* confirmed the high activity of these C_7 *sites* on the (111) and (211) faces of a large Fe crystal [102]. Further work on adsorption of CO and NO confirmed that sites for dissociative adsorption consist of a rather large *"ensemble"* of five or more atoms [103,104].

- Dissociation of CO is the first and crucial step in Fischer-Tropsch catalysis of hydrocarbons [105,106,107,]. Parallel Schulz-Flory lines for alkanes and oxygenates indicate identical chain growth kinetics [108]. CH_x groups are responsible for chain initiation and chain growth, but attachment of undissociated CO to a growing chain is the first part of a termination step. It results in the formation of an aldehyde or a primary alcohol.

- Even the adsorption of olefins includes rupture of C-H bonds. The groups of Somorjai [109] and Ibach [110] showed that ethylidene and ethylidyne groups are formed when ethylene is adsorbed on platinum [111]. However, as the reaction of such groups with chemisorbed H atoms is sluggish, it is uncertain whether they are intermediates or mere spectators in hydrogenation reactions.

- Strong chemisorption on open crystal faces is *"corrosive",i.e.,* surface reconstruction takes place at high coverage [75]. The chemisorption induced disorder is clearly visible in the field ion microscope, as was shown by Holscher *et al.* [91] "Peeling off" of the outermost layers of metal atoms by field evaporation shows that the defects generated by chemisorption have penetrated several atomic layers deep into the crystal. Likewise, Brill *et al.* observed surface reconstruction of Fe crystals by chemisorbing nitrogen [102].

• Reactions such as CO oxidation on platinum show the phenomenon of *catalytic oscillation.* As was shown by the group of Ertl in Berlin, Germany. The analysis of this phenomenon is of general validity to reaction of strongly adsorbing molecules on metals.

Most research on metal single crystals was done by studying one adsorbate on one face. Co-adsorption of two molecules at the same face was initiated by the group of M. Wyn Roberts in Cardiff. These authors developed a UHV compatible multi-photon source photoelectron spectrometer; they found that O_2 chemisorption on metals is not necessarily dissociative in one single step, but the precursor stages of an adsorbed superoxide-like $O_2^{\delta-}$ and an $O^{\delta-}$ species were identified. These adsorbates were found to be more reactive than the final state of O^{2-}, which is rather inert towards many reactants. Co-adsorption of O_2 and NH_3 on a Mg (0001) surface revealed that $O^-_{(s)}$ induces H abstraction from co-adsorbed ammonia, leading to the formation of surface amide, $NH_2(s)$ and $OH(s)$ groups [112,113]. On Zn(0001) co-adsorption of O_2 with ammonia or pyridine indicates formation of a charge transfer complex of an adsorbed superoxide ion, $O_2^-(s)$, and the nitrogen base [114]. In the case of ammonia, the complex $(O_2^-..NH_3)_{(s)}$ is shown to decompose to an amide, a hydroxyl and an oxide ion

$$(O_2^-.. NH_3)_{(s)} ==> NH_{2(a)} + OH_{(a)} + O^-_{(a)} \qquad (8)$$

and the $O^-_{(a)}$ moiety is transformed to an oxide ion. The ammonia molecule which moves rather freely over the Zn (0001) surface thus acts as a chemical trap for the adsorbed superoxide ion. No amide formation is observed if O_2 is admitted first and permitted to form O^{2-} ions. On copper crystals the result of co-adsorption of O_2 and NH_3 depends on the crystal surface. On Cu(110) oxydehydrogenation is similar to that observed for Zn(0001), but on Cu(111) the oxygen transients $O^{\delta-}$ and $O_2^{\delta-}$ have been characterized spectroscopically [115].

On Ag single crystals, the mode of non-dissociative O_2 adsorption is strongly dependent on the crystal face. On the (110) face the adsorbed O_2 molecule displays an O-O vibration frequency of 640 cm^{-1}. It is attributed to an $O_2^=$ ion bonded parallel to the surface in the troughs of the (110) grooves, as follows from a comparison of the spectroscopic data with the first principle calculations by Upton *et al* [116]. This is the favored position for optimum interaction with the $2\pi^*$ orbitals. On the (111) surface, the O_2 behaves rather isotropically because the spatial extension of the $2\pi^*$ and the 2p orbitals is rather similar. Electron back donation is higher for oxygen atoms having two silver atoms as nearest neighbors than only one [117] (see Chapter 3 of this reference).

The identification of a diatomic oxygen adsorbate is remarkable, as it was often assumed that spontaneous dissociation of O_2 is facile on metal surfaces. The findings of the Cardiff group suggest that an $O_2^{\delta-}$ intermediate is even more likely on catalysts exposing transition metal ions that are supported on an electrically insulating carrier, such as a zeolite. As we shall see below, superoxide ions, O_2^-, and peroxide ions, O_2^{2-}, have recently been implicated in catalysis by zeolite supported Co and Fe ions. In these systems the adsorbing metal atoms are not each other's close neighbors as they are in metal surfaces. Also several metal ions, which are held in fixed positions on a zeolite, are required to deliver the four electrons needed for the transformation of an O_2 molecule into two oxide ions.

5.   ADSORPTION AND CATALYSIS BY ALLOYS; ENSEMBLE AND LIGAND EFFECT

In the 1950s many scientists hoped that the electron band model of metal physics could be applied to predict catalysis. The obvious fact that the best hydrogenation catalysts, Pt, Rh, Pd etc., were transition metals with incompletely filled *d*-bands suggested a causal relation of catalytic activity and the *collective* parameters describing electric conductivity and magnetism. Much attention was paid on alloys, such as CuNi, as they seemed to permit a *gradual* change in the parameters used in the band model. Nickel is ferromagnetic, but alloying Cu to Ni lowers magnetic susceptibility and a CuNi alloy containing 60% Cu was reported to be diamagnetic. In that period such observations were interpreted as meaning that each Cu atom in the alloy donates one electron to its *d*-band; so if the *d*-band of pure nickel metal contains 60% holes, no holes were left in the *d*-band of the $Cu_{0.6}Ni_{0.4}$ alloy. As also the hydrogenation activity of CuNi alloys is lower than that of pure nickel, the holes in the *d*-band were supposed to be the real cause of hydrogenation activity.

In hindsight these models appear strongly oversimplified. They were, however, quite valuable in inspiring experimental scientists to study the catalytic activity of metals and alloys. Hydrogenation of ethylene and benzene were used to probe the catalytic activity of transition metals and their alloys [118]. An even wider range of metals and alloys could be probed with the decomposition of formic acid. G.M. Schwab developed an elegant all-glass reactor for the study of the rate of this reaction at various temperatures. With this device, activation energies were measured in a simple and reproducible manner [119].

The idea of a simple causal relationship between catalytic activity and the mentioned *collective parameters* of the band model was, however, strongly challenged when the surface of alloy films, prepared by deposition of the metal vapors under a good ultra-high vacuum, was studied. Even a small addition of Cu to Ni causes a sharp decrease in hydrogen adsorption per unit surface area, which can be expressed as the ratio of chemisorbed H atoms to physisorbed Xe atoms. This ratio remains constant over a wide range of compositions before decreasing to the very low value typical of pure Cu [120]. The observations were rationalized by applying two basic thermodynamic principles of general validity [56]: (1) Formation of Cu-Ni alloys from the metals is a slightly endothermic process. The Gibbs Free Energy-vs.-composition-plot has two minima. Clearly, two phases coexist in equilibrium, one rich in Cu, the other rich in Ni. (2) The principle of minimizing the surface energy of a solid in equilibrium favors an arrangement of these particles such that the copper-rich phase should envelop particles of the nickel-rich phase. The result is that in a wide range of overall compositions the surface of CuNi films exposes a surface of constant composition, *viz.* that of the Cu-rich phase. This model became known as the "*cherry-model*" of alloy films [121,122,123,124]. It was confirmed for PtAu alloys for which the phase diagram of two coexisting phases has been well known [72]. The most important conclusion from this work is that catalysis is overwhelmingly controlled by the surface and depends little on the electronic state of the metal or alloy two or more atomic layers below the surface.

For an alloy consisting of atoms of almost equal size, such as Cu and Ni, the enveloping phase will become enriched with the element of lowest heat of sublimation, *i.e.*, the copper. The same criterion predicts that in PtAu or PdAg alloys the outermost surface tends to be enriched with Au or Ag in equilibrium under uhv. This holds for slightly endothermic alloys and alloy components of nearly equal size. For exothermic AB_x alloys, such as PtSn, however, the tendency of the A atoms to surround themselves with B atoms becomes a dominant principle [125].

Cherries not only have a kernel surrounded by pulp, there is also a skin. Indeed, it was found that for any homogeneous alloy and thus also for the "pulp" phase of the "cherry", the surface composition differs from that of the interior. The composition of the outermost atomic layers depends on the chemistry of the gas phase, as chemisorption becomes a new factor affecting the Gibbs free energy of the system. For a PtAu alloy in contact with carbon monoxide, strong chemisorption bonds are formed of CO with Pt, but not with Au atoms. The Gibbs free energy will thus be lower for a Pt-rich surface than for an Au-rich surface. Indeed, the process of *"chemisorption induced surface aggregation"* of Pt is observed even at room temperature. This process is reversible: upon desorbing the chemisorbed CO molecules, the surface returns to the Au-rich composition state [125,126,127,128,129].

Of particular interest to catalysis is, of course, understanding the cause of the catalytic *selectivity* of alloys. This problem is illustrated by the results of two papers that appeared in the same issue of *J. Catalysis*. Both made use of the Cu Ni alloy system. Roberti *et al.* studied the hydrogenolysis and the H/D exchange of methylcyclopentane over CuNi films [130]. At the same time, Sinfelt compared hydrogenation and hydrogenolysis of benzene over supported CuNi catalysts [131]. Both groups found that small additions of Cu to Ni induce a dramatic lowering of the selectivity for hydrogenolysis, while the propensity of the alloy to catalyze hydrogenation or H/D exchange of the carbon hydrogen bonds changed much less. The two groups gave different interpretations for their findings. Sinfelt proposed an interpretation in terms of collective electronic properties of the alloy, whereas Roberti *et al.* interpreted the results in terms of surface composition and local site geometry.

As numerous field emission results had shown that most chemisorbed atoms on a metal surface prefer *"hollow sites"* over *"atop sites"*, Roberti *et al.* reasoned that two adjacent hollow sites between Ni, not Cu atoms, are needed for the fission of the C-C bond. Even on the close-packed (111) face of the fcc alloy, a cluster of five Ni atoms is required to create a pair of adjacent hollow sites. Such Ni_5 *ensembles* are, of course, ubiquitous on a nickel surface, but on a CuNi alloy surface enriched with Cu, very few large Ni ensembles will be found. The dramatic change in selectivity upon alloying Cu with Ni is thus attributed to the different *ensemble requirements* of the competing reactions: hydrogenation and hydrogenolysis. This geometry favors the reaction which requires only small ensembles of Ni atoms, whereas hydrogenolysis, requiring much larger Ni ensembles, is suppressed.

The intuitive assumption that two adjacent "hollow" sites are required for dissociative adsorption on metals has been confirmed by theoretical calculations [117]. Dissociation is favored with the dissociated atoms ending in high-coordination sites and sharing the fewest surface metal atoms. This is a consequence of the Bond Order Conservation principle. For the dissociation of CO, the interaction between the C and the O atom becomes repulsive when they share the same metal atom.

This model is of general validity: the *ensemble effect* has become a unifying concept for the rationalization of numerous phenomena of catalysis. It is also not confined to alloys. Van Hardeveld and Hartog [132] call an ensemble of n surface atoms a B_n site and show that the relative concentration of B_5 sites for truncated fcc particles in equilibrium depends on the size of the particle [133]. On Ni surfaces the IR data of van Hardeveld and van Montfort indicate that N_2 [105] molecules are adsorbed on B_5 sites without dissociation [102]. Dissociative chemisorption of N_2 molecules does, however, take place during ammonia synthesis on iron catalysts. As mentioned above, Brill *et al.* showed the high crystal face specificity of this process with the (111) face of bcc Fe being the most active plane. The site responsible for this high activity consists of seven Fe atoms, including not only Fe atoms of the outermost plane but also of two crystal planes below it. While dissociative adsorption of N_2 is crucial for NH_3 synthesis, dissociation of CO is the intial step in the Fischer Tropsch process, as shown by Araki and Ponec [105]. As the best Fischer Tropsch catalysts, Fe and Ru, are also excellent ammonia synthesis catalysts, one may assume that Brill's conclusion also applies to the Fischer-Tropsch process. For the dissociative chemisorption of NO on a Rh (111) face van Santen showed that at least five surface atoms are required, forming two triangles with one common corner. Each of the fragments of the diatomic molecule ends up in a hollow site in the center of one of the two triangles.

Where dissociative adsorption is rate limiting, the effect of topography will be more important than in the non-dissociative adsorption of a molecule such as CO. In general, it is, quite convenient to distinguish "structure-sensitive" and "structure-insensitive" reactions in catalysis This definition is very useful, as it permits classification of observations without making specific assumptions as to the cause of the "structure-sensitivity" or its absence.

Stressing the importance of topography does not mean that electronic effects should be ignored. On the contrary, all "chemical" effects are, of course, a consequence of the shape and occupation of the electronic orbitals and their occupation with electrons. Moreover, the strength of the chemical bond between a surface atom and an adsorbate will, of course, be affected by the other atoms to which that surface atom is bonded. This influence is strongest for the direct neighbors of the metal atom and much smaller for more distant atoms. To distinguish this electronic effect due to near neighbors from the previous concept of *collective* electronic effects, the term *ligand effect* has been proposed [128,134].

As a concept of metal catalysis the *ensemble effect* is helpful in understanding modern bimetallic "platforming catalysts" such as PtSn or PtGe. In these cases it is in particular the large ensemble requirement of the formation of coke precursors which is thought to be responsible for the much extended catalyst life of bimetallic systems in comparison to platinum-only catalysts. Also, PtRe can be understood under the umbrella of the ensemble effect, provided that an additional fact is considered: Pure PtRe alloys have a high selectivity for undesired hydrogenolysis; but in the H_2S containing atmosphere of industrial platforming, the Re atoms of the surface are capped with sulfur atoms, whereas the Pt atoms are not. These Pt atoms are thus interdispersed within sulfur capped Re atoms and therefore present as very small ensembles, ideal for (de-)hydrogenation but not for hydrogenolysis and formation of coke precursors [135,136].

In summarizing this section, we conclude that the physics of the delocalized collective phenomena, electric conductivity or magnetism do not dominate catalytic activity and selectivity of metals and alloys. Instead, the catalytic phenomena are controlled by two chemical causes, namely, the chemical composition of the surface and the atomic geometry of the sites.

It should be mentioned that these are the very same principles which A.A. Balandin used in his model of the atomic *multiplets* and the *volcano-shaped curves* [137].

6. ADSORPTION AND CATALYSIS ON ZEOLITE BASED SYSTEMS

Zeolites are still a largely untapped reservoir of highly promising catalysts. Previous research was largely confined to zeolites in their H form, as these materials are used at huge scale as catalysts in petroleum conversion processes. Their acid strength is higher than that of amorphous silica-alumina, and their well defined channel structure leads to shape-selectivity. Large molecules that cannot enter the channel system are not converted under conditions where smaller molecules are present. Large reaction products formed inside a cage may be unable to leave it unless they first isomerize to a molecule with smaller kinetic diameter. For instance, among the dialkyl-benzenes, the *para-* isomer leaves the channel system easily, whereas the *meta* isomer is trapped [138,139]. In the methanol conversion to gasoline over H-MFI catalysts, the cut-off in the product is given by the 1,2,4,5 tetramethyl benzene which is the largest substituted benzene molecule capable of leaving the zeolite (21). Finally, two transition states leading to different products may have different space requirements. Transition state selectivity is, therefore, a third parameter contributing to the shape-selectivity of zeolite catalysts. As many competent reviews have appeared in the literature on the science and technology of hydrocarbon conversions over acid zeolites, we shall not re-discuss this subject here. The reader is referred, for instance, to the paper by W.O. Haag [140] and references therein.

Zeolites can, however, do much more for catalysis than merely act as strong acids with well defined channel or cage structure. Of interest is their ability to host inside their cavities species such as bare metal clusters, ligated metal clusters such as metal carbonyl clusters, and even single metal atoms. Some of these species are electroneutral and others carry a positive charge, including bare metal ions that are ligated only to zeolite oxygens and oxo-ions, such as $(GaO)^+$ that carry an oxygen atom that is not part of the zeolite lattice and therefore called an "extra-lattice oxygen" or ELO. The positive charge of bare ions and protons compensates the negative charge of the Al-centred tetrahedra. But since oxo-ions have a smaller charge than that given by their valency, the maximum loading of a zeolite with oxo-ions is significantly higher than that of bare ions. For instance, the maximum loading of a given zeolite with $(RhO)^+$ ions is three times higher than the maximum loading with Rh^{3+} ions. $(GaO)^+$ and $(RhO)^+$ are examples for mononuclear oxo-ions. Other oxo-ions, such as $[Cu-O-Cu]^{2+}$, are binuclear and connected *via* an oxygen bridge [141,142,143]. All species mentioned in this paragraph have peculiar propensities as catalytic sites.

Ligated metal clusters can be introduced into zeolite cages or they can be formed there from smaller building blocks. This *ship-in-a-bottle* technique permits the synthesis of large complexes which fit into certain cages but are too big to leave them through any of the cage windows. We refer to the reviews by M. Ichikawa [144] Sachtler and Zhang [145] and B.C. Gates [146]. Figure 3 shows the FTIR spectrum which is obtained when Pd ions are exchanged into zeolite Y, followed by reduction with H_2 and exposure to carbon monoxide gas at room temperature [147,148]. Analysis of the spectrum indicates a cluster of 13 Pd atoms with icosahedral structure, surrounded by CO ligands in linear, bridging, triple bridging, and butterfly coordinations. No such cluster has ever been synthesized outside a zeolite cavity. After similar observations with Pt in Y zeolites, Stakheev *et al.* obtained IR evidence for the formation of neutral Pt carbonyl clusters inside cavities of zeolite KL; these authors showed that formation of such clusters took place only at pressures of 5 mbar or higher [149]. The mechanism of Pd carbonyl cluster genesis has been analyzed by Zhang et al [150]. When Rh/NaY zeolites are used as catalysts for CO hydrogenation, it is found that the reaction product depends crucially on the reaction conditions. At low pressure, the reaction product is mainly hydrocarbons, but at high CO pressure when Rh clusters and Rh-carbonyl clusters are expected to co-exist inside the zeolite cavities. Acetic acid has been found to become the major product of CO hydrogenation over Rh/Y catalysts carrying Rh metal and Rh-carbonyl clusters in its cavities [151].

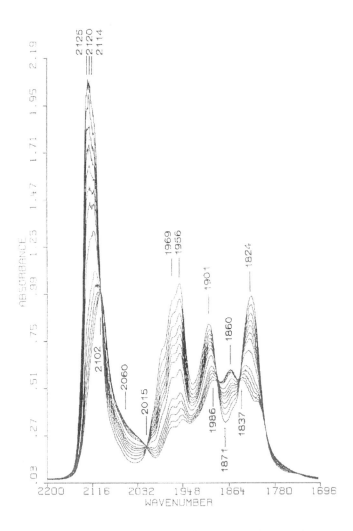

Figure 3. IR Spectrum of $Pt_{13}(CO)_x$ in supercages of NaY zeolite (from *Catalysis Lett.* 2 **1989** 132).

If this tentative interpretation is confirmed, transition metal/zeolites are candidates as catalysts for high yields of oxygenates from syngas at CO pressures that favor formation of carbonyl clusters inside the zeolite cavities. When reduced transition metal clusters share a zeolite cage with one or several acid protons, they form an adduct. As the positive charge of the proton is shared with the cluster, XPS shows it to be electron deficient [152].

As the negative charge of the Al centered tetrahedra in the zeolite matrix is fixed, the metal clusters forming an adduct with protons are less mobile than in the absence of protons, and they coalesce less. This opens the possibility to keep reduced metal atoms apart from each other and study the catalysis by single atoms. As follows from the ensemble model, isolated Pt atoms do not catalyze hydrogenolysis of hydrocarbons. When catalyzing the H/D exchange with cyclopentane the product pattern differs significantly from that obtained over multiatomic Pt particles [153].

The combination of hydrogenolysis by metal clusters and the special geometry of zeolites has been combined to obtain selective ring-opening catalysis by orienting methyl-cyclopentane molecules before they hit a metal particle. The C_2-C_3 bond will be selectively opened if the molecule is directed to hit the metal cluster with this part of the ring. Conversely, if the molecule diffuses through the main channel of mordenite carrying Pt clusters in the side pockets of these channels, the molecule will be slit open on its sides [154].

Zeolite supported transition metal ions have been identified as unusually powerful and selective catalysts for the reduction of nitrogen oxides from the emission of combustion engines working with lean air/fuel mixtures. Cu/MFI, Fe/MFI, Co/MFI, and Pd/MFI catalyze traces of NO and NO_2 with hydrocarbons present at concentrations of 1000 ppm or below. Over Co/MFI catalysts NO or NO_2 the reduction to N_2 with isobutane is 100% [155]. This catalyst even performs appreciable NO_x reduction with methane. NO_x reduction over zeolite based catalysts has recently become a field of very intensive research, because lean burn gasoline and diesel engines have a higher energy conversion efficiency than traditional engines working with stoichiometric or rich fuel/air mixtures. The "Three Way Catalyst" which is used in traditional passenger cars, is unable to catalyze NO abatement in the presence of oxygen which is, of course, an important component of the emission from lean-burn or diesel engines. This situation obviously presents a great challenge to research on heterogeneous catalysis. The finding by Iwamoto and Held [156] that zeolite supported transition metal ions or oxo-ions are able to catalyze NO_x reduction has opened a new field of intensive research. Some of these catalysts lose their high performance in an atmosphere rich in water vapor, but others show an even higher performance in a wet atmosphere similar to that in emission coming out of the tail pipe of a passenger car. The reader interested in this dynamic field of research is referred to recent papers by Iwamoto [157] Misono [158] Cant and Liu [159] and Chen *et al* [160,161].

Although this research is still in progress, it appears that binuclear oxo-ions such as $[HO-Fe-O-Fe-OH]^{2+}$ in the cavities of the MFI zeolite display an exceptionally high activity for NO_x reduction with hydrocarbons in atmospheres rich in oxygen and water vapor [162,163,164,165,166]. Formation of such oxo-ions in zeolites is favored by the fact that the positions of the negative charges, *i.e.*, the Al centered tetrahedra, are fixed. It is therefore impossible for an Fe^{3+} ion to be simultaneously close to the three Al centered tetrahedra of which it compensates the charge, and much better Coulomb interaction between positive and negative charge carriers is obtained when the effective positive charge of cations is lowered by extra-lattice oxygen. Binuclear, oxygen bridged oxo-ions have the additional advantage of being larger than bare ions, so they can easily connect with two Al centered tetrahedra in zeolites with moderate Si/Al ratios, as was shown by B.R. Goodman *et al* [167]. This chemistry provides zeolite based systems with an additional dimension for optimizing catalytic performance that is not available to catalysts with amorphous supports.

Oxo-ions can be prepared by hydrolysis of bare ions or by sublimation of a volatile compound such as $GaCl_3$ or $FeCl_3$ onto a zeolite in its H-form;

$$FeCl_3 \text{ (vapor)} + H^+\text{(zeolite)} \rightleftharpoons HCl \text{ (gas)} + [FeCl_2]^+\text{(zeolite)} \qquad (10)$$

followed by hydrolytic treatment which converts the $[FeCl_2]^+$ ions into oxo-ions.

For catalysis an important difference between mononuclear and binuclear oxo-ions is in their reducibility. All ions can be reduced with hydrogen, but only oxo-ions can be reduced with CO. A mononuclear ion such as $[GaO]^+$ or $[RhO]^+$ is reduced to Ga^+ and Rh^+, *i.e.*, its valency is lowered from 3+ to 1+. In contrast, a binuclear oxygen-bridged ion will be reduced by <u>one</u> unit, so that $[Cu-O-Cu]^{2+}$ becomes $[Cu- -Cu]^{2+}$ and $[HO-Fe-O-Fe-OH]^{2+}$ becomes $[HO-Fe- -Fe-OH]^{2+}$.

Obviously, the energy difference of the reduced and the oxidized state will be of importance for all catalytic oxidation and reduction reactions. Two research groups have recently confirmed the binuclear model for Fe/MFI by a combination of techniques, including EXAFS [168,169]. These data support the hypothesis that the extraordinary catalytic performance of catalysts such as Cu/MFI, Fe/MFI, or Co/MFI is due to the presence of binuclear oxygen-bridged transition metal ions.

A surprising and still incompletely understood phenomenon is the isothermal catalytic oscillation which is observed for N_2O decomposition over Cu/MFI [170,171] Fe/MFI [172] and Co/MFI [173] catalysts. It shows that the ions in different zeolite cages are not totally independent of each other but their chemical reactions can be synchronized. It appears difficult to apply the concepts that are valid for oscillatory reactions on metal surfaces [174] to zeolites. Both the gas phase and the zeolite lattice are potential coupling media. It is well documented that transition metal ions and oxo-ions become attached to the cage walls in a manner that gives rise to discrete perturbations of the lattice vibrations, resulting in a new IR band [175,143]. Further research will be necessary to clarify the mechanism of the oscillation process with zeolite catalysts.

A conceptual problem for active sites that are encaged in different zeolite cavities is the mechanism of dissociative adsorption. On a macroscopic metal crystal where every metal atom is surrounded by other metal atoms, it is not difficult to imagine that an impinging O_2 molecule will dissociate as sketched above in the section on metals. Even N_2 dissociates fairly easily on a C_7 site. But the metal ions in a zeolite are not each other's first neighbors, so they do not form "ensembles". It appears more likely that diatomic adsorbates are formed, for instance superoxide ions, O_2^-, or peroxide ions, O_2^{2-}. Dissociation of these precursors to monoatomic O^- or O^{2-} ions will require a high activation energy. It is also possible that zeolite oxygen participates in this process, as is indeed suggested by the results of isotope exchange between $^{18}O_2$ and Cu/MFI [176] and Fe/MFI [177].

Wyn Roberts has succeeded in identifying undissociated diatomic O_2 precursors on Zn and Mg surfaces and Madix showed the presence of diatomic O_2 entities on Ag. For Fe/MFI recent work at Northwestern University indicates that both peroxo- and superoxo- groups are formed on Fe/MFI [178]. Superoxide ions have been identified on Fe/MFI and Co/MFI at low temperature by EPR [173]. We mentioned above that the ion pairs of O_2^- and Co^{3+} show the phenomenon of super-hyperfine structure caused by the interaction of the nuclear spin of Co ($I = 7/2$) with the unpaired electron in the O_2^- ion. With uv Raman spectroscopy, peroxo groups have been found; superoxo groups are indicated by weak signals [179]. Binuclear oxo complexes of iron and Fe-O-O-Fe configurations are not unique to zeolites but are also assumed to be crucial in the catalysis of enzymes of the MMO class that are able to catalyze the oxidation of methane to methanol [180]. In this chemistry a double-oxygen bridged di-iron complex has been discussed, but Siegbahn and Crabtree showed by DFT calculations that the Fe-O-O-Fe structure is energetically more favorable [181].

A different form of active oxygen has been identified on Lewis sites of a zeolite. A high concentration of such sites is created, for instance, by calcining the H-form of MFI at 900°C. Kustov *et al.* [182] showed that exposure of such material to nitrous oxide results in the formation of *singlet oxygen*, while N_2 is released. Together with the Al^{3+} ion, the singlet oxygen forms a Lewis base- Lewis acid pair, which can be probed by volumetric H_2 adsorption at 77K . The polarized H_2 molecule at the Lewis base-Lewis acid pair gives rise to an IR absorption band which is shifted to lower frequency compared to the gas phase (4163 cm^{-1}). The singlet oxygen can be determined quantitatively by volumetric measurement of the adsorbed H_2, a value of 5-7 x10^{19}/g has been reported after N_2O adsorption on an H/MFI sample that was calcined at 1170K. The singlet oxygen is quite reactive; for instance it converts benzene to phenol. As this oxygen can be regenerated from N_2O, a catalytic cycle is possible, *i.e.*, benzene is oxidized with N_2O to phenol with a selectivity of 98-100%. Previously, Panov *et al* [183] had demonstrated catalytic oxidation of benzene to phenol with N_2O as the oxidant over an Fe/MFI catalyst; these authors ascribe this catalytic action to so-called "δ-oxygen" The same species was claimed to also lead to catalytic oxidation of methane with nitrous oxide to methanol. It is not clear at present whether this "δ-oxygen" is identical or even related to the singlet oxygen identified by Kustov *et al*. Clearly, much more research will be required to fully exploit the potential for oxidation catalysis of the three forms of active oxygen, superoxide ions, peroxide ions and singlet oxygen, that have been identified recently on the surface of zeolites.

7. CONCLUSIONS

Research on industrial catalysis using solid acids and academic work with labeled molecules and on metal single crystals with modern physical techniques have not only produced a flood of valuable data but also led to important new concepts on the nature of heterogeneous catalysis. Geometric principles such as structure sensitivity, ensemble effects and transition state shape selectivity are no longer uncertain hypotheses. The formation of unsaturated reaction intermediates on surfaces, including metal alkylidynes or nitrenes or cyclopropane derivatives, can be used with confidence. The role of intermolecular H transfer, familiar in acid catalysis begins to enter the family of concepts in metal catalysis. Zeolites, besides being strong acids with peculiar three-dimensional geometry, are showing their unique ability to form multinuclear oxo-ions and encaged carbonyl clusters leading to new catalytic challenges, inaccessible to amorphous carriers.

8. REFERENCES

1. JW. Goethe, who had promoted Döbereiner's professorship, wrote in his Faust a scene where the devil, disguised as a university professor, teaches a young student: "Dann eben wo Begriffe fehlen, da stellt ein Wort zur recthen Zeit sich ein."
2. H. Pines, and V.N. Ipatieff, J. Am. Chem. Soc., 70 (1948) 531.
3. E.K. Rideal, "Concepts in Catalysis." Academic Press: London, New York. 1968.
4. S. Brunauer, P.H. Emmett, E. Teller, J. Am. Chem. Soc., 60 (1938) 309.
5. H.S. Taylor, Proc. Royal Soc. London. A108 (1925) 105.
6. H.S. Taylor, Adv. Catal., 1, (1948) 1

7. H. Leidheiser, A. Gwathmey, J. Am. Chem. Soc., 70 (1948) 1206.
8. E.W. Müller, Z. Physik, 106 (1937) 541.
9. M.W. Roberts, Catal. Lett., 67 (2000) 1.
10. H. Pines and R.C. Wakher, J. Am. Chem. Soc., 68 (1946) 599.
11. V.N. Ipatieff and L. Schmerling, Advances in Catalysis, 1 (1948) 27-64.
12. H. Pines, Adv. Catal., 1(1948) 201.
13. H. Pines, Aristoff, and V.M. Ipatieff, J. Am. Chem. Soc., 71 (1949) 749.
14. H. Pines, F.J. Pavlik and V.N. Ipatieff, J. Am. Chem. Soc., 73 (1959) 5738.
15. R.C. Hansford, Adv. Catal., 4 (1952) 1.
16. P.B. Weisz, Chemtech. (1973) 498.
17. V. Haensel, Adv. Catal., 3 (1951) 179.
18. R.A. Shankland, Adv. Catal., 6 (1954) 271.
19. H. Heinemann, Chapter 1, Volume 1, Catalysis, Science and Technology. Springer Verlag: Heidelberg, New York. 1981.
20. H. Heinemann, "Development of Industrial Catalysis," Handbook of Heterogeneous Catalysis. Wiley-Verlag Chemie: Weinheim. 1997.
21. N.Y Chen, R.L. Gorring, H.R. Ireland, and T.R. Stein, Oil and Gas Journal 15 (1977) 165.
22. C.D. Chang, A. Silvestri, J.Catal. 47 (1977) 249.
23. A.G. Oblad, H. Heinemann, L. Friend, and A. Gameron, Proc. 7th World Petroleum Congress, 5 (1967) 197.
24. F.C. Whitmore, J. Am. Chem. Soc., 54 (1932) 3274.
25. D.M. Brouwer, and H. Hoogeveen, Progr. Phys. Org. Chem. 9 (1972) 179.
26. D.M. Brouwer, "Chemistry and Chemical Engineering of Catalytic Processes." Sijthoff & Noordhof, Alphen and. Rijn: Netherlands. 1980.
27. M.R. Guisnet, Acc. Chem. Res. 23 (1990) 392.
28. C. Bearez, F.Avendano, F. Chevalier and M. Guisnet, Bull. Soc. Chim. France (1985) 346.
29. V. Adeeva, G.D Lei and W.M.H. Sachtler, Appl. Catal. A: General, 118 (1994) L11.
30. V. Adeeva, H.-Y. Liu, B-Q. Xu and W.M.H. Sachtler, Topics in Catalysis 6 (1998) 61.
31. V. Adeeva, J.W. de Haan, J. Jänchen, G.-D. Lei, V. Schünemann, L.J.M. van de Ven, W.M.H. Sachtler and R.A. van Santen, J. Catal. 151 (1995) 364.
32. J.W. Otvos, D.P. Stevenson, C.D. Wagner and O. Beeck, J. Am. Chem. Soc. 73 (1951) 5741.
33. G.B. McVicker, G.M. Kramer and J.J. Ziemiak, J. Catal. 83 (1983) 286.
34. M. Zaki and H. Knözinger, J. Catal., 119 (1989) 311.
35. K. M. Neyman, P. Strodel, S. Ph. Ruzankin, N. Schlensorg, H. Knözinger, and N. Rösch, Catal. Lett., 31 (1995) 273.
36. K.M. Neyman, P. Strodel, R.A. van Santen, J.Jänchen, and E. Meijer, J. Phys. Chem., 97 (1993) 11071.
37. G.A. Olah, O.Farooq, A.Hussain, N.Ding, N.J. Grivedi, and J.A. Olah, Catal. Lett., 10 (1991) 239.
38. J.M. Sommer, Hachoumi, F. Garin, and D. Barthomeuf, J. Am. Chem. Soc. 116 (1994) 5401.
39. J.M. Sommer, M. Hachoumi, F . Garin, and D. Barthomeuf, and J. Védrine, J. Am. Chem. Soc., 117 (1995) 1135.
40. W.O. Haag, R.M. Dessau, Proceedings Intern. Congress Catal. in Berlin, Germany, Vol II, p. 305. Dechema-Verlag: Frankfurt am Main, Germany. 1984.
41. G.J. Kramer, R.A. van Santen, C.A. Emeis, and A.K. Novak, Nature, 363 (1993) 529.

42. R.A. van Santen, G.J. Kramer, Chem. Rev., 95 (1995) 637.
43. I.N. Senchenya and V. B. Kazansky, Kinetika i Kataliz, 28 (1987) 566.
44. V.B. Kazansky and I.N. Senchinya, J. Catal., 119 (1989) 108.
45. Kazansky, VB., Accounts Chemical Research, 24 (1991) 379.
46. J.F. Haw, J.B. Nichols, T. Xu, L.W. Beck, D.B. Ferguson, Acc. Chem. Res., 29 (1996) 259.
47. S. Kotrel, M.P. Rosynek, J.J. Lunsford, J. Catal., 191 (2000) 55.
48. V. Haensel, US Patent 2,611,736, September 23, 1952.
49. Boudart, M. Adv.Catal., 20 (1969) 153.
50. W.C. Conner, S.J. Teichner and G.M. Pajonk, Adv. Catal., 34 (1986) 1.
51. W.C. Conner and J.L. Falconer, Chem. Rev., 95 (1995) 759.
52. H-Y. Liu, W.-A. Chiou, G. Fröhlich and W.M.H. Sachtler, Top. Catal., 10 (2000) 49.
53. O.E. Lebedeva, W.-A. Chiou and W.M.H. Sachtler, J. Catal., 188 (1999) 365.
54. O.E. Lebedeva and W.M.H. Sachtler, J. Catal., 191 (2000) 364.
55. F.G. Gault, Adv. Catal., 30 (1981) 1.
56. V. Ponec, Adv. Catal., 32 (1983) 149.
57. Y-Y. Huang and W.M.H. Sachtler, J. Catal., 184 (1999) 247.
58. Y-Y. Huang and W.M.H. Sachtler; J. Catal., 190 (2000) 69.
59. T. Yamada and K. Tamaru, J. Phys. Chem., 144 (1985) 195.
60. J. Horiuchi and M. Polanyi, Nature, 132 (1933) 931.
61. O. Beeck, Rev. Mod. Phys., 17 (1945) 61.
62. O. Beeck, Disc. Faraday Soc., 8 (1950) 118.
63. R. Suhrmann and W.M.H. Sachtler, Z. f. Naturforschung, 9a (1954) 14.
64. R. Suhrmann, Adv. Catal., 7 (1955) 303.
65. J.C.P. Mignolet, Disc.Faraday Soc., 8 (1950) 326.
66. C. Kemball, "Advances in Catalysis". Academic Press: New York and London. 1959.
67. R.L.Jr. Burwell, J. Catal., 138 (1992) 761.
68. J. Addy and G. C. Bond, Trans Faraday Soc., 53 (1957) 368, 383, 388.
69. V. Ponec and G.C. Bond, "Catalysis by Metals and Alloys," Studies in Surface Science and Catalysis. Elsevier Science B.V. Amsterdam etc.
70. J.J. Rooney J. Catal., 2 (1963) 53.
71. R. Jongepier and W.M.H. Sachtler, J. Res. Inst. Catalysis, Hokkaido University, 16 (1968) 69 .
72. F.J. Kuyers, R.P. Dessing, and W.M.H. Sachtler. J. Catal., 33 (1974) 316.
73. L.B. Xu, Marshik, Z. Zhang and W.M.H. Sachtler; Catalysis Letters, 10 (1991) 121.
74. O. Beeck, J.W. Otvos, D.P. Stevenson and C.D. Wagner, J. Chem. Phys., 16 (1948) 255.
75. W.M.H. Sachtler and L.L. van Reijen. J. Res. Inst. Catalysis, Hokkaido University, 10 (1962) 87.
76. W.M.H. Sachtler and G.J.H. Dorgelo, Proc. Intern. Congress of Electron Microscopy, 2 (1958) 801.
77. W.M.H. Sachtler, Surf. Sci., 22 (1970) 468.
78. W.J.M. Rootsaert, L.L. van Reijen, and W.M.H. Sachtler., J. Catal., 1 (1962) 416.
79. W.J.M. Rootsaert and W.M.H. Sachtler, Z. Phys. Chem., 26 (1960) 16.
80. B.E. Nieuwenhuys, D.Th. Meijer and W.M.H. Sachtler, Surf. Sci., 40 (1973) 125.
81. A.A. Balandin, Adv. Catal. Rel. Subj., 19 (1969) 1.
82. J. Fahrenfort, L.L. van Reijen and W.M.H. Sachtler, Ber.Bunsenges Phys. Chem., 64 (1960) 216.
83. J. Fahrenfort, L.L. van Reijen and W.M.H. Sachtler. "The Mechanism of Heterogeneous Catalysis." Elsevier: Amsterdam. 1960.

84. W.M.H. Sacthler and J. Fahrenfort, Actes du 2ieme Congres Intern. De Catalyse 1960. Editions Technip. Paris. 831-863. 1961.
85. R.J. Gomer, Chem. Phys., 21 (1953) 1869.
86. G. Ehrlich, Advan. Catal., 14 (1963) 255.
87. W.J.M. Rootsaert, L.L. van Reijen, and W.M.H. Sachtler., J. Catal., 1 (1962) 416.
88. I.N. Stranski, Z. Phys. Chem., 136 (1928) 259.
89. I.N. Stranski and R. Kaischew, Annalen Physik, 25 (1935) 330.
90. E.W.Müller, Z. Physik, 131 (1951) 136.
91. A.A. Holscher, W.M.H. Sachtler, Disc.Faraday Soc., 41 (1966) 29.
92. G.A. Somorjai, "Chemistry in Two Dimensions: Surfaces". Cornell University Press: Ithaca, London. 1981.
93. J.A. Rodriguez and D.W. Goodman, Surf. Sci. Rep., 14 (1991) 108.
94. M.W. Roberts, Adv. Catal., 29 (1980) 55.
95. M.W.Roberts, Surf. Sci., 299/300 (1994) 769.
96. D.A. King, "The Chemical Physics of Solid Surfaces and Heterogeneous Catalysis. Elsevier". Amsterdam. 1995.
97. G. Ertl, Ber. Bunsenges. Phys. Chem., 86 (1982) 425.
98. Ertl, G, Adv. In Catal., 37 (1990) 213.
99. H. Ibach, H. Hopster and B. Sexton, Appl. Surf. Sci., 1 (1977) 1.
100. C.N. Panja, N. Saliba, and B. E. Koel, Surf. Sci., 395 (1998) 248.
101. R. Brill, E.L. Richter and E. Ruch, Angew. Chemie, Intern. Ed. Eng, 6 (1967) 882.
102. N.D. Spencer, R. C. Schoonmaker, and G.A. Somorjai, J. Catal., 74 (1982) 129.
103. R.A. van Santen, "Theoretical Heterogeneous Catalysis". World Scientific: Singapore, New Jersey, London, Hong Kong. 1991.
104. H. Topsoe, N.Topsoe, H. Bolbro and J. Dumesic Proc. 7th Intern Congr Catalysis, Part A. Kodansha: Tokyo. 1981.
105. M. Araki and V. Ponec, J. Catal., 44 (1976) 438.
106. P. Biloen and W.M.H. Sachtler. Adv. Catal. 30 (1981) 165.
107. W.M.H. Sachtler, Chem. Ing. Tech., 54 (1982) 901.
108. W.M.H. Sachtler, in Proceedings Intern. Congress Catal.,Berlin, Germany, 1984. Dechema- Verlag: Frankfurt am Main, Germany.
109. P. Stair and G.A. Somorjai, J. Chem. Phys., 66 (1977) 2036.
110. Lehwald, S, H. Ibach, Surf. Sci., 89 (1979) 425.
111. G.A. Somorjai, "Chemistry in Two Dimensions". Cornell University Press: Ithaca, London. 1981.
112. C.T. Au, X-Ch. Li, J-A. Tang, and M. W. Roberts, J. Catal., 106 (1987) 518.
113. C. T. Au and M. W. Roberts, J. Chem. Soc. Trans., I 83 (1987) 2047.
114. A.F. Carley, M. W. Roberts and S. Yan, Catal. Lett., 1 (1988) 265.
115. P.R. Davies, M. W. Roberts, N. Shukla, and D. Vincent, Surface Sci., 325 (1995) 50.
116. T.H. Upton, P. Stevens and R.J. Madix, J. Chem. Phys., 88 (1988) 3988.
117. R.A. van Santen, "Theoretical Heterogeneous Catalysis". World Scientific: Singapore, New Jersey, London, Hong Kong. 1991.
118. D. Eley, Catalysis, 3 (1955) 49.
119. G. Schwab, Adv. Catal. 2 (1950) 251.
120. W.M.H. Sachtler, J. Catal., 12 (1968) 35.
121. W.M.H. Sachtler, G. Dorgelo and R. Jongepier, J. Catal., 4 (1965) 100.
122. W.M.H. Sachtler, G. Dorgelo and R. Jongepier, J. Catal., 4 (1965) 654.
123. W.M.H. Sachtler and R. Jongepier. J. Catal., 4, (1965) 665.

124. W.M.H. Sachtler, G. Dorgelo and R. Jongepier, "Grundprobleme dünner Schichten," Basic Probl. Thin Film. Phys., Proc. Int. Symp. 218 (1966).
125. R.A. van Santen and W.M. H. Sachtler, J. Catal., 33 (1974) 202.
126. R. Bouwman and W.M. H. Sachtler, Surface Science, 24 (1971) 350.
127. R. Bouwman and W.M. H. Sachtler, J. Catal., 26 (1972) 63-69.
128. W.M.H. Sachtler, Le Vide, 19 (1973) 163-165.
129. W.M.H. Sachtler, J. Vac. Sci. Tech., 9 (1972) 828.
130. A. Roberti, V. Ponec, and W.M. H. Sachtler. J. Catal. 29 (1973) 381.
131. J. Sinfelt, J. Catal., 29 (1973).
132. R. van Hardeveld and F. Hartog, Surf. Sci., 15 (1969) 184.
133. R. van Hardevelt and A. van Montfort, Surf. Sci., 15 (1969) 90.
134. W.M.H. Sachtler and R. A. van Santen, Adv. Catal., 26 (1977) 69.
135. P. Biloen, J. Helle, H. Verbeek, F. Dautzenberg, and W.M. H. Sachtler., J. Catal., 63 (1980) 112 .
136. V. Shum, J. Butt and W.M.H. Sachtler, Appl. Catal., 11 (1984) 151.
137. A. Balandin, Actes du 2ieme Congres Intern. De Catalyse 1960. Editions Technip., Paris. 1135 (1960).
138. S. Csicsery, "Shape-Selective Catalysis", ACS Monograph. 171 (1976) 680.
139. W.C. Kaeding, W, C. Chu, L. Young and S. Butler, J. Catal., 69 (1981) 392.
140. W. Haag, "Zeolites and Related Microporous Materials, State of the Art, 1994," Volume 84 of Studies in Surface Science and Catalysis. Elsevier Science B.V. Amsterdam. 1994.
141. W. Delgass, R. Garten and M. Boudart, J. Phys. Chem., 73 (1969) 2970.
142. J. Valyon and W. Hall, J. Catal., 143 (1993) 520.
143. T. Beutel, J. Sarkany, J. Yan and W.M. H. Sachtler, J. Phys. Chem, 100 (1996) 845.
144. M. Ichikawa, Adv. Catal., 38 (1992) 283.
145. W.M. H. Sachtler and Z. Zhang, Adv. Catal. 39 (1993) 129.
146. B. Gates, "Supported Metal Cluster Catalysts", Handbook of Heterogeneous Catalysis. Wiley-Verlag Chemie: Weinheim. 1997.
147. L. Sheu, H. Knözinger and W.M. H. Sachtler, Catal. Letters, 2 (1989) 129.
148. L. Sheu, H. Knözinger and W.M. H. Sachtler, J. Am. Chem. Soc., 111 (1989) 8125.
149. A. Stakheev, E. Shpiro, N. Jaeger and G. Schulz-Ekloff, Catal. Lett., 34 (1995) 293.
150. Z. Zhang, W.M. H. Sachtler and H. Knözinger, J. Phys. Chem., 97 (1993) 3579.
151. B. Xu, and W.M. H. Sachtler, J. Catal., 180 (1998) 194.
152. W.M. H. Sachtler and A. Stakheev, Catal. Today, 12 (1992) 283.
153. V. Zholobenko, G. Lei, B. Carvill, B. Lerner and W.M. H. Sachtler, J. Chem. Soc. Faraday Trans., 90 (1994) 233.
154. B. Lerner, B. Carvill and W.M. H. Sachtler, J. Molec. Catal., 77 (1992) 99.
155. X. Wang, H-Y. Chen and W.M. H. Sachtler, Appl. Catal. B. 29 (2001) 47.
156. M. Iwamoto, H. Yahiro, K. Tanda, N. Mizuno, Y. Mine, S. Kagawa J. Phys. Chem., 95 (1991) 3727.
157. M Iwamoto, Proc. 10 th Int. Zeolite Conference. Garmisch-Partenkirchen, July 1994. Elsevier: Amsterdam. 1994.
158. M. Misono, CatTech, 4 (1998) 183.
159. N. Cant and I. Liu, Catal. Today, 63 (2000) 133.
160. H-Y. Chen, T. Voskoboinikov and W.M. H. Sachtler, J. Catal., 186 (1999) 91.
161. H-Y. Chen, X. Wang, W.M. H. Sachtler, Appl. Catal. A, General, 194 (2000) 159.
162. H-Y. Chen and W.M. H. Sachtler, Catal. Today, 42 (1998) 73.
163. H-Y. Chen, T. Voskoboinikov and W.M. H. Sachtler, Catal. Today, 54 (1999) 483.

164. T. Voskoboinikov, H.-Y. Chen and W.M. H. Sachtler, Appl. Catal. B (Env), 19 (1998) 275.
165. H-Y. Chen, X. Wang and W.M.H. Sachtler, Phys. Chem. Chem. Phys. 2 (2000) 3083.
166. H-Y. Chen, H-Y., El M. El Malki, X. Wang and W.M.H. Sachtler. NATO Sci. Ser., II 13 (2001) 75.
167. B. Goodman, K. Hass, W. Schneider and J. Adams, Catal. Lett., 68 (2000) 85.
168. P. Marturano, L. Drozdova, A. Kogelbauer and R. Prins, J. Catal., 192 (2000) 236.
169. A. Battiston, J. Bitter and D. Koningsberger, Cat Lett., 66 (2000) 75.
170. T.J. Turek, J. Catal., 174 (1998) 98.
171. P. Giambelli, E. Garufi, R. Pirone, G. Russo and F. Santagata, Appl. Catal. B, 8 (1996) 331.
172. E.M. El Malki, R. van Santen and W.M.H. Sachtler, Mircopor. Mesopor. Mat., 35- 36 (2000) 235.
173. E.M. El-Malki, D. Werst, P. Doan and W.M.H. Sachtler, J. Phys Chem. B, 104 (2000) 5924.
174. F. Schüth, Adv. Catal., 39 (1993) 51.
175. G. Lei, B. Adelman, J. Sarkany and W.M.H. Sachtler, Appl. Catal.B, 112 (1995) 245.
176. J. Valyon, W.K. Hall, J. Catal., 143 (1993) 520.
177. T. Voskoboinikov, H.-Y. Chen and W.M.H. Sachtler, J. Mol. Catal. A, 155 (2000) 155.
178. Z. Gao, Z, Q. Sun, W. Sachtler, Appl. Catal., (subm.).
179. H. Kim, Z. Gao, Q. Sun, P. Stair and W. Sachtler (in preparation).
180. L. Que and R. Ho, Chem. Rev., 96 (1996) 2607.
181. P. Siegbahn and R. Crabtree, J. Am. Chem. Soc., 119 (1997) 3103.
182. L. Kustov, A. Tarasov, V. Bogdan, A. Tyrlov and J. Fulmer, Catal. Today, 61 (2000) 123.
183. G. Panov, A. Uriarte, M. Bodkin and V. Sobolev, Catal. Today, 41 (1998) 365.

Chapter 10

ENANTIOSELECTIVE REACTIONS USING MODIFIED MICROPOROUS AND MESOPOROUS MATERIALS

Graham J. Hutchings

Key words: Enantioselective heterogeneous catalysis, zeolites, mesoporous materials, alcohol dehydration, acid catalysis, aziridination of alkenes, epoxidation

Abstract: Homochiral molecules are important chemical intermediates for pharmaceuticals and agrochemicals and, at present, most are synthesised using either non-catalytic methods or using homogeneous catalysts. There is a significant interest in the design of immobilised heterogeneous catalysts that are capable of high enantioselection. An approach based on the immobilisation of cations within microporous (zeolite Y) and mesoporous (MCM-41) materials will be described. Three enantioselective reactions will be discussed: (a) the dehydration of butan-2-ol over sulfoxide modified zeolite H-Y; (b) the aziridination of alkenes over bis(oxazoline) modified Cu-exchanged zeolite H-Y and (c) the epoxidation of cis-stilbene using chiral salen modified Mn-exchanged AlMCM-41. The main feature of the approach used to design enantioselective heterogeneous catalysis concerns the modification of cations that are ion-exchanged with extra-framework sites of zeolites and mesoporous materials. These charge balancing cations can function as heterogeneous catalysts and, under appropriate conditions (*e.g.* non-aqueous solvents), are highly stable. Furthermore, these cations can be modified by the addition of chiral ligands and, consequently, can function as enantioselective catalysts. The zeolite or mesoporous framework can be considered as part of the coordination shell of the cation, which can enhance the enantioselection of these catalysts.

1.   INTRODUCTION

In many catalysed reactions, a large number of reaction pathways may be possible. The purpose of the catalyst researcher, or designer, is to try to select a formulation that will promote the rate of the desired reactions over those of the non-desired pathways. This is referred to as selectivity and is, today, considered to be the most important feature of catalyst design. The reason for the emphasis on selectivity in modern chemical processes has been encouraged by environmental considerations. The manufacture of non-desired by-products is wasteful of expensive reactants but, also, these by-products have to be disposed of in some way (*e.g.* by combustion, deposited in landfill sites, or vented to the environment directly). The costs associated with the disposal of these by-products have escalated markedly in recent years and, hence, selectivity control for catalytic processes is now a key issue. Of course, for some reactions, it must be recognised that selectivity is not an issue, *e.g.* ammonia synthesis, and such processes can be operated under thermodynamic control. However, it must be recognised that the need to control selectivity is not a new issue. For example, the selectivity control for the catalysed reactions of CO/H_2 mixtures is well known and documented, since methanol is the least thermodynamically preferred product, yet with a sophisticated Cu/ZnO catalyst, this chemical can be manufactured with a selectivity of >99.5% for CO/H_2 mixtures, *via* CO_2 as an intermediate. In fact, in the petrochemical and bulk chemical industries, the control of product selectivity has been exercised extremely well for several decades. These processes, operated on huge scales (*e.g.* >100,000 tons product per annum) produce relatively few unwanted by-products. However, this is not the case in the synthesis of fine chemicals. Often these materials are synthesised using stoichiometric, rather than catalytic, reactions and generate vast quantities of waste material relative to the desired product. Unfortunately, the synthesis of catalysts, which themselves can be considered as fine chemicals, can produce significant waste material. Hence, the emphasis for some research in the design of new heterogeneous catalysts concerns the control of selectivity in the synthesis of fine chemicals. In this respect, the synthesis of a pure enantiomer represents the most significant challenge in the control of selectivity and this chapter will introduce this topic and describe the design approach we have developed using modified zeolite and mesoporous materials. This chapter is dedicated to Professor Wyn Roberts, who has been a significant pioneer concerning the reactivity of molecules on surfaces which underpin the factors that lead to the control of reaction selectivity.

2. DESIGN APPROACH FOR ASYMMETRIC HETEROGENEOUS CATALYSTS

Interest in asymmetric synthesis continues to increase, and this has highlighted the need for the design of highly selective asymmetric catalysts. Most of this research activity concerns homogeneous catalysts and has led, for example, to the design of catalysts for the synthesis of L-Dopa [1] and to the Sharpless epoxidation and dihydroxylation processes [2]. More recently, increased attention has been given to the identification of suitable heterogeneous asymmetric catalysts [3-7], since such catalysts readily overcome the problems typically encountered with homogeneous systems, namely product recovery and catalyst separation. A heterogeneous asymmetric catalyst must not only activate the substrates but must exert stereochemical control, and this requires the preferential formation of a particular diastereoisomeric transition state. To date, three approaches have been taken in the design of heterogeneous enantioselective catalysts : (i) the use of a chiral support for an achiral metal catalyst; (ii) modification of an achiral heterogeneous catalyst using a chiral cofactor, and (iii) the immobilisation of a homogeneous catalyst. Early studies in this field adopted the first of these approaches. In 1932, Schwab *et al.* [8,9] demonstrated that supporting Cu, Ni, Pd and Pt on the enantiomers of quartz gave a catalyst capable of the enantioselective dehydration of butan-2-ol, and the best results were obtained from a catalyst with sub-monolayer coverage of the metal. Subsequently, natural fibres were used as catalyst supports, including silk and cellulose. Synthetic chiral polymers were also evaluated as supports, but the early work experienced a range of problems including lack of reproducibility and low enantioselection. These studies have, however, recently led to the eventual development of polypeptides as catalysts, for example, in the epoxidation of chalcones with enantiomeric excesses of *ca.* 99% [10].

The creation of a chiral catalyst surface by the adsorption of a chiral modifier onto an achiral catalyst has been successful in a number of studies, particularly for enantioselective hydrogenation. For example, the modification of platinum catalysts with cinchona alkaloids for the hydrogenation of prochiral α-ketoesters [11-13], and the modification of Raney nickel catalysts with diethyl tartrate, used for the hydrogenation of prochiral β-dicarbonyl compounds [14], have been extensively studied. The immobilisation of an enantioselective homogeneous catalyst onto an achiral support, such as a zeolite, has also proved to be a viable approach. For example, Wan and Davis [15] have immobilised a homogeneous hydrogenation catalyst with retention of both the high activity and the enantioselection of the non-immobilised catalyst. Recently, Bein and coworkers [16] have encapsulated chiral Mn (III) salen complexes in zeolite Y and have reported high enantiomeric excesses for epoxidation of *cis*-β-methylstyrene with sodium hypochlorite.

These early studies demonstrate that very high levels of enantioselection can be achieved using heterogeneous asymmetric catalysts. Most of these catalysts tend to be very specific, however, and small changes to the nature of the catalyst, the modifier or the substrate can lead to a loss of enantioselectivity. We, therefore, consider that a generic approach to the design of heterogeneous asymmetric catalysts would have an immense benefit for this field. We believe that this can be provided by heterogeneous catalysts with well-defined structures. For this reason, we have concentrated our initial research [17-30] on the chiral modification of zeolite catalysts, since these microporous materials are well-defined crystalline solids and the nature of the modified surface can be probed with a range of spectroscopic techniques. The central aspect of this design approach concerns the use of the charge balancing extra-framework cations as heterogeneous catalysts. These cations can be readily introduced into the intra-crystalline pore structure of zeolites using standard ion-exchange techniques from aqueous solutions. The cations are held within the structure by electrostatic forces and, when used in non-aqueous solvents or with gas phase reactants/products these cations are stable within the structure and are not readily removed from the zeolite. It is the careful control of the reaction conditions that ensures the stability of these catalysts. Furthermore, the cations present within the zeolite micropores can readily be modified by chiral ligands. Indeed, the zeolitic framework should be viewed as part of the coordination shell of the cation, and chiral ligands can be accommodated as a further part of this coordination shell. We have now used this approach to design heterogeneous asymmetric catalysts for three reactions : (i) the enantioselective dehydration of butan-2-ol [17-22]; (ii) the enantioselective aziridination of alkenes [23-27] and (iii) the enantioselective epoxidation of alkenes [28-30]. The results of these studies will be described in this chapter. It should be noted that the enantioselective dehydration of butan-2-ol can be viewed as the initial proof of concept stage for this approach. This reaction deals with the selective dehydration of one enantiomer in the presence of the second and, for the catalyst system described, this effect was observed only for a very short time under very specific conditions. However, the examples dealing with the enantioselective aziridination and epoxidation of alkenes demonstrate that this approach can produce stable, highly enantioselective catalysts and, hence, confirms the generic nature of this design approach.

3. ENANTIOSELECTIVE DEHYDRATION OF BUTAN-2-OL USING ZEOLITE Y MODIFIED WITH CHIRAL DITHIANE OXIDES – A PROOF OF CONCEPT STUDY

Zeolites are extensively used as catalysts in the refinery and petrochemical industries [31] and, in recent years, the use of microporous materials has been explored in the field of fine chemical synthesis [32]. In most of these applications, use is made of the ion-exchange properties of zeolite and they are used as either Brønsted acidity/basicity has been demonstrated as a mechanism for controlling product selectivity in zeolite catalysed reactions [33,34]. Although ion exchange in zeolites is well studied, the role and formation of other non-framework species has received less attention. Until recently, most attention has been given to the formation of non-framework aluminium species during the dealumination of zeolite Y by steaming. The combination of the octahedrally coordinated, non-framework aluminium species acting as a Lewis acid together with the tetrahedrally coordinated, framework aluminium acting as a Brønsted acid has been shown to lead to the formation of a new site with enhanced catalysed reactions [35-37]. In this section, the modification of zeolite Y with chiral dithiane oxides will be described and it will be shown that the combination of Lewis and Brønsted acidity is important with respect to the nature of the active site for this enantioselective aziridination catalyst.

Two types of chiral modifier have been contrasted in this study. First, substituted 1,3-dithiane 1-oxides I, which are chiral at sulfur, have been used. Samples where R=H, CH_3 and Ph were utilised in both racemic and enantiomerically enriched form. Where R=CH_3, or Ph, a second chiral centre is present but the relative configuration at this centre is always controlled and, in the materials described, only the *trans* configuration was utilised. These materials were compared with 1,3-dithiane II, and cystine III as alternative modifiers.

3.1 Conversion of Butan-2-ol over Zeolite Y Modified with Dithiane Oxide during Synthesis

Samples of zeolite Y were prepared in the presence and absence of a racemic sample of 1,3-dithiane 1-oxide. The samples were characterised by powder X-ray diffraction and both were found to be crystalline; ^{27}Al MAS NMR spectroscopy confirmed that all aluminium atoms were tetrahedrally coordinated. ^{29}Si MAS NMR spectroscopy was used to determine the framework Si/Al ratio of the zeolite. For the zeolite Y made in the absence of the dithiane oxide, the Si/Al ratio was 2.7, which is similar to commercial samples of zeolite Y. For the zeolite synthesised in the presence of dithiane oxide, the framework Si/Al ratio was significantly lower, at 2.0, indicating that the presence of the dithiane in the synthesis gel increases the incorporation of aluminium into the framework. ^{13}C MAS NMR spectroscopy of a sample prepared using 1,3-dithiane 1-oxide in the synthesis gel showed signals characteristic of the 1,3-dithiane 1-oxide confirming that the modifier was incorporated in the zeolite. These materials were found to be inactive for the dehydration of racemic butan-2-ol, and conversions typically of <8% were observed even at temperatures as high as 200°C (Table 1). The presence of the dithiane oxide modifier leads to a decrease in activity, and no effect on the chemoselectivity was observed. In this case, we consider that the zeolite is present in the sodium ion-exchanged form, and hence the concentration of Brønsted acid sites is very low. The materials were then ion-exchanged with ammonium nitrate and calcined for a short time at 400°C to give the proton exchanged forms. The dehydration of racemic butan-2-ol was then examined using the ion-exchanged materials (Table 1). The proton forms are much more active than the Na$^+$ forms, and the proton form of the dithiane oxide-modified zeolite is much more active than the unmodified zeolite, and also gives a somewhat higher selectivity for production of but-1-ene.

Table 1. Catalytic Performance of Zeolite Y Modified by the Addition of 1,3-Dithiane 1-Oxide to the Synthesis Gel for the Dehydration of Butan-2-ol[a]

Catalyst[b]	Temperature (°C)	Conversion (%)	Product Composition (%)		
			But-1-ene	c-But-2-ene	t-But-2-ene
Na-Y	190	7.1	16.1	36.4	47.5
H-Y	140	1.0	15.7	28.7	55.6
	160	4.0	17.0	36.0	47.0
	170	16.0	17.4	37.4	45.2
Na-SO-Y	190	3.7	17.4	36.3	46.3
H-SO-Y	140	1.8	25.9	27.7	46.4
	160	16.4	22.6	29.8	47.6
	200	42.4	22.0	33.2	44.8

[a] Catalyst (0.3 g) reacted with butan-2-ol (3.2 x 10^{-3} mol h^{-1}) prevaporised in diluent nitrogen (4.3 x 10^{-2} h^{-1}).

[b] Na- denotes that the zeolite is in the sodium-exchanged form. H- denotes that the zeolite is in the proton-exchanged form, and SO- denoted that 1,3-dithiane 1-oxide has been added to the synthesis gel.

These results show that modified catalysts can be prepared by the incorporation of dithiane oxide modifiers in the synthesis gel of zeolites. However, only low levels of the modifier are incorporated by this method, as evidenced by the integrated intensities of the ^{13}C MAS NMR spectra. In addition, to prepare catalytically active materials, it is necessary to carry out ion exchange of the Na^+ form of the zeolite with NH_4^+ cations and to form the active proton form by a calcination procedure. During this process, it can be expected that some of the protonated dithiane oxide will also be exchanged as the cation and an equilibrium will be established between the concentrations of Na^+, NH_4^+, and $C_4H_9S_2O^+$ as counter-cations in the zeolite. In addition, the calcination procedure required for the conversion of the NH_4^+ form to the H^+ form could lead to thermal decomposition of the dithiane oxide. For these reasons, we decided not to pursue this method of catalyst synthesis, and instead a postsynthesis modification procedure was investigated.

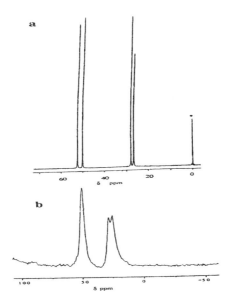

Figure 1. ^{13}C NMR spectra (a) 1,3-dithiane-1-oxide dissolved in $CHCl_3$; *TMS; (b) 1,3-dithiane 1-oxide adsorbed in zeolite Y (LZY 82, Union Carbide; one molecule/supercage).

3.2   Conversion of Butan-2-ol over Zeolite Y Modified with Dithiane Oxide Postsynthesis

Catalyst stability. Samples of zeolite H-Y modified by adsorption of 1,3-dithiane 1-oxides at one molecule per supercage were prepared and characterised by powder X-ray diffraction and ^{13}C MAS NMR spectroscopy. Powder X-ray diffraction showed that the modification process had not caused any significant loss of crystallinity. The ^{27}Al MAS NMR spectra of both the unmodified zeolites indicated the presence of non-framework aluminium, but this was not significantly increased by the adsorption of the dithiane oxide. ^{13}C MAS NMR spectroscopy of the modified zeolite, when compared with the solution NMR spectra of the equivalent dithiane oxide (Figure 1), indicated that the dithiane oxide was molecularly adsorbed within the zeolite.

The stability of the modified zeolites was then investigated with a range of techniques. First, the modified zeolites were examined with EIMS. The samples were gradually heated at a constant rate (2°C/min), and the evolved gases were analysed. The results for 2-phenyl-1,3-dithiane 1-oxide as modifier are shown in Table 2, and the results are compared with the pure modifier and the unmodified zeolite.

Table 2. EIMS Analysis for 2-Phenyl-1,3-dithiane 1-oxide-Modified Zeolite Y

Temperature (°C)	*m/z*			
	40-100	200-300	300-400	400-500
Zeolite-HY	46, 44	-	ND	-
2-Phenyl-1,3-dithiane 1-oxide	46, 44	212,121, 90,45	ND	-
Zeolite HY 2-Phenyl-1,3-dithiane 1-oxide	46, 45, 44	-	ND	106,91,78 64,45

Note : ND, not detected

The pure dithiane oxide decomposed in the region 200-300°C, whereas the zeolite-modified sample did not show any decomposition products below 400°C; at this temperature there was no evidence for the molecular ion (*m/e* 212), which indicates that the dithiane oxide decomposes at this temperature rather than simply desorbing. The unmodified zeolite, as expected, did not show any desorption or decomposition products in this temperature region. Secondly, the modified samples were heated (180°C, 2 h) in a fixed-bed microreactor under flowing nitrogen. The samples were cooled and then extracted with deuteriated chloroform using a soxhlet procedure.

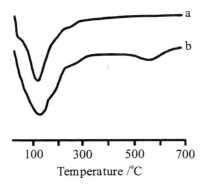

Figure 2. TGA of (a) zeolite Y and (b) 1,3-dithiane 1-oxide adsorbed in zeolite Y (LZY) 82, Union Carbide, one molecule/supercage)

Analysis of the extract by ^{1}H NMR spectroscopy confirmed that the dithiane modifier could be recovered unchanged by this procedure. Thirdly, thermogravimetric analysis (heating rate 10°C/min) was employed to assess further the stability of the modifier in the zeolite. A comparative analysis of the unmodified zeolite H-Y and zeolite H-Y modified with 1,3-dithiane 1-oxide is shown in Figure 2. Both samples show similar weight loss below *ca.* 250°C due to water loss. The modified zeolite does show an additional weight loss at *ca.* 450°C. This behaviour is consistent with the results of the EIMS study.

To assess whether the modifier was evenly distributed throughout the microporous structure of the zeolite, an XPS study was conducted for zeolite H-Y modified with 1,3-dithiane 1-oxide at one molecule per supercage. Analysis of the peak areas of the S_{2p} and the Si_{2p} peaks were used to determine the S/Si ratio, which was found to be 0.07. The calculated value of the ratio expected for the bulk of the sample for uniform distribution is 0.07, and the agreement between these values indicates that the dithiane oxide modifier is indeed uniformly distributed.

Zeolite H-Y and zeolite H-Y modified with cystine were analysed by thermal gravimetric analysis and the results are shown in Figure 3. The unmodified and Na^+ exchanged zeolites showed similar weight loss up to *ca.* 250°C due to water loss. The cystine modified zeolites exhibited similar weight loss for water removal but an additional weight loss at *ca.* 280-300°C was observed. Separate experiments using L-cystine showed that this weight loss is coincident with the decomposition of the amino acid. This indicates that there is only a weak interaction between the cystine and the zeolite framework.

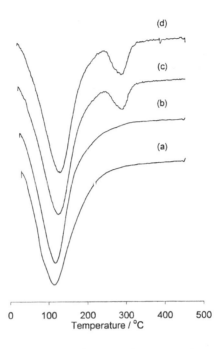

Figure 3. TGA profiles of : (a) H-Y, (b) Na-Y, (c) H-Y-D-cystine, and (d) H-Y-L-cystine

3.3   Catalytic Performance of Modified Zeolites for the Dehydration of Butan-2-ol

A range of catalysts was prepared by the adsorption of racemic 2-substituted-1,3-dithiane 1-oxides (substituent R=H, CH_3, C_6H_5; one molecule/supercage), and these were used as catalysts for the dehydration of racemic butan-2-ol. The results are shown in Table 3, together with those for the unmodified zeolites. It is apparent that modifications of zeolite H-Y (LZY 82, Union Carbide) with 1,3-dithiane 1-oxide gives considerably higher catalytic activity than zeolite H-Y (Crosfield) modified in the same way. However, both modified samples displayed an enhanced catalytic activity when compared with their respective unmodified zeolite. This enhanced activity was maintained for many hours, whereas the unmodified zeolites were rapidly deactivated under the reaction conditions. Turnover numbers of >1200 based on the dithiane 1-oxide were observed for the conversion of butan-2-ol over the modified zeolite, indicating that the enhanced activity is a sustained catalytic effect. Modification with 2-methyl-

1,3-dithiane 1-oxide gave similar results, but 2-phenyl-1,3-dithiane 1-oxide did not lead to such a pronounced rate enhancement, although it still gave a higher activity than the unmodified zeolite. Following reaction, all the catalysts were analysed by powder X-ray diffraction, and some loss of crystallinity was apparent. The zeolites were found to have coked during the reaction, but the dithiane oxide modifier could still be recovered by soxhlet extraction using deuteriated chloroform.

Table 3. Catalytic Performance of Dithiane Oxide-Modified Zeolite Y[a]

Catalyst[b]	Temperature (°C)	Conversion (%)	Product Composition (%)		
			But-1-ene	*c*-But-2-ene	*t*-But-2-ene
US-Y	115	0	-	-	-
	140	6.0	16.0	47.4	36.6
	150	18.2	15.6	51.2	33.2
SOUS-Y	100	11.6	24.4	51.3	24.3
(R=H)	140	95.4	15.0	46.0	39.0
	180	97.3	13.3	44.2	42.5
SOUS-Y	115	10.0	15.9	33.6	50.5
(R=CH₃)	150	98.8	13.7	38.5	47.8
	190	100	12.4	35.3	52.3
H-Y	115	0	-	-	-
	140	2.0	17.0	36.0	47.0
	170	5.0	17.0	37.0	46.0
SOH-Y	110	1.7	14.6	28.3	57.0
(R=H)	140	10.6	12.5	29.2	58.3
SOUS-Y	120	0.5	17.2	31.0	51.7
(R=C₆H₅)	153	3.5	15.9	34.5	49.6
	170	9.1	16.5	36.5	47.0

[a]Catalyst (0.3 g) reacted with butan-2-ol (3.2 x 10⁻³ mol h⁻¹) prevaporised in diluent nitrogen (4.3 x 10⁻² mol h⁻¹).
[b]Key to catalysts: US-Y, LZY 82, Union Carbide; H-Y, ion-exchanged NaY, Crosfield; SO denotes the zeolite has been modified by addition of 2-*R*-1,3-dithiane 1-oxide (one molecule per supercage, R=H, CH₃, C₆H₅).

A range of control experiments was carried out to probe the origin of this rate enhancement (Table 4). Modification of zeolite Y with 1,3-dithiane (compound II) using an analogous procedure led to a small rate enhancement, significantly lower than that observed for modification by the equivalent dithiane oxide, indicating the importance of the sulfoxide oxygen atom as a vital component of the catalytically active site. Investigation of a non-microporous SiO₂/Al₂O₃ catalyst indicated that, in this case, the dithiane oxide modifier acted as a poison. In addition, the same reaction carried out over dithiane oxide supported on an inactive support, BN, indicated that the dithiane oxide by itself does not exhibit any catalytic activity for this reaction. These control experiments demonstrate that it is the combination of the microporous zeolite, together with the dithiane oxide, that is important for the creation of the enhanced activity catalytic site observed with the modified zeolites.

The conversion of racemic butan-2-ol over zeolites modified with the amino acid cystine was carried out at 160°C and the results for the effect of time-on-line are shown in Figure 4. The highest initial conversion for the dehydration reaction was observed using the unmodified zeolite Y. The sodium exchanged zeolite Y (two Na$^+$ per supercage) also gave a high conversion under these conditions and, after 180 min time-on-line, very similar conversions were observed with the unmodified and Na$^+$ exchanged zeolites. In contrast, the cystine modified zeolites gave much lower conversions and similar activity was observed for both D- and L-modified zeolites.

Table 4. Reaction of Racemic Butan-2-ol over Modified Catalysts[a]

Catalyst :	Y[b]		Y-SO[c]	Y-S[d]		SiO$_2$/Al$_2$O$_3$[e]	SiO$_2$/Al$_2$O$_3$[f]	BN-SO[g]	
Temperature (°C)	115	225	115	110	175	200	200	115	225
Conversion (%)	0	90	90	3.6	35.6	31.5	24.5	0	0
Selectivity (%)									
But-1-ene	-	17.7	8.3	15.4	14.7	16.5	21.6	-	-
trans-But-2-ene	-	40.6	53.8	53.8	46.5	31.7	33.9	-	-
cis-But-2-ene	-	41.6	37.8	30.7	38.8	51.8	46.5	-	-

[a] Catalysts (0.3 g) tested in a conventional glass microreactor using on-line GC analysis with butan-2-ol (3.3 x 10^{-3} mol h^{-1}) prevaporised in a nitrogen diluent (3.7 x 10^{-2} mol h^{-1}).
[b] Zeolite Y (ultrastabilised LZY 82, Union Carbide)
[c] Zeolite Y modified with (±)-1,3-dithiane 1-oxide, one molecule per supercage (7.6 wt%).
[d] Zeolite Y modified with 1,3-dithiane, one molecule per supercage (7.2 wt%).
[e] Non-microporous silica alumina, SiO$_2$/Al$_2$O$_3$ = 5.7.
[f] Non-microporous silica alumina, SiO$_2$/Al$_2$O$_3$ = 5.7, modified with 1,3-dithiane 1-oxide (6.9 wt%).
[g] Boron nitride modified with 1,3-dithiane 1-oxide (7.4 wt%) under these conditions unmodified BN was also inactive.

This was also observed when the conversion was decreased to *ca.* 1.0% by decreasing the temperature to 120°C. It is clear from these results that the Na (average 2 per supercage) present in the NaHY are not responsible for the loss of dehydration activity and that it is the cystine (average 1 per supercage) that causes this effect.

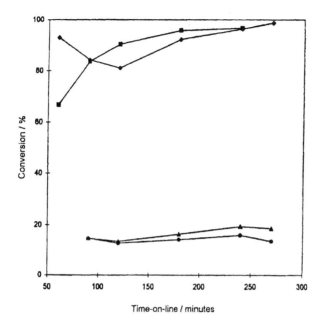

Figure 4. Conversion *vs.* time-on-line plots for : (◆) H-Y, (■) Na-Y, (σ) H-Y-D-cystine, and (●) H-Y-L-cystine

3.4 In Situ FTIR Spectroscopic Investigation of Butan-2-ol Dehydration over Dithiane Oxide-modified Zeolite H-Y

The adsorption of racemic butan-2-ol on unmodified zeolite H-Y (Crosfield) at 50°C leads to the formation of hydrogen-bonded species. Figure 5 shows the IR spectra of H-Y at 50°C, before and after exposure to butan-2-ol vapour. Two hydroxyl stretching vibrations are observed at 3632 and 3542 cm^{-1}, denoted high (HF) and low (LF) frequency bands, respectively. van Santen *et al.* have shown that the protons of the LF band are hydrogen bonded to other lattice oxygen atoms [38]. Adsorption of butan-2-ol leads to the formation of hydrogen bonds between the adsorbate and the Brønsted acidic zeolite hydroxyl groups, as evidenced by the HF zeolite hydroxyl band decreasing in integrated intensity with respect to the LF band. Direct comparison of band intensity is not valid in this case as the extinction coefficients of bands originating from hydrogen-bonded species are significantly increased compared to those of the free molecular species

[39]. The broad bands centred around 2400 and 1700 cm^{-1} are two of the three characteristic bands formed by species strongly hydrogen bonded to zeolites [40-42].

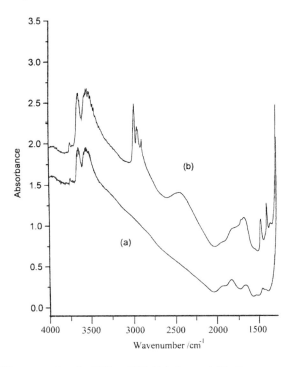

Figure 5. FTIR spectra of zeolite H-Y at 50°C (a) before and (b) after exposure to butan-2-ol (133 N m^{-2})

On heating under a static vacuum, some of the adsorbed butan-2-ol desorbs. A small amount of butanone is also formed, indicated by the presence of a C=O stretch band centred at 1705 cm^{-1}. However, at 90°C water is eliminated, signified by the large increase in intensity of the water deformation band at 1645 cm^{-1}. At the same time, a shift occurs in the CH$_3$ and CH$_2$ band positions (Figure 6). For hydrogen-bonded molecules, the asymmetric and symmetric CH$_3$ stretch bands of adsorbed butanol appear at 2976 and 2888 cm^{-1}; above 90°C, these bands shift to 2960 and 2873 cm^{-1}. This is indicative of the formation of butoxide species [43].

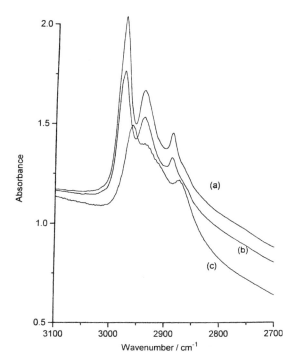

Figure 6. FTIR spectra of zeolite H-Y after exposure to butan-2-ol (133 N m^{-2}) at (a) 50°C, (b) 80°C, and (c) 90°C

When butan-2-ol was adsorbed onto the zeolite modified with 1,3-dithiane 1-oxide (one molecule/supercage), the same initial hydrogen-bonded species was formed. However, on heating, the elimination of water and the band shift characteristic of butoxide formation, described above, occurred at 70°C, rather than 90°C. Modification with higher loading of dithiane-1-oxide did not lead to a further decrease in the temperature of butoxide formation. A blank experiment in which the zeolite H-Y was treated with water at 50°C for 2 h in a procedure analogous to that used in the preparation of the dithiane oxide modified zeolite did not give this effect. For the water treated zeolite H-Y, the butoxide species was still formed at 90°C, indicating that the observed decrease in the temperature of formation of the butoxide species was due to the modification of zeolite Y by the 1,3-dithiane 1-oxide.

3.5   Reaction of R-butan-2-ol and S-butan-2-ol over Zeolite Y Modified with Enantiomerically Enriched Dithiane Oxides

A catalyst was prepared by adsorbing enantiomerically enriched (R)-1,3-dithiane 1-oxide (enantiomeric excess 83%) in zeolite H-Y (Crosfield, one molecule/supercage). A series of control experiments were carried out in which enantiomerically pure (R)- and (S)-butan-2-ol were reacted over the modified zeolite, and it was determined that, in the temperature range investigated (110-150°C), no racemisation of the butan-2-ol occurred. In addition, following reaction with either enantiomerically pure (R)- or (S)-butan-2-ol, or racemic butan-2-ol, the chiral modifier was extracted with deuteriated chloroform, as described above, and was investigated by ^{1}H NMR spectroscopy in the presence of a chiral shift reagent. In this way, it was determined that these treatments had not induced racemisation of the chiral modifier.

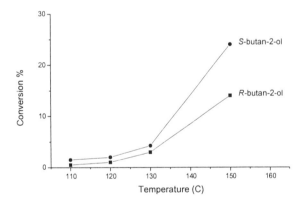

Figure 7. Effect of temperature on the conversion of (R)-butan-2-ol and (S)-butan-2-ol over zeolite H-Y (Crosfield) modified with (R)-1,3-dithiane 1-oxide (one molecule/supercage); reaction conditions as in Table 6.

A series of experiments were then carried out at increasing temperatures for the separate reactions of (R)- and (S)-butan-2-ol over the chirally modified zeolite; the results for conversion and selectivity are shown in Figure 7. It is clear that, when the zeolite is modified with (R)-1,3-dithiane 1-oxide, (S)-butan-2-ol is more reactive over the entire temperature range compared with (R)-butan-2-ol. In a further set of experiments, the temperature was maintained at 110°C, and the flow rates of the (*R*)- and (*S*)-butan-2-ols were separately varied (Table 5). Enhanced reactivity of (*S*)-butan-2-ol was again observed.

Table 5. Effect of reactant feed rate on the conversion of (R)- and (S)-butan-2-ol to alkenes over Zeolite Y modified with (R)-1,3-dithiane 1-oxide[a]

Reactant flow rate	Conversion (%)	
(10^{-2} mol h^{-1})	(*R*)-Butan-2-ol	(*S*)-Butan-2-ol
2.1	5.7	12.2
6.6	4.5	8.0
13.0	1.48	4.75
18.0	0.01	1.05

[a] Zeolite Y (ion-exchanged Na-Y ex Crosfield modified with one molecule per supercage of (R)-1,3-dithiane-1-oxide, 0.3 g) reacted separately with (R)- and (S)-butan-2-ol prevaporised in diluent nitrogen (5.6×10^{-2} mol h^{-1}) at 110°C.

A similar set of experiments was conducted with zeolite Y (LZY, 82, Union Carbide) modified with enantiomerically enriched ((*S*)-2-phenyl-1,3-dithiane 1-oxide (enantiomeric excess 99%) at a loading of one molecule/supercage; the results for the separate reactions of (*R*)- and (*S*)-butan-2-ols with increasing temperature are shown in Figure 8. In this case, it is apparent that (*R*)-butan-2-ol is more reactive than (*S*)-butan-2-ol over the temperature range. Over both modified zeolites, the activation energies for the separate rate conversions of (*R*)- and (*S*)-butan-2-ols are very similar (105 ± 5 kJ mol^{-1}).

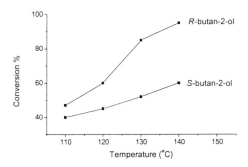

Figure 8. Effect of temperature on the conversion of (R)-butan-2-ol and (S)-butan-2-ol over zeolite H-Y (LZY 82, Union Carbide) modified with (S)-2-phenyl-1,3-dithiane 1-oxide (one molecule/supercage); reaction conditions as in Table 6

3.6 Reaction of Racemic Butan-2-ol over Zeolite Y Modified with Enantiomerically Enriched Dithiane Oxides

The difference in reaction rate observed for the reaction of the enantiomerically pure (*R*)- and (*S*)-butan-2-ol suggests that this catalyst system should be able to discriminate between the two enantiomers of racemic butan-2-ol. Hence, a series of experiments were carried out for the reaction of racemic butan-2-ol over zeolite H-Y modified with enantiomerically enriched dithiane oxides; the results are given in Table 6. As observed previously with the enantiomerically pure (*R*)- and (*S*)-butan-2-ols, when the zeolite is modified with (*R*)-1,3-dithiane 1-oxide, it is apparent that (*S*)-butan-2-ol reacts in preference.

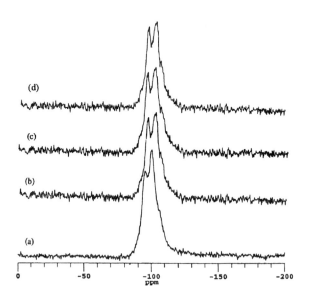

Figure 9. ²⁹Si MAS NMR spectra: (a) zeolite H-Y, (b) after modification with 1,3-dithiane-1-oxide, (c) after modification with 1,3-dithiane, and (d) after modification with D-cystine

Table 6. Reaction of Racemic Butan-2-ol over Zeolite Y Modified by Enantiomerically Enriched Dithiane Oxides

Catalyst		Temp.	Conversion	Product composition[b] ($\times 10^{-3}$ mol h^{-1})				Butan-2-ol conversion (%)		
zeolite	modifier	(°C)	(%)[a]	But-1-ene	But-2-ene	R-Butan-2-ol	S-Butan-2-ol	R	S	Relative rate
Y[c]	(R)-**I**, R=H	110	0.5	-	0.037	3.673	3.640	0.002	0.035	1:17.5
Y[c]	(R)-**I**, R=H	120	1.3	0.015	0.085	3.669	3.581	0.006	0.094	1:15.7
Y[c]	(R)-**I**, R=H	150	9.9	0.105	0.620	3.657	2.968	0.018	0.707	1:39.3
Y[d]	(S)-**I**, R=Ph	110	4.2	0.020	0.271	3.399	3.660	0.276	0.015	18.4:1
Y[d]	(S)-**I**, R=Ph	120	7.5	0.044	0.507	3.260	3.594	0.415	0.081	5.1:1

[a] Total conversion of R- and S-butan-2-ol
[b] Flow rate, 10^{-3} mol h^{-1}.
[c] Zeolite Y (Crosfield NaY, ion exchanged with NH$_4$NO$_3$ and calcined at 550°C), modified with R-1,3-dithiane 1-oxide, **one molecule per supercage** (0.1g), tested in a conventional glass microreactor with racemic butan-2-ol (7.35 x 10^{-5} mol h^{-1}), prevaporised in a nitrogen diluent (6.7 x 10^3 mol h^{-1}).
[d] Zeolite Y (ultrastabilised LZY 82, Union Carbide) modified with S-2-phenyl-1,3-dithiane 1-oxide, **one molecule per supercage (0.1 g), tested in a** conventional glass microreactor with racemic butan-2-ol (7.35 x 10^{-3} mol h^{-1}), prevaporised in a nitrogen diluent (6.2 x 10^3 mol h^{-1}) to (R)-butan-2-ol. In addition, when the zeolite is modified with (S)-2-phenyul-1,3-dithiane 1-oxide, (R)-butan-2-ol reacts in preference to (S)-butan-2-ol. **In both these sets of** experiments, enantioselection was observed for a number of turnovers, although the effect was **short-lived at the higher temperatures** (turnover numbers, 17 and 2, respectively (based on dithiane 1-oxide) at 110 and 150°C, respectively).

This is in agreement with molecular simulation pore access calculations using the Insight II code [44] that show that both 1,3-dithiane-1-oxide and cystine can enter and diffuse within the zeolite pores with ease. However, calculated protonation energies for cystine were found to be 110 kJ/mol less than those calculated for dithiane 1-oxide using the same basis set [45]. These molecular simulations confirm the weak interaction between cystine and zeolite Y compared to the interaction of dithiane-1-oxide and zeolite Y.

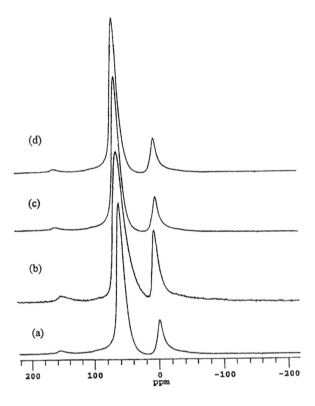

Figure 10. ^{27}Al MAS NMR spectra: (a) zeolite H-Y, (b) after modification with 1,3-dithiane-1-oxide, (c) after modification with 1,3-dithiane, and (d) after modification with D-cystine

^{29}Si and ^{27}Al MAS NMR spectra were recorded for the unmodified zeolite H-Y and for the zeolite modified with 1,3-dithiane-1-oxide, 1,3-dithiane and cystine (one molecule/supercage), and these are shown in Figures 9 and 10. From the ^{29}Si MAS NMR spectra, it is apparent that no differences are induced when the zeolite is modified. In particular, no dealumination of the zeolite lattice is observed. The ^{27}Al spectra show that, when the zeolite is modified with 1,3-dithiane-1-oxide, an increase is observed in intensity of the resonance at 0 ppm, which is associated with the extra-framework aluminium. This is not observed when the zeolite is modified with either 1,3-dithiane of D-cystine. Since the ^{29}Si NMR spectra do not show evidence of dealumination of the zeolite, this increase in intensity is considered to result from a specific interaction between the extra-framework aluminium and the 1,3-dithiane-1-oxide. Further evidence for this interaction with sulfoxides has been obtained using model modifiers such as DMSO [46]. Figure 11 shows ^{13}C CP MAS NMR spectra of : (a) 1,3-diethiane-1-oxide modified H-Y, and (b) cystine modified H-Y. The spectra were obtained under the same conditions, each sample being subjected to 2500 pulses. Comparisons of the two ^{13}C CP spectra indicate that there is significantly greater heteronuclear dipolar coupling in the case of the sulfoxide modified zeolite than the cystine modified sample, there being only a broad carbon resonance in the dithiane oxide case compared to the sharp signals derived from the cystine modified zeolite. This is in agreement with the molecular simulation studies that indicate a significantly lower protonation energy for cystine and the ^{27}Al MAS NMR data that shows that there is little interaction between the cystine and zeolite framework. It can be concluded, therefore, that the cystine is relatively mobile within the pore structure of the host zeolite and does not form the rigid active site that occurs on modification with 1,3-dithiane-1-oxide.

Modification of zeolite H-Y with dithiane oxides leads to the creation of a catalyst that has enhanced activity for the acid-catalysed dehydration of butan-2-ol when compared with the unmodified zeolite. This effect is observed both for samples prepared by addition of the modifier to the synthesis gel and for the postsynthesis modification of two different commercial samples of zeolite Y (Crosfield and Union Carbide). When the modifier is added to the synthesis gel, the resulting zeolite contains only a small quantity of the modifier, but the effect is still apparent.

The interaction of alcohols with the acid forms of zeolites has been particularly well studied in recent years [47,48], and it is generally considered that the adsorbed species that is formed initially is an alkoxide. *In situ* FTIR spectroscopy confirmed that, in the present study, an alkoxide species is formed on reaction of the butan-2-ol with both the unmodified zeolite and the dithiane oxide-modified zeolite. Consistent with the enhanced reactivity observed with the flow reactor studies, the *in situ* FTIR spectroscopy experiments reveal that the alkoxide species is formed at a significantly lower temperature for the dithiane oxide modified zeolite than for the unmodified zeolite.

Dehydration of butan-2-ol leads to the formation of but-1-ene, *cis*-but-2-ene and *trans*-but-2-ene. At equilibrium, but-1-ene becomes the major product at temperatures >90°C and, for the conditions used in this study, the equilibrium compositions are 47.5:35:17.5, for *trans*-but-2-ene, *cis*-but-2-ene, and but-1-ene, respectively. When the modified zeolites are used as catalysts, product compositions that are very close to the equilibrium values are observed and, consequently, it is not possible to draw any mechanistic information from the observed chemoselectivity.

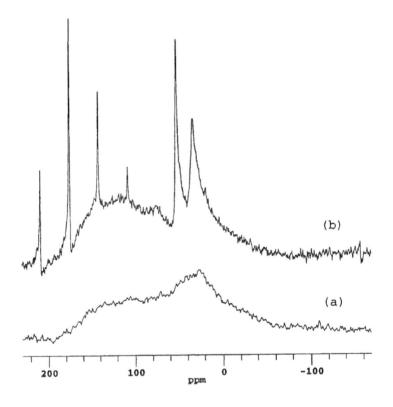

Figure 11. ¹³C CP MAS NMR spectra of: (a) H-Y after modification with 1,3-dithiane-1-oxide, (b) H-Y after modification with D-cystine

All these data indicate that, when the dithiane oxide is added to the zeolite, a new high activity site is formed. It is apparent that the enhancement in activity is greatest for the sample of zeolite Y supplied by Union Carbide (LZY 82). This is an ultrastabilised zeolite Y that has been steamed and, consequently, contains significant amounts of nonframework aluminium. The rate enhancement is less pronounced, but still significant, with the modification of the zeolite Y supplied by Crosfield. The proton form of this zeolite was obtained by ion exchange of the commercially supplied Na-Y sample which originally contained no extraframework aluminium. The ion-exchange procedure adopted led to dealumination of the framework, but the extent was much less marked than with zeolite H-Y (LZY 82). Hence, the degree of enhancement observed may be related to the presence of extraframework aluminium. This must, however, act in combination with the Brønsted acid site of the bridging hydroxyl group associated with the zeolite framework aluminium, since the dehydration reaction investigated is catalysed by Brønsted acidity. A number of studies have indicated that a combination of extraframework (Lewis acid sites) and framework (Brønsted acid sites) aluminium can give enhanced acidity in zeolite catalysts. Haag and Lago [35] showed that steaming zeolites at 500°C with low levels of water vapour can lead to an increase in the activity of the zeolite for acid-catalysed reactions. Mirodatos and Bartemeuf [36] showed that superacid sites could be created in mordenite by a steaming procedure. These effects were subsequently explained by Fritz and Lunsford [37] in terms of the initial dealumination of the zeolite to form nonframework aluminium that imparts, presumably through an electrostatic effect, strong acidity of the remaining framework Brønsted acid sites. This leads to an increase in acidity of the zeolite and, for simple acid-catalysed reactions such as cracking, an enhanced activity is observed. In this catalyst, the high activity site is considered to be formed by the specific interaction of the dithiane oxide with both the extra frame work aluminium and the Brønsted acid site associated with the framework aluminium (Figure 12).

The formation of high activity acid sites enables these modified zeolites to be used as enantioselective catalysts since, although only one active site per supercage is modified, it is many orders of magnitude more active than the remaining unmodified sites. When the dithiane oxide is used in the enantiomerically enriched form, the active sites are able to discriminate between the enantiomers of butan-2-ol.

This important effect is achieved by enantioselective rate enhancement, *i.e.* both enantiomers react faster in the chiral environment than in the absence of the chiral modifier, but one reacts faster than the other. The observation that modification leads to the formation of a high activity site has also been observed for the cinchona alkaloid modification of platinum for the enantioselective hydrogenation of prochiral α,β-ketoesters. In the case of cinchona modification, rate enhancements of three orders of magnitude are observed and, hence, the observation of a rate enhancement effect may be a prerequisite for the design of effective enantioselective catalysts.

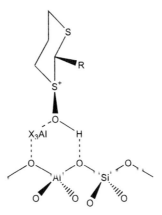

Figure 12. Schematic representation of the structure of the high activity site formed on addition of 1,3-dithiane-1-oxide to zeolite H-Y (Al^+X_3 denotes extra-framework aluminium species)

It is observed that the activation energies for the dehydration of the separate enantiomers over the zeolites doped with the enantiomerically enriched dithiane oxides are identical within experimental error (105 ± 5 kJ mol^{-1}). In addition, it should be noted that the enantiomeric discrimination obtained from the reaction of racemic butan-2-ol is higher than that based on the conversions of the separate enantiomers. The apparent inconsistency can be resolved by regarding the modified zeolite as if it were an enzyme (a zeozyme?) operating with its active sites saturated with substrate. The conversions of the separate enantiomers then yield the relative reactivities in the catalytic step, whereas the results using the racemate reflect the enantioselection not only of the catalytic step but also of the competitive binding of the substrates. Present indications, supported by molecular modelling studies, are that the latter effect is the larger.

It is important to note that the reactants and products in this reaction are gas phase. However, the reactant butan-2-ol is present at high partial pressures and, at the lower temperature used in this study, which is only 12°C above the boiling point of butan-2-ol, it is possible that condensation may occur in the micropores of the zeolite. This effect would not, however, be expected to contribute significantly to the results obtained at the higher temperatures, where enantioselection is still observed. Although the initial approach has been to design a catalyst that consumes chiral molecules, the study is, in effect, a proof of the concept that a zeolite can be modified so that it preferentially catalyses reaction of one enantiomer of a chiral substrate in the presence of both. In view of the immense range of microporous zeolites available, together with the large number of enantiomerically pure modifiers, we consider that this approach is capable of providing the basis for the generic design of a new type of enantioselective catalysts, and demonstrate this in the subsequent sections.

4. CATALYTIC HETEROGENEOUS ASYMMETRIC AZIRIDINATION OF ALKENES USING Cu-EXCHANGED ZEOLITE Y AND Al-MCM-41

In the preceding section, the active site of the catalyst comprised the modification of the proton in zeolite Y. We wished to extend this approach to other cations ion-exchanged into a zeolite. At this stage, zeolite Y was retained as the zeolite since it could be obtained in high purity and, as the framework aluminium concentration was relatively high compared to other commercially available zeolites, it permitted the study of relatively high cation concentrations. We decided to select a reaction which was known to be homogeneously catalysed by a cation in solution and of which, furthermore, the heterogeneously catalysed counterpart was unknown. There are clearly a number of reactions that fit these criteria, but we selected the aziridination of alkenes, as Evans *et al.* [49,50] had previously shown that copper cations in solution could readily catalyse the aziridination of alkenes using (*N*-(*p*-tolylsulfonyl)imino)phenyliodinane (PhI=NTs) as the nitrene donor. Furthermore, it was known that use of chiral bis(oxazoline) [50] ligands could effect enantioselective aziridination on pro-chiral alkenes. With this homogeneously catalysed reaction, it was known that high concentrations of Cu^{2+} were required (3000-8000 ppm) and this limited the efficacy of the methodology for the synthesis of chiral aziridines, which are important intermediates for the synthesis of pharmaceuticals. With this in mind, we investigated the aziridination of alkenes using Cu-exchanged zeolite H-Y (*ca.* 50% of the protons were exchanged for Cu^{2+}) using PhI=NTs as the nitrene donor and the results are shown in Table 7.

These initial results confirmed our contention that cations within zeolites could be used as heterogeneous counterparts of known homogeneous catalysts. In order to confirm that this process was truly heterogeneous, after reaction the catalyst was removed by filtration, another aliquot of reactants, without the catalyst, was added to the solvent and the reaction monitored. No further product was observed. Further, the removed catalyst was reused with fresh reagents and solvent, and the zeolite demonstrated similar activity to fresh catalyst. Evans noted in earlier studies that, in the homogeneously catalysed reaction [49], the yield of aziridine decreased to 37% when a 1:1

styrene:PhI-NTs molar ratio was employed; this was considered to be due to the competing breakdown of the PhI=NTs reagent.

Table 6. CuHY-catalysed Aziridination of Representative Alkenes

Entry	Alkene[a]	Cu (mol %)	Yield (5)[b]
1	Styrene	25	90 (92)
2	Styrene[c]	25	87 (35)
3	Styrene	5	62
4	α-Methylstyrene	25	33
5	p-Chlorostyrene	25	76
6	p-Methylstyrene	25	66
7	Cyclohexene	25	50 (60)
8	Methyl cinnamate	25	84 (73)
9	*trans*-Stilbene	25	0 (52)
10	*trans*-2-Hexene	25	44

Values in parentheses indicate yields obtained from homogeneous reactions
[a] Unless otherwise specified, reaction conditions were : solvent MeCN, 25°C; styrene:PhI=MTs=5:1 molar ratio
[b] Isolated yield of aziridine based on PhI=NTs
[c] Styrene:PhI=NTs=1:1 molar ratio

This decrease was found to be less significant when using the heterogeneous CuHY catalyst, for which a yield of 87% was obtained with 1.0 equivalent of styrene (Table 7, entry2); this compares favourably to a yield of 90% based on PhI=NTs when a 5:1 styrene:PhI=NTs molar ratio was used (Table 7, entry 1). CuHY is an effective catalyst for the aziridination of a range of alkenes (Table 7). The catalyst gives best results with phenyl-substituted alkenes, and lower yields are observed with cyclohexene and *trans*-hex-2-ene. Interestingly, in the aziridination of *trans*-stilbene, no product could be observed when the CuHY catalyst was utilised. However, in previous studies, Evans *et al* [49] demonstrated that the homogeneous copper catalysts were effective for the aziridination of this bulky alkene. The ineffectiveness of the heterogeneous CuHY catalyst is thought to be due to the relatively bulky aziridine product being too large to diffuse from the pores of CuHY. In a further set of experiments, a range of solvents were investigated and the results are given in Table 8. For the heterogeneously catalysed reaction, the best results are obtained using acetonitrile as solvent. This is in contrast to the homogeneously catalysed reaction, for which high yields were obtained using benzene as solvent. The three experimental observations (a) that the reactivity was not dependent on the styrene:PhI=NTs molar ratio, (b) the lack of reaction for the bulky substrate *trans*-stilbene, and (c) the differences in reactivity observed with different solvents, are clear indications that the heterogeneously catalysed reaction displays some significantly different characteristics from the homogeneously catalysed reaction previously described by Evans *et al.* [49,50]. To date, the only major disadvantage with the heterogeneous catalyst is that the reaction time for 100% conversion of the PhI=NTs reagent is somewhat longer than that for the homogeneous catalyst under equivalent experimental conditions; *e.g.* for the reaction of styrene (entry 1, Table 7), the heterogeneous catalyst required 2 h reaction time compared to 0.5 h for the homogeneous catalyst, although the yields of the aziridine product were the same within the experimental error.

Table 7. Effect of Solvent on the Aziridination of Styrene[a]

Solvent	Yield (%)
DMF	17
MeCN	86
THF	5
DCM	5
Benzene	32
Toluene	2
Hexane	10

[a] Isolated yields based on PhI=NTs. Reaction conditions : 25°C; styrene:PhI=NTs = 5:1 molar ratio

Zeolites are known to be affected by the adsorption of water and it, therefore, appeared possible that traces of water might act as an inhibitor for the aziridination reaction either by physically blocking the zeolite pores or by coordination to the Cu^{2+} cation. In view of this, the effect of water on the aziridination reaction was investigated for both the homogeneous and heterogeneous systems. Comparison of the homogeneously catalysed reaction for styrene under the conditions used in Table 7 in the presence and absence of added water (1 wt%) showed that the addition of water decreased the reaction rate by an order of magnitude and that the yield was decreased to only 50% at 100% PhI=NTs conversion due to the competing degradation of this reagent. The effect of water is more marked in the heterogeneously catalysed reaction. We have shown that the rate of the heterogeneous reaction is generally lower than the homogeneously catalysed reaction. When acetonitrile containing water (1.0 wt%) was used as solvent, the reaction proceeded very slowly such that, after 5 h, only a 5.0% yield of the aziridine product was observed. Similar effects were observed if the water was added to the zeolite by preadsorption prior to the reaction. These studies show that the dry solvents must be used when CuHY is used as a catalyst for the aziridination of alkenes.

We have successfully recovered and reused the catalyst for 13 consecutive experiments and have found that only traces of Cu^{2+} are lost from the catalyst and that the yield of products is retained. We have already noted that adsorbed water can build up within the pores of the zeolite on continued use and this can lead to some loss of activity. However, full yield can be recovered if the catalyst is simply recalcined in air prior to reuse. We are, therefore, confident that this catalyst system can form the basis of a commercial heterogeneous catalyst for the aziridination of alkenes.

IV

Evans *et al.* [50] have shown that the modification of the homogeneous copper catalysts using chiral bis(oxazoline) ligands induces enantioselectivity in the aziridination reaction. In view of this, we have extended the above studies by examining the modification of the CuHY catalyst with the chiral bis(oxazoline) 2,2,-bis[2-((4*R*)-(1-phenyl-1,3-oxazolinyl)propane **IV**, and have observed *N*-tosylaziridine products with up to 42% enantiomeric excess. The optimum conditions for racemic heterogeneous aziridination of alkene, in the absence of bis(oxazoline), were observed to be 25°C using acetonitrile as solvent. As expected, for the enantioselective reaction, the use of lower reaction temperatures was found to give the highest enantioselectivities. Initially, we found that a temperature of −10°C provides the highest enantioselectivity without compromising yield when using acetonitrile solvent and the bis(oxazoline) **IV** as the chiral modifier (Table 9). It would be reasonable to assume that one chiral modifier would be required per zeolite supercage to obtain maximum enantioselectivity. We have, however, found that very low levels of the expensive modifier can be used without resulting in decreases in yield or enatioselectivity (Table 10). An excess of bis(oxazoline) significantly reduces the yield of aziridine, due to pore-blocking, but both yield and enantioselectivity were maximised with a molar ratio of PhI-NTs:bis(oxazoline) of only 1:0.05. This corresponds to a molar ratio of Cu^{2+}:bis(oxazoline) of 2:1, indicating that not all the Cu^{2+} cations are modified in our experiments. In a subsequent experiment, an excess of bis(oxazoline) **IV** was stirred with CuHY (Cu^{2+}:bis(oxazoline)=1:5 mole ratio) in acetonitrile. The zeolite was filtered and washed with acetonitrile, then used as the catalyst in fresh solvent and reactants. Both yield and enantioselectivity observed were identical to those obtained when bis(oxazoline) was added directly to the reaction mixture (molar ratio of PhI=NTs:bis(oxazoline)=1:0.05). It is clear that very low levels of the modifier are required to obtain the enantioselectivities reported.

Table 8. Effect of Temperature on the Asymmetric Aziridination of Styrene[a]

Temperature (oC)	Yield (%)	e.e. (%)
25	80	18
0	72	25
-10	75	33
-20	84	34
-35	22	32

[a] Isolated yields based on PhI=NTs, e.e. determined by chiral HPLC.
Reaction conditions: solvent MeCN, styrene:PhI=NTs:bis(oxazoline)=5:1:0.05 molar ratio.

Table 9. Effect of Amount of Bis(oxazoline) on Aziridination of Styrene[a]

Bis(oxazoline):PhI-NTs mol ratio	Yield (%)	ee (%)
0.5	18	29
0.2	45	35
0.2	50	35
0.05	84	34
0.025	77	32
0.01	64	28

[a] Dependence of the aziridination of styrene on the amount of bis(oxazoline) measured as molar equivalents based on PhI-NTs. Isolated yields based on PhI=NTs, e.e. determined by chiral HPLC. Reaction conditions: solvent MeCN, -20°C; styrene:PhI=NTs=5:1 molar ratio

We have subsequently extended these studies to investigate a range of chiral bis(oxazoline) ligands and also an alternative nitrene donor, (*N*-(*p*-nitrobenzylsulfonyl)imino)phenyliodinane (PhI=NNs), and these results are shown in Table 11. Careful experimentation using a new nitrene donor was found to give very high enantioselectivities with styrene as substrate. Most recently, we have improved upon these results and have obtained e.e. of 87% for an isolated yield of the aziridine product of >90%.

Table 10.   CuHY-catalysed aziridination of representative alkenes

Bis-oxazoline	Alkene[a]	Temp °C	PhINTs Yield[b] %	PhINTs e.e.[c] %	PhINNs Yield[b] %	PhINNs e.e.[c] %
None	Styrene	25	90 (92)	-	93 (97)	-
	α-Methylsturene	25	33	-		
	Cyclohexene	25	50 (60)	-		
	Methyl cinnamate	25	84 (73)	-		
	trans-Stilbene	25	0 (52)	-		
	trans-Hex-2-ene	25	44			
(Me, Me bis-oxazoline; Ph, Ph)	Styrene	25	87	29	69	52
	Styrene	-10	82	44	100[e]	75[e]
	trans-β-Methylstyrene	-10	74	36		
	trans-β-Methylcinnamate	-10	8 (21)	61 (70)	30[f]	59[f]
(Me, Me bis-oxazoline; CMe₃, CMe₃)		25			87	64
	Styrene	-20	64	0		
	Styrene[d]	-20	15 (89)	18 (63)		
		0			100	34
(bis-oxazoline; Ph, Ph)	Styrene	25	78 (75)	10 (10)		
(bis-oxazoline; Ph···Ph, Ph, Ph)	Styrene	25	73 (74)	0 (15)		
(pyridine bis-oxazoline; iPr, iPr)	Styrene	-10	4	61		

[a] Solvent CH₃CN, alkene : PhI=NTs=5:1 molar ratio, [b] Isolated yield of aziridine based on PhI=NTs. Values in parentheses indicate yields obtained from homogeneous reactions, [c] Enantioselectivity determined by chiral HPLC, [d] styrene was used as solvent, [e] 0°C, [f] 25°C. Absolute configurations of major products, determined by optical rotation, are (*S*) for trans-β-methylstyrene and trans-β-methylcinnamate, (*R*) for styrene.

To demonstrate that alternative types of silicate framework can be used for this reaction, experiments were carried out with copper-exchanged MCM-41. Yields of up to 87% of the aziridine with e.e. of 37% were obtained. Using this type of mesoporous catalyst system greatly enhances the versatility of the heterogeneous aziridination reaction.

5.   ASYMMETRIC EPOXIDATION USING MODIFIED Mn-EXCHANGED MATERIALS

To demonstrate the flexibility of the approach to catalyst design described in this paper, the epoxidation of alkenes using iodosylbenzene has also been studied. Initial studies focused on MnHY:salen catalysts for the epoxidation of styrene, however, the reaction was slow, and low yields of styrene oxide were observed. Analysis of the reaction mixture revealed the breakdown of the salen ligand within a few turnovers. Subsequently, Mn-AlMCM-41 was used with iodosylbenzene as the oxygen donor and *cis*-stilbene was used as substrate, and the results, together with those of control experiments, are shown in Table 12. $Mn(OAc)_2$ in the absence of AlMCM-41 or salen ligand is not particularly reactive, and only 1.5% yield of the epoxide was formed after reaction for 24 h at 25°C. Modification of Mn^{2+} in solution by the salen ligand, as expected, leads to a significant rate enhancement, and both the *cis*-epoxide and the *trans*-epoxide is formed with an enhanced *cis/trans* ratio to the homogeneously catalysed Mn:salen catalyst. This effect suggests that the Al-MCM-41 is occupying part of the Mn^{2+} coordination sphere, restricting the *cis* → *trans* transformation. Further modification of the Mn-exchanged Al-MCM-41 with salen leads to a further enhancement in reactivity, and the *trans* epoxide is formed with a 70% e.e., very similar to that observed for the equivalent homogeneous reactions. *trans*-Stilbene is found to be a significantly less reactive substrate, and the e.e. of the resultant *trans*-epoxide is significantly decreased with the salen modified Mn-exchanged Al-MCM-41. The use of Mn-exchanged Al-MCM-41:salen catalyst for this epoxidation does not result in the formation of significant levels of by-products as has been observed when manganese bypiridyls have been used as catalysts, and typically only deoxybenzoin is observed at low levels (*ca.* 5-10%).

A further set of experiments was carried out to examine the reusability of the Mn-exchanged Al-MCM-41:salen catalyst. Following the reaction, the Mn-exchanged Al-MCM-41:salen catalyst was recovered by filtration and the solid was reused in a new catalytic reaction; although the reactivity and enantioselectivity had declined, epoxide was still formed and the *cis/trans* ratio was unchanged. Recalcination of the recovered material and addition of new salen ligands essentially restored both the reactivity and the enantioselection. Use of the solution following the filtration did not give any activity and, furthermore, this solution contained no Mn^{2+}. These experiments demonstrate that the reaction occurring with Mn-exchanged Al-MCM-41:salen is wholly heterogeneously catalysed. At this stage, we have made no attempt to optimise the catalytic performance, but we anticipate that appropriate modification of the chiral salen ligand in the reaction conditions will lead to enhanced reactivity and e.e.

Table 11. Epoxidation of Stilbene at 25°C using Mn exchanged MCM-41

Entry	Catalyst	Time h[a]	Conv.[b]	Epoxide Yield (%)[f]	Selectivity (%)		e.e. *trans* (%)[g]
					cis	*trans*	
1	none	25	0	0	0	0	0
2	Mn(OAc)$_2$[c,d]	24	100	1.5	0	100	0
3	Mn(salen) complex[c]	1	100	86	29	71	78
4	AlMCM-41[c]	24	0	0	0	0	0
5	MnMCM-41[c]	2	45	3	0	100	0
6	MnMCM-41+salen[c]	2	100	69	58	42	70
7	MnMCM-41+salen[c,e]	26	100	35	0	100	25
8	solution[c]	2	0	0	0	0	0
9	MnMCM-41 reused[c]	2	37	18	61	39	30
10	MnMCM41 recalcined +salen[c]	2	100	52	63	37	54

[a] reaction time, [b] as determined by decomposition of iodosylbenzene to iodobenzene, using HPLC, [c] reactions were conducted in DCM unless otherwise noted with molar ratio of *cis*-stilbene:catalyst:iodosylbenzene=7:1:0.13, [d] reaction conducted in CH_3OH, [e] *trans*-stilbene used as substrate, [f] Conversions, yields and selectivity determined by HPLC, using APEX ODS reverse-phase column, [g] enantiomeric excess determined by chiral HPLC using (*R,R*) Whelk-O 1 column, 92% hexane/*i*PrOH.

6.   CONCLUDING COMMENTS

In this paper, we have described a design approach for heterogeneous enantioselective catalysts. The approach is based upon modification of the counter-cation of zeolites or mesoporous alumino-silicates with a suitable chiral ligand. We have demonstrated the approach with three examples : (a) enantioselective dehydration, (b) enantioselective aziridination of alkenes using Cu^{2+}-exchanged zeolite Y modified with chiral oxazolines and (c) the modification of manganese-exchanged Al-MCM-41 by a chiral salen ligand for the enantioselective epoxidation of alkenes. Since there is a broad range of zeolites and mesoporous materials available as catalytic materials, it is anticipated that the approach described in this paper can form the basis of a generic design of new enantioselective catalysts.

7. REFERENCES

1   W.S. Knowles, M.J. Sabacky, B.D. Vineyard and D.J. Weinkauf, J. Am. Chem. Soc., 97 (1975) 2565.
2   T. Katsuki and K.B. Sharpless, J. Am. Chem. Soc., 102 (1980) 5974.
3   Y. Izumi, Adv. Catal., 32 (1983) 215.
4   H. Brunner, Topics Stereochem., 18 (1988) 129.
5   M. Bartock, "Stereochemistry of Heterogeneous Metal Catalysts", p. 511, Wiley, New York, 1985.
6   J.D. Morrison, "Asymmetric Synthesis", Vol. 5, Academic Press, New York, 1985.
7   H.U. Blaser, Tetrahedron Asymmetry, (1991) 843.
8   G.M. Schwab and L. Rudolph, Naturwisserschaft, 20 (1932) 362.
9   G.M. Schwab, F. Rost and L. Rudolph, Kolloid Z., 68 (1934) 157.
10 W. Kroutil, P. Mayon, M.E. Lasterra-Sanchez, S.J. Maddrell, S.M. Roberts, S.R. Thornton, C.J. Todd and M. Tuter, J. Chem. Soc., Chem. Commun., (1996) 845.
11 H.U. Blaser, H.P. Jalett, D.M. Monti, A. Baiker and J.T. Wehrli, Stud. Surf. Sci. Catal., 67 (1991) 147.
12 G. Webb and P.B. Wells, Catal. Today, 12 (1992) 319.
13 A. Baiker, Stud. Surf. Sci. Catal., 101 (1996) 51.
14 M.A. Keane and G. Webb, J. Catal., 136 (1992) 1.
15 K.T. Wan and M.E. Davis, Nature, 370 (1994) 449.
16 T. Bein and S.B. Ogunwumi, Chem. Commun., (1997) 901.
17 S. Feast, D. Bethell, P.C.B. Page, M.R.H. Siddiqui, D.J. Willock, F. King, C.H. Rochester and G.J. Hutchings, J. Chem. Soc., Chem. Commun., (1995) 2409.
18 S. Feast, D. Bethell, P.C. Bulman Page, F. King, C.H. Rochester, M.R.H. Siddiqui, D.J. Willock and G.J. Hutchings, J. Mol. Cat. A, 107 (1996) 291.
19 S. Feast, D. Bethell, P.C. Bulman Page, M.R.H. Siddiqui, D.J. Willock, G.J. Hutchings, F. King and C.H. Rochester, Stud. Surf. Sci. Catal., 101 (1996) 211.
20 S. Feast, M.R.H. Siddiqui, R.P.K. Wells, D.J. Willock, F. King, C.H. Rochester, D. Bethell, P.C. Bulman Page and G.J. Hutchings, J. Catal., 167 (1997) 533.
21 R.P.K. Wells, P. Tynjälä, J.E. Bailie, D.J. Willock, G.W. Watson, F. King, C.H. Rochester, D. Bethell, P.C. Bulman Page and G.J. Hutchings, Appl. Catal. A, 182 (1999) 75.
22 G.J. Hutchings, Chem. Commun., (1999) 301.
23 C. Langham, P. Piaggio, D. Bethell, D.F. Lee, P. McMorn, P.C. Bulman Page, D.J. Willock, C. Sly, F.E. Hancock, F. King and G.J. Hutchings, Chem. Commun., (1998) 1601.
24 C. Langham, S. Taylor, D. Bethell, P. McMorn, P.C. Bulman Page, D.J. Willock, C. Sly, F.E. Hancock, F. King and G.J. Hutchings, J. Chem. Soc., Perkin Trans. 2, (1999) 1043.
25 C. Langham, D. Bethell, D.F. Lee, P. McMorn, P.C. Bulman Page, D.J. Willock, C. Sly, F.E. Hancock, F. King and G.J. Hutchings, Appl. Catal. A, 182 (1999) 85.

26   G.J. Hutchings, C. Langham, P. Piaggio, S. Taylor, P. McMorn, D.J. Willock, D. Bethell, P.C. Bulman Page, C. Sly, F.E. Hancock and F. King, Stud. Surf. Sci. Catal., 130 (2000) 521.

27   C. Langham, P. Piaggio, D. Bethell, D.F. Lee, P.C. Bulman Page, D.J. Willock, C. Sly, F.E. Hancock, F. King and G.J. Hutchings, "Catalysis of Organic Reactions", ed. F.E. Herkes, Marcel Dekker, Inc., (1998), Vol. 3, p. 25.

28   P. Piaggio, P. McMorn, C. Langham, D. Bethell, P.C. Bulman Page, F.E. Hancock and G.J. Hutchings, New J. Chem., (1998) 1167.

29   P. Piaggio, C. Langham, P. McMorn, D. Bethell, P.C. Bulman Page, F.E. Hancock, C. Sly and G.J. Hutchings, J. Chem. Soc., Perkin Trans. 2, (2000) 143.

30   P. Piaggio, P. McMorn, D. Murphy, D. Bethell, P.C. Bulman Page, F.E. Hancock, C. Sly, O.J. Kerton and G.J. Hutchings, J. Chem. Soc., Perkin Trans. 2, (2000) 2008.

31   D.A. Whan, Chem. Di., 17 (1981) 532.

32   W. Hölderich, M. Hesse and F. Neumann, Angew. Chem., Int. Ed. Engl., 27 (1988) 226.

33   R.M. Dessau, Zeolites, 10 (1990) 205.

34   G.J. Hutchings and D.F. Lee, Catal. Lett., 34 (1995) 115.

35   W.O. Haag and R.M. Lago, US Patent 4 326994.

36   C. Mirodatos and D. Bartemeuf, J. Chem. Soc., Chem. Commun., (1981) 39.

37   P.O. Fritz and J.H. Lunsford, J. Catal., 118 (1989) 85.

38   M.J.P. Brugmans, A.W. Kleyn, A. Lagendijk, W.P.J.H. Jacobs and R.A. van Santen, Chem. Phys. Lett., 217 (1994) 117.

39   G.C. Pimentel and A.L. McClellan, "The Hydrogen Bond", Freeman, San Francisco, 1960.

40   S. Bratos and H. Ratajczak, J. Chem. Phys., 76 (1982) 77.

41   R.A. van Santen, Rec. Trav. Chimiques Pays-Bas, 113 (1984) 423.

42   M.F. Claydon and N. Sheppard, J. Chem. Soc., Chem. Commun., (1969) 1431.

43   C. Williams, M.A. Makarova, L.V. Malysheva, E.A. Paukshtis and Zamaraev, J. Chem. Soc., Faraday Trans. 86 (1990) 3473.

44   Insight II software package version 4.0 from Molecular Simulations Inc., 1996.

45   D.J. Willock, D. Bethell, S. Feast, G.J. Hutchings, F. King and P.C.B. Page, Topics Catal., 3 (1996) 77.

46   R.P.K. Wells, PhD Thesis, University of Dundee, 1996.

47   C.D. Chang, Stud. Surf. Sci. Catal., 61 (1991) 393.

48   G.J. Hutchings, P. Johnson, D.F. Lee, A. Warwick, C.D. Williams and M. Wilkinson, J. Catal., 147 (1994) 177.

49   D.A. Evans, K.A. Woerpel, M.M. Hinman and M.M. Faul, J. Am. Chem. Soc., 113 (1991) 726.

50   D.A. Evans, M.M. Faul and M.T. Bilodeau, J. Am. Chem. Soc., 116 (1994) 2742.

Chapter 11

MOLECULAR DESCRIPTION OF TRANSITION METAL OXIDE CATALYSTS

Jerzy Haber

Key words: Transition metal oxides, selective oxidation, oxide surfaces, cobalt films, Co_3O_4, conversion electron mössbauer spectroscopy, cyclic voltammetry, surface oxygen species.

Abstract: The nature of the surfaces of bulk and supported transition metal oxides, important as catalysts for the selective oxidation of hydrocarbons, is explored. It is emphasized that the oxide surface is in dynamic interaction with the gas phase and adapts itself to reaction conditions. The importance of the transfer of electrons between the oxide and the hydrocarbon/oxygen gas mixture as a redox system, and the influence of surface defects, is illustrated. Reactions of oxidation of hydrocarbons may start from the activation of either the hydrocarbon molecule or oxygen. The catalytic behaviour of the oxide is determined by the relative importance of nucleophilic and electrophilic oxidation pathways. Monolayer transition metal oxides supported on a second oxide phase exhibit properties which are different from the bulk material. These materials may be systematically modified to investigate the factors which are important in selective oxidation reactions and to tailor catalysts to the requirements of the reactions

"That is the essence of science: ask an impertinent question, and you are on the way to pertinent answer."
J.Bronowski, The Ascent of Man

1.   INTRODUCTION

The spectacular progress of catalysis, both science and technology, was in the last twenty years driven by the development in four fields:

- new materials, in particular those based on the principle of molecular imprinting (use of templates) [1-4];
- application of new surface science techniques to identify active sites, in particular the *in situ* techniques [5-7];
- quantum chemical modelling of catalyst surface and elementary steps of catalytic reactions [8-10];
- design of new reactors [11].

A prominent place in science of catalysis and catalysis-based modern chemical industry is occupied by reactions of catalytic oxidation [12-14]. They have vastly contributed to the development of modern society, for their products are incorporated into an amazingly large proportion of the materials and commodities in daily use. Selective oxidation is the easiest route for the functionalization of hydrocarbon molecules. Today catalytic oxidation is the basis of the production of almost all monomers used in manufacturing of synthetic fibres, plastics and many other products [15]. Understanding of the mechanism of chemo-, regio- and enantioselective introduction of oxygen into complex organic molecules may open new perspectives on the energy saving and wasteless production of many important chemicals, creating the basis for new green technologies.

The essential components of catalysts for selective oxidation of hydrocarbons are transition metal atoms. They are used in different speciations:

- organic complexes of transition metals in liquid phase,
- organic complexes of transition metals deposited at surfaces of oxide supports, as discussed in interfacial coordination chemistry,
- bulk transition metal oxides or oxysalts,
- transition metal oxy-ions deposited at surfaces of oxide supports to form oxide monolayer catalysts,
- transition metal ions incorporated into frameworks of zeolites, pillared clays or polymers.

2. BULK TRANSITION METAL OXIDES AND OXYSALTS

One of the fundamental conceptual developments of the science of interfaces in recent years is the understanding that the surface of a solid is not a rigid static structure, on which various phenomena occur, involving molecules adsorbed from the fluid phase, but is always in a dynamic interaction with the latter. The surface "lives" and adapts itself to the "living conditions". Adaptability to the change of these conditions is one of the important properties of solid surfaces, resposible for many phenomena of great theoretical interest and practical importance. When properties of the fluid phase are altered, the fluid/solid interface immediately responds, very often by changes of the composition and structure of the outermost layer of the solid. This in turn may result in changes of the physical and chemical properties of the solid surface, which become different from the bulk. This is particularly important for its behaviour in catalytic reactions.

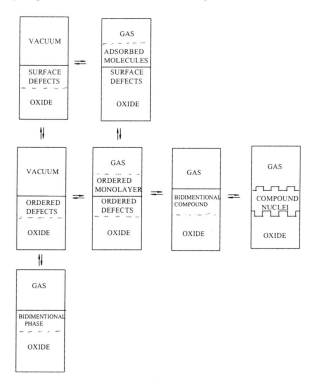

Figure 1. Dynamic phenomena at oxide surfaces

Dynamic interactions which may take place at the surface of a crystal of such chemical compound as transition metal oxide, oxysalt, sulphide etc. are summarized in Fig.1. An equilibrium is always established between the surface of the transition metal oxide crystal and the gas phase. A certain number of metal cations and oxide anions pass into the gas phase, where they form the metal vapour, composed of atoms or small clusters, and O_2 molecules, leaving behind in the crystal lattice point defects such as vacancies and interstitials or extended defects (shear planes). This leads to the change of stoichiometry, these oxides belonging to the class of nonstoichiometric solids showing broader or narrower range of nonstoichiometry. The equilibrium concentration of defects *i.e.* the nonstoichiometry is determined by oxygen pressure and temperature. At higher temperatures the defects diffuse into the bulk and the whole crystal becomes equilibriated with the gas phase. Such crystals may be considered as solid solutions of defects in the crystal lattice. As the appearance of a defect at the surface of an oxide crystal changes its surface free energy, adsorption of defects may take place at the solid side of the gas/solid interface, the surface layer becoming enriched in those defects, whose presence decreases the surface tension. By changing the oxygen pressure, the concentration of defects in the oxide may be altered, which is followed by an appropriate change of the surface enrichment. When the latter attains a certain critical value, ordering of these defects in the surface layer may take place. Futher changes of the surface composition may cause the shifts of the positions of ions resulting in surface reconstruction and formation of new bidimensional phases.

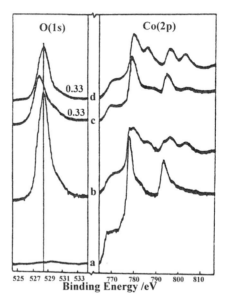

Figure 2. XPS spectra of O1s electrons and Co2p electrons of: a) metallic cobalt film, b) after exposure to 10^{-4} Torr of oxygen for 5 min at 773 K, c) after subsequent exposure to 10^{-1} Torr of oxygen for 5 min at 873 K, d) after outgassing for 30 min at 873 K [16].

All these phenomena may be strongly influenced by the presence of adsorbates in the gas phase. Adsorbed molecules interact with the surface of the oxide shifting the defect equilibria, changing the surface composition and loosening the bonds between the lattice constituents, what may lead to the formation of an ordered adsorbed monolayer, followed by migration of surface ions and formation of new bidimensional compounds. In appropriate conditions nucleation may take place at various surface defects and crystallites of a new bulk compound may appear.

The fact that the surface of a given transition metal oxide may be covered by a layer of a different oxide phase of this transition metal may be illustrated by the results of an experiment, in which a thin layer of metallic cobalt was exposed to oxygen in different conditions and its surface was analysed by XPS (Fig.2). Spectrum (a) registered for the initial film after deposition shows only the two Co2p peaks. After exposing the film at 773K to 1×10^{-4} Torr of oxygen for 5 min O1s peak of high intensity appears at 529.2 eV. Simultaneously the two new Co2p peaks become visible, whose intensity is comparable to the peaks in the initial metallic film and shifted relatively to their position by 2.4 eV towards higher B.E. values. The presence of the satellites indicates that paramagnetic Co^{2+} ions have been formed. After raising the temperature to 873K and prolonging the exposition to oxygen (curve c) the Co2p peaks of initial sample disappear completely indicating that now the surface of the sample is totally covered with an oxide layer thicker than the electron escape depth. As the satellites of the Co2p peaks are no longer present in the spectrum, it may be concluded that oxidation now resulted in the formation of trivalent cobalt ions. When the sample was then outgassed at 600°C for 30 min (curve d) the satellites of Co2p peaks appeared again. This indicates that the sample, consisting of bulk Co_3O_4, stable in the conditions of preparation – depending on the oxygen pressure in the gas phase – contains at the surface either Co^{2+} or Co^{3+} ions. This may be easily visualized on the basis of structural model. Co_3O_4 contains Co^{3+} ions in one half of the octahedral sites and Co^{2+} ions in one in eight of the tetrahedral sites. On removal of oxygen by outgassing the Co^{3+} ions are reduced to Co^{2+} and shifted, filling all octahedral sites and forming a surface structure analogous to CoO. On annealing in oxygen a reverse process takes place. Thus, irrespectively of the composition of the bulk a surface layer exists characterized by the composition which depends on the oxygen pressure in the gas phase.

Similar behaviour was quite recently observed [17] in the case of thin films of magnetite Fe_3O_4 (001) epitaxially grown on MgO (001) using molecular beam of ^{57}Fe in the presence of oxygen. LEED pattern indicated a perfect structure with a p(1x1) reconstructed surface, as labeled with respect to the bulk unit cell of magnetite. *In situ* UHV Conversion Electron Mössbauer Spectroscopy (CEMS) revealed the stoichiometry corresponding to magnetite. When however the sample was exposed to air at atmospheric pressure, a drastic change of the intensity of spectral components was oberved, indicating the surface oxidation. It could be estimated that a 15Å thick layer of γ–Fe_2O_3 was formed at the surface of the 100Å Fe_3O_4 film. Annealing the sample at 600K for one hour at UHV restored the spectra of the "as prepared" state. The reducing condition of UHV reversed the oxidation. The short time and relatively low annealing temperature confirm that the oxidation-reduction processes are limited to the surface.

Very often the transition metal oxysalts may form different polymorphic modifications, showing sometimes completely different catalytic properties. A question arises, which modification is in fact present at the temperature of the catalytic reaction. It should be remembered that due to the contribution from the surface free energy the value of the Gibbs free energy, at which a given phase transformation occurs at the surface may be different from that required for this transformation to take place in the bulk of the crystal. Thus, on heating, the phase transformation may take place at the surface long before the temperature is attained at which this transformation is observed in the bulk of the crystallites. The crystallites are then composed of the bulk having the structure of one polymorphic modification, enveloped by a surface layer of the structure characteristic for another modification. As an example, the case of cobalt molybdate $CoMoO_4$, an important catalyst for selective oxidation of hydrocarbons, working usually in the temperature range 573-673K may be quoted [18]. Under normal pressure this compound forms two polymorphic modifications: the low temperature green modification "b", in which molybdenum is in tetrahedral coordination, and the high temperature violet modification "a" with molybdenum in octaheral sites. In normal conditions the transition takes place at 693K. XPS studies of this system revealed however that on heating modification "b", its surface becomes reconstructed into modification "a" already at 373K, this reconstruction being fully reversible. This is illustrated in Fig.3, in which the photoelectron spectra of Mo3d electrons from modification "b" of $CoMoO_4$ are shown at different stages of heating and cooling the sample. At ambient temperature the Mo3d doublet visible at 234.2 and 237.2 eV is characteristic of tetrahedrally coordinated hexavalent molybdenum. On heating the sample a shift of this doublet to lower values of binding energy started to be visible around 373K, and at 473K the doublet assumed a new position at 232.2 and 235.3 eV, near that of hexavalent molybdenum in octahedral position, which did not change on further heating to higher temperatures. When the sample was cooled again to room temperature, the doublet returned to its initial position. It should be mentioned that on heating above 663K reduction of the sample began to be noticeable, as indicated by the shift of the doublet to lower binding energy values. It may be assumed that the shift of Mo3d peak reflects the reversible polymorphic transformation, which at the surface takes place at the temperature of 373K, more than 300K lower than in the bulk. A general conclusion may be thus formulated that not only the chemical composition of the surface layer, but also its phase structure is very often different from that of the bulk.

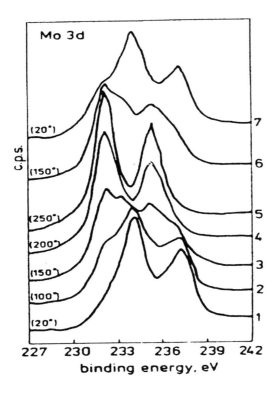

Figure 3. Changes of the photoelectron spectra in the range of the Mo3d electron energies of the sample of $CoMoO_4$ on heating and subsequent cooling [19].

Reactions of the oxidation of hydrocarbons proceed through several elementary steps [20]. They start by the cleavage of the C-H bond, resulting in the formation of a H-O and C-O bonds with the surface oxide ions of the catalyst . The alkoxy-group loses the second hydrogen atom and desorbs as oxygenated species or undergoes further transformations, oxygen vacancies being simultaneously formed. These steps, in which the cations of the catalyst are reduced, are followed by adsorption of oxygen from the gas phase, transfer of electrons from the solid to adsorbed oxygen molecule and incorporation of oxygen ions into the lattice of the oxide with simultaneous oxidation of the cations to their initial state, which completes the redox cycle. The cycle involves thus in the simplest case two adsorbed redox couples:

$$RH + 2O^{2-} \rightarrow R\text{-}O^- + HO^- + 2e^-$$
$$\tfrac{1}{2} O_2 + 2e^- \rightarrow O^{2-}$$

of which the first injects electrons into the oxide, the second one extracts them from the oxide. The injection of electrons from an adsorbed redox pair into an oxide can take place spontaneously only if the redox potential of this pair is situated above the Fermi level, and the extraction of electrons from the solid – when the redox potential is located below this level. The probability of these processes is a function of the density of states at the level corresponding to the redox potential of the adsorbed species. This density is equal to zero in the band gap, therefore the potential of the adsorbed species must lie above or below the band gap (Fig.4).

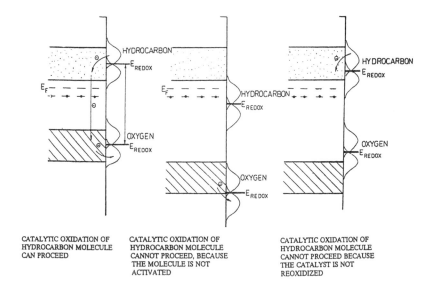

Figure 4. The transfer of electrons across the adsorbate/oxide interface in the course of the heterogeneous catalytic oxidation of a hydrocarbon molecule by gas phase oxygen [14]

To this end, the relative positions of the energy bands in the solid and the redox potential of the adsorbed molecules may be adjusted by:
- generation of surface defects, which would create surface electronic states or a surface band participating in the exchange of electrons with the reacting adsorbed molecules,
- doping of the oxide with altervalent ions, which will shift the Fermi level,

deposition of transition metal ions or their oxide monolayer on an oxide suppport to form an oxide/oxide interface with such value of the contact potential that the positions of the energy bands will shift to the optimum position.

The role of surface defects in mediating the electron transfer between the solid catalyst and the reacting molecule may be illustrated by the results obtained with a monocrystal of TiO_2. Its (110) surface was characterized by UHV Scanning Tunneling Microscopy [21]. At negative sample bias voltage the occupied electronic states are probed, thus surface oxygen atoms could be imaged, because the highest occupied states are the oxygen 2p levels. After annealing in UHV oxygen vacancies become visible at the surface (Fig.5), their concentration increasing on prolonging the time of annealing and decreasing on exposing to oxygen.

Figure 5. STM image of the clean 1x1 TiO_2 surface taken at the sample bias voltage -1.7V. 8.5x 8.5 nm² [21]

The surface properties of the TiO_2 (110) crystal plane of the monocrystal were then studied by the electrochemical methods [22]. Fig.6 shows the cyclic voltammetry curves after reduction in hydrogen and evacuation. The strong anodic current flowing for potentials higher than approximately +1.2 V indicates that oxidation of water to molecular oxygen takes place. The reversible potential for the reaction:

$$2H_2O \rightarrow O_2 + 4H^+ + 4e^-$$

at pH=4 is 0.992 V, whereas the conduction band edge is located at approximately –0.3 V. Hence, it is rather improbable that this reaction proceeds *via* the conduction band. Evidently, surface oxygen vacancies created during reduction provide energy states located in the band gap, which mediate the process of charge transfer.

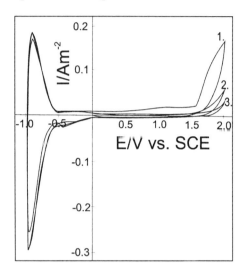

Figure 6. Voltammetric curves (first three cycles) in Na_2SO_4 solution for rutile electrode after reduction in hydrogen and annealing under vacuum

It can be seen from Fig.6 that the anodic current in consecutive sweeps diminishes. When cycling was continued for a longer time, the anodic current for potentials higher than 1.2 V dropped almost to zero, which means that the process of oxygen evolution was accompanied by the process of surface oxidation, with the annihilation of defects, present at the surface. Etching in H_2SO_4 at the temperature of 510K causes a complete disappearance of the current for potentials higher than 1.2 V. Etching in concentrated sulfuric acid is non-oxidative, causing only the removal of the outermost surface layer of rutile, together with the surface defects, without changing the stoichiometry of the bulk of the sample. Hence, it may be concluded that oxygen vacancies at the surface of transition metal oxide may play an important role in catalytic reaction mechanism, mediating the charge transfer through the interface.

3. THE ROLE OF DIFFERENT SURFACE OXYGEN SPECIES

Reactions of oxidation of hydrocarbons may start either from activation of the hydrocarbon molecule or activation of oxygen. In the first case the first step consists in activation of the hydrocarbon molecule by abstraction of hydrogen from a given carbon atom, which becomes prone to a nucleophilic addition of the oxide ion O^{2-}. It should be emphasized that the latter has no oxidizing properties, but is a nucleophilic reactant. The consecutive steps of hydrogen abstraction and oxygen addition may be then repeated to obtain selectively more oxygenated molecules. These reactions are classified as nucleophilic oxidation. The role of oxidant in these steps of the reaction sequence is played by cations of the catalyst lattice. After the addition of the surface lattice oxide ion the oxygenated product is desorbed, leaving at the surface an oxygen vacancy, which is then replenished by oxygen from the gas phase.

In the second case activated oxygen species: singlet oxygen, molecular or atomic ion radicals O_2^- or O^- play the role of oxidant. All these forms are strongly electrophilic reactants. They may abstract hydrogen from the hydrocarbon molecule generating alkyl radicals, which may start a chain reaction leading, in the conditions of heterogeneous catalysis, to total oxidation. In reactions with olefins or aromatics they attack the regions of highest electron density – the π–bond system, resulting in C-C bond cleavage to give oxygenated fragments or undergo total oxidation. These reactions may be classified as electrophilic oxidation.

Ample experimental evidence indicates that in the case of transition metal molybdates as catalysts the oxy-polyhedra of transition metal ions play the role of active sites, activating the hydrocarbon molecules. Quantum chemical calculations of the interaction of methane with oxy-vanadium clusters V_2O_9 as model of the interaction of hydrocarbons with transition metal oxide catalysts, carried out by DFT method, showed that on cleavage of the C-H bond both partners – hydrogen and alkyl group – become linked to oxide ions forming hydroxyl and alkoxy-group respectively, two electrons being simultaneously injected onto the d-orbitals of vanadium [23]. However, at the surface of a transition metal oxide or oxysalt the situation is more complex.

As already mentioned, transition metal oxides are nonstoichiometric compounds, their composition depending on the equilibrium between the lattice and its constituents in the gas phase. This is a dynamic equilibrium, in which the rate of dissociation of the oxide lattice and evolution of oxygen in the form of O_2 molecules is equal to the rate of its incorporation from the gas phase. In the process of dissociation the oxide ion must be extracted from the surface of the solid, its two electrons injected into the solid, oxygen atoms recombined to form a molecule and the latter desorbed as dioxygen. The reverse series of elementary steps must take place upon incorporation. They result in surface equilibria:

$$2O^{2-} \; \rightleftarrows \; 2O^{-}_{chem} + 2V_O \; \rightleftarrows \; O_2^{2-}{}_{chem} \; \rightleftarrows \; O_2^{-}{}_{chem} \; \rightleftarrows \; O_{2ads} \; \rightleftarrows \; O_{2gas}$$

so that the surface is always covered by different oxygen species, as illustrated in Fig.7 [24]. The surface coverage by these species depends on the oxygen pressure in the gas phase, the rate constants of adsorption and surface oxygen transformations (transfer of electrons between adsorbed oxygen species and the solid), the rate of incorporation into the vacancies and the dissociation pressure of the oxide.

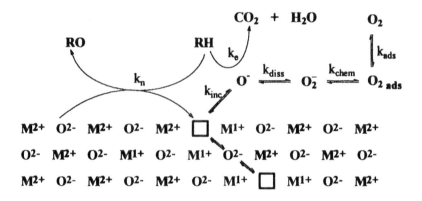

Figure 7. Mechanism of the oxidation of a hydrocarbon molecule over an oxide catalyst [24]

At higher temperatures the defects, created at the surface, diffuse into the bulk and equilibrium is established between the gas phase and the solid oxide. O^{2-} ions, exposed at the surface, are nucleophilic reagents, whereas transient $O_2^{-}{}_{chem}$, $O^{-}{}_{chem}$ and O_{2ads} are electrophilic species. When such a surface is contacted with hydrocarbon molecules, nucleophilic surface oxide ions O^{2-} react with activated hydrocarbon molecules to form selective oxidation products. In the parallel reaction pathway transient electrophilic oxygen species attack the hydrocarbon molecules to generate radicals which lead to total oxidation.. Thus, different oxygen species, present at the oxide surface, compete for hydrocarbon molecules. Simultaneously, competition exists between hydrocarbon molecules and surface oxygen vacancies for the electrophilic $O^{-}{}_{chem}$ species which can either be captured by hydrocarbon molecules to form products of total oxidation or react with oxygen vacancies to become incorporated into the surface. The nucleophilic and electrophilic pathways are described by two different rate equations and the ratio of their rates determines selectivity of the reaction. The two rate equations are coupled by the equation, describing the generation and annihilation of surface oxygen vacancies:

$$O_o^{2-} = V_o^- + O^-_{chem}$$

Catalytic properties of transition metal oxides, particularly the selectivity in oxidation reactions, are thus strongly dependent on the dissociation pressure of the oxide and the properties of its defect structure.

It should be remembered that Roberts postulated the formation of similar transient oxygen species at the surface of metals and showed that they can be responsible for reactions of catalytic oxidation [25].

Another factor which may be important in determining the reaction pathway of a hydrocarbon molecule at the surface of an oxide catalyst is the electrostatic field. As revealed by quantum-chemical calculations of the surface structure of such oxides as V_2O_5 or MoO_3, an electrostatic field extends above the surface [10] the gradient varying considerably on moving along the surface. This field may polarize the reacting molecules and determine their orientation in respect of the surface.

4. TRANSITION METAL IONS DEPOSITED AT OXIDE SUPPORTS

When a metal oxide phase is dispersed on a second oxide phase, present in large excess and playing the role of the support, a number of phenomena may take place in the course of thermal treatment, which may be summarized by scheme 1.

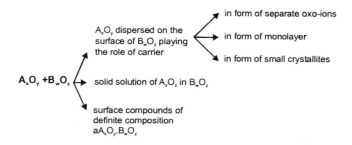

Scheme 1. Processes taking place on depositing active oxide phase on oxide support.

The type of the process taking place depends on chemical and crystal properties of both phases, their surface free energies relation, method of preparation, temperature and the atmosphere of the pretreatment. Irrespective of whether obtained by impregnation, precipitation-deposition, grafting or thermal spreading in the solid oxide mixture, only when the temperature of annealing is low enough and the miscibility and chemical affinity of the two compounds is very limited, will the minority phase A_xO_y almost entirely accumulate at the surface of the support. In the submonolayer range, isolated transition metal-oxygen polyhedra are anchored at the surface of the oxide support. They may be located on the surface of the oxide support (extraframework), or may become incorporated into its outermost surface layer (surface framework), or may penetrate into the subsurface layers (surface solid solution). Depending on whether the energy of cohesion of the minority phase is smaller or larger than the energy of its adhesion to the support, this phase will form at first the bidimensional islands, then a monolayer and finally small three-dimensional crystallites.

Thermodynamic considerations of the system, composed of crystallites of a transition metal oxide (active phase "a") supported on another oxide (support "s"), indicate that when the energy of cohesion of the supported phase is smaller than the energy of adhesion of this phase to the oxide, playing the role of a support, spontaneous spreading of the former over the surface of the latter will take place until a uniform coverage is attained [18.26]. This is a manifestation of the phenomenon of wetting, described by the equation:

$$\Delta F = \gamma_{ag}\Delta A_a - \gamma_{sg}\Delta A_s + \gamma_{as}\Delta A_{as} < 0$$

where γ_{ij} denote the specific interface free energy between phases i and j, the subscripts a, s and g denote active phase, support and gas phase respectively, and ΔA are changes in interface areas between respective phases. For spreading of the active phase over the surface of the support to occur, ΔF must be negative. In the opposite case, when ΔF assumes a positive value, the active phase has a tendency to form crystallites on the surface of the support, and when it is dispersed into a monolayer *e.g.* by grafting, coalescence takes place and the system transforms into a heterogeneous mixture of crystallites of active phase and the support. Thus, wetting of the support by the active phase is a condition of the stability of oxide monolayer catalysts. Indeed, experiments showed that, when titania is mixed with vanadia, spreading of vanadia occurs and a monolayer of vanadium oxides supported on titania is formed indicating that wetting takes place. Such catalysts are stable when working in the catalytic oxidation of hydrocarbons. Conversely, vanadium oxides supported on silica form crystallites, because silica is not wetted by vanadium oxides.

Results of an investigation of spreading vanadia over the surface of different samples of titania are shown in Fig.8, in which the dependence of the extent of vanadia disappearance (α), as determined by XPS, is plotted versus time in parabolic law coordinates [27]. A linear relationship indicates that the process is diffusion controlled. It may be noted that spreading proceeds in an oxidizing atmosphere, whereas in helium, when surface reduction of V_2O_5 occurs, the spreading is inhibited. Apparently, VO_2 being isomorphous with TiO_2, has similar surface free energy and the system lacks the necessary driving force of spreading. This indicates that redox properties of the gas phase are of great importance for the process of spreading.

It has been shown with the aid of EXAFS that increase in the BET surface area of TiO_2 results in a progressive change from long range order in external planes of crystalllites in case of low surface area samples to that characteristic of the amorphous oxide [28]. Thus, the decrease of the rate of V_2O_5 spreading with increase of the titania surface area as indicated by the data in Fig.8 can be explained in terms of increasing difficulty in formation of two dimensional vanadium oxide sheet on progressively disordered TiO_2 surface.

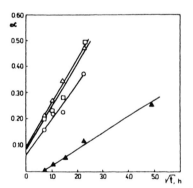

Figure 8. Kinetic curves of V_2O_5 spreading over the surface of TiO_2 samples of different surface areas in terms of the parabolic law. Anatase surface area in dry air: ($\bullet$) 101 m^2/g, ($\circ$) 52 m^2/g, ($\triangle$) 46 m^2/g, ($\square$) 41 m^2/g; in helium: ($\blacktriangle$) 46 m^2/g.

An attempt to induce spreading of vanadia over titania previously covered with a monolayer of vanadia failed completely showing that after the surface has been saturated with vanadia monolayer, migration of vanadia is no more possible, because the surface free energy of the system has approached that of vanadia itself, thus removing the driving force, required for migration.

Electrochemical studies of the interaction of various transition metal ions with the surface of a rutile monocrystal revealed that insertion of these metal ions into the surface and subsurface layers of the crystal takes place, the degree of insertion depending strongly on the metal in question and the conditions of experiment. [29].

The interaction of rutile surface with metal oxides changes dramatically the conditions of charge transfer at the rutile surface, which may influence the course of catalytic reactions occuring at the surface. Indeed, incorporation of vanadium ion into subsurface layers of V_2O_5/TiO_2 oxidation catalysts has been observed [30]

The rationale behind the preparation of monolayer catalysts is the fact that their properties are different from those of the bulk phase and this makes possible the tailoring of the catalyst to the requirements of the reaction. This may be achieved by taking advantage of following factors:

- in the monolayer, collective properties of the crystal may not yet dominate over the atomic properties of the component atoms, which permits tailoring the monolayer properties using both the properties of isolated transition metal complexes and those of a solid phase;
- surface transition metal ions and the lattice defects, generated by their presence, create local energy levels or bands which can mediate the transfer of electrons between the catalyst and the reacting molecules;
- the transfer of electrons between the support and the monolayer influences the occupancy of molecular orbitals of the transition metal oxy-ions and may change the type of orbitals involved in their HOMO and LUMO. This may profoundly influence the course of the interaction of the active sites with reacting molecules;
- the oxide monolayer exposes only one crystal face, which may be different at surfaces of various supports, due to the effect of epitaxy. As majority of catalytic reactions are structure-sensitive, this enables the preparation of catalysts with the surface structure optimal for the given reaction and simultaneously the elimination of adverse effects, due to parasitic reactions at other crystal faces, present in the case of bulk crystallites.
- properties of the monolayer are controlled by the surface free energy relationships at the interfaces. As the surface free energy of oxides is strongly influenced by adsorption and by incorporation of foreign ions into the surface layers of the lattice, the behaviour of the monolayer may be modified by doping of the support or the monolayer with foreign ions or by thermal treatment in the presence of various gases, in particular water vapour.

Supported transition metal oxide catalysts have been extensively studied because of their industrial importance and theoretical interest. They are ideal model systems to study the fundamental questions, concerning the mechanism of selective oxidation, because the active sites can be characterized molecularly with *in situ* Raman, IR, solid state NMR, DRS, EXAFS/XANES, XPS as well as measurements of chemisorption and catalytic activity. Moreover, one can study the dependence on the surface coverage, the type of support and the character of additives, introduced to the bidimensional overlayer or the support. Ideal molecular dispersion can be achieved by chemical vapour deposition (grafting), in which surface hydroxyl groups are titrated. The coverage may be thus controlled by changing the degree of surface hydroxylation. Other methods of preparation as adsorption, ion exchange or spreading, result in molecular dispersion only at very low coverages, polymeric species being formed at higher coverages. Isotopic oxygen exchange experiments and *in situ* Raman spectroscopy indicated that terminal M=O bonds of the supported species are not involved in oxidation reactions on the monolayer type catalysts. The question, which surface oxygen species take part in the oxidation: the bridging M-O-M oxygen atoms of the polymeric forms of the overlayer, or bridging M-O-support oxygen atoms still remains the subject of discussion. The type of support has a dramatic influence on the catalytic activity. In the case of oxidative dehydrogenation of methanol, requiring only one surface active site, on a series of catalysts, composed of VO_x deposited in the same amount on different supports, the TOF varies by 5 orders of magnitude [31]. It is interesting that the chemistry of the oxy-ions in the two-dimensional monolayer is analogous to their chemistry in aqueous solutions, the point-of-zero-charge of the support instead of pH being the critical factor, determining the type of oxy-ions formed.

"The true scientist never loses the faculty of amazement. It is the essence of his being."
Hans Seyle

5.   REFERENCES

1   Davis, M.E., 2000, Molecular Design of Heterogeneous Catalysts, *Stud.Surf.Sci.Catal.*, **130**,1.
2   Davis, M.E.,1997, Catalytic Materials *via* Molecular Imprinting, *CATTECH,* **1**, 19
3   Thomas, J.M., 1996, Catalysis and Surface Science at High Resolution, *Faraday Disc.*, **105**, 1
4   Ying, J.Y., Mehnert, C.P., Wong, M.S., 1999, Synthesis and Applications of Supramolecular-Templated Mesoporous Materials, *Angew.Chem.,Intern.Ed.*, **38**, 56.

5   Topsoe, H., 2000, In-Situ Characterization of Catalysts, *Stud.Surf.Sci. Catal.*, **130**, 1
6   Somorjai, G. A. 1994 Introduction to Surface Chemistry and Catalysis, John Wiley and
     Sons Inc, New York.
7   Weiss, W., Schlogl, R., 2000, An Integrated Surface Science Approach Towards Metal
     Oxide Catalysis, *Topics Catal.*, **13**, 75.
8   van Santen, R.A., 1995, Concepts in Theoretical Heterogeneous Catalytic Reactivity,
     Catal.Rev.-Sci.Eng., **37**, 557
9   Catlow, C.RF.A., Bell, R.G., Gale, J.D., Lewis, D.W., Jayle, D.L., Sinclair, P.E., 1998,
     An Introduction to Molecular Heterogeneous Catalysis, in *Catalytic Activation and
     Functionalization of Light Alkanes* (E.G.Derouane,J.Haber,F.Lemos,F.R.Ribeiro,
     M.Guinet, eds), Kluwer Acad.Publ., Dordrecht, p.189.
10   Hermann, K., Witko, M., 2001, Theory of Physical and Chemical Behavior of
     Transition Metal Oxides. Vanadium and Molybdenum Oxides, in *The Chemical Physics
     of Solid Surfaces, vol.9, Oxide Surfaces*, (D.P.Woodruff, ed), Elsevier Science,
     Amsterdam, chapter 4
11   Eigenberger, G., 1997, Reaction Engineering, in *Handbook of Heterogeneous Catalysis*
     (G.Ertl, H.Knozinger, J.Weitkamp, eds) VCH, Weinheim, p.1399.
12   Bielanski, A., Haber, J., 1991, *Oxygen in Catalysis*, Marcel Dekker Inc., New York.
13   Delmon, B., 1997, The Future of Industrial Oxidation Catalysis Spurred By
     Fundamental Advances, *Stud.Surf.Sci.Catal.*, **110**, 43
14   Haber, J., 1997, Molecular Mechanism of Heterogeneous Oxidation – Organic and
     Solid State Chemist's View, *Stud.Surf.Sci.Catal.*, **110**, 1
15   Haber, J., 1992, Catalytic Oxidation: State of the Art and Prospects,
     Stud.Surf.Sci.Catal., **72**, 279.
16   Haber, J., 1982, Reactivity of Solids – A Rapidly Advancing Field of Research, in
     Reactivity of Solids (K.Dyrek, J.Haber, J.Nowotny, eds) PWN – Polish Scientific Publ.
     Warszawa, p.3
17   Handke, B., Haber, J., Slezak, T., Kubik,M., Korecki, J., 2001, Magnesium
     Interdiffusion and Surface Oxidation in Magnetite Epitaxial Films Grown on
     MgO(100), Vacuum **63**, 331.
18   Haber, J., 1984, The Role of Surfaces in the Reactivity of Solids, *Pure Appl.Chem.*, **56**,
     1663.
19   Haber J., Stoch J., Unger L., 1984, unpublished results
20   Haber J., 1996, Selectivity in Heterogeneous Oxidation of Hydrocarbons, in
     Heterogeneous Hydrocarbon Oxidation, ACS Symp.Series 638 (B.K.Warren, S.Ted
     Oyama, eds), American Chemical Society, Washington D.C., p.20.
21   Spiridis, N., Haber, J., Korecki, J., 2001, STM Studies of Au-Nanoclusters on TiO_2
     (110), *Vacuum* **63**, 99.
22   Haber, J., Nowak, P., 1999, The Interaction of Rutile (TiO_2) Surface with some
     Catalytically Active Transition Metal Oxides During Heating, Studied by
     Electrochemical Methods, *Topics Catal.*, **8**, 199.
23   Broclawik, E., Haber, J., Piskorz, W., 2001, Molecular Mechanism of C-H Bond
     Cleavage at Transition Metal Oxide Clusters, *Chem.Phys. Lett.* **333**, 332.
24   Haber, J., Turek, W., 2000, Kinetic Studies as a Method to Differentiate between
     Oxygen Species Involved in the Oxidation of Propene, *J.Catal.*, **190**, 320.
25   Roberts, M.W., 1996, The Role of Short-lived Oxygen Transients and Precursor States
     in the Mechanisms of Surface Reactions: a Different View of Surface Catalysis,
     Chem.Soc.Reviews, 437.

26   Knozinger, H. Taglauer, E., 1993, Toward Supported Oxide Catalysts *via* Solid-Solid Wetting, in *Specialist Periodical Report, Catalysi, Vol.10* Royal Society of Chemistry 1

27   Haber, J., Machej, T., Serwicka, E.M. Wachs, I.E., 1995 Mechanism of Surface Spreadng in Vanadia-Titania System, *Catal.Lett.*, **32**, 101.

28   Haber, J., Kozlowska, A., Kozlowski, R., 1988, Nature of the Support-Active Phase Enhancement in Vanadium Oxide Monolayer Catalysts, in *Proc. 9th Intern. Congress on Catalysis, Calgary 1988*, (M.J.Phillips, M.Ternan, eds), Chemical Institute of Canada, Ottawa, p.1481

29   Haber, J., Nowak, P., 1995, b. Catalysis related Electrochemical Study of the V_2O_5/TiO_2 (rutile) System, *Langmuir*, **11**, 1024

30   Centi, G., Giamello, E., Pinelli, D., Trifiro, F., 1991, Surface Structure and Reactivity of V-Ti-O Catalysts prepared by Solid State Reactions. 1. Formation of a V^{IV} Interacting Layer, *J.Catal.*, **130**, 220.

31   Wachs, I.E., Deo, G., Jehng, J-M., Kim, D.S., Hu, H., 1996, The Activity and Selectivity Properties of Supported Metal Oxide Catalysts During Oxidation Reactions, in *Heterogeneous Hydrocarbon Oxidation, ACS Symp.Series 638*, (B.K.Warren, S.Ted Oyama, eds), American Chemical Society, Washington D.C., p.292.

Chapter 12

SELECTIVITY IN METAL-CATALYSED HYDROGENATION

Peter B. Wells

Key words: Catalytic hydrogenation; selectivity; enantioselectivity; alkyne
              hydrogenations; alkadiene hydrogenations; α,β-unsaturated aldehyde
              hydrogenations; pyruvate ester hydrogenation; alkene isomerisations;
              supported metal catalysts; evaporated metal films; D-tracer studies; butadiene;
              crotonaldehyde; promotion of selectivity by non-metals; electronic effects on
              activity and selectivity; hydrogen occlusion in metals; defective metallic
              structure and its effect on selectivity; molecular congestion and its effect on
              selectivity; alkyne polymerisation; alkyne isomerisation; selective enantioface
              adsorption; modifiers; cinchona alkaloids; alkaloid conformations; solvent
              effects on enantioselectivity; strychnos alkaloids; vinca alkaloids; morphine
              alkaloids; synthetic modifiers; chiral metallic surfaces.

Abstract: The many factors that influence selectivity in metal-catalysed reaction are
              reviewed. Alka-1,3-diene hydrogenation occurs by 1:2- and 1:4-addition; the
              extent of 1:2-addition is determined by the electronegativity of the metal but
              electronegative additives promote 1:4-adition at the expense of 1:2-addition. In
              α,β-unsaturated aldehyde hydrogenation saturation of the carbon-carbon
              double bond is normally favoured over that of the carbon-oxygen double bond,
              but additives that cause the formation of $M^{\delta+}$ promote carbonyl hydrogenation
              and selectivity to the (desired) unsaturated alcohol. Detailed evidence is
              presented to show that selectivity is influenced by molecular congestion in the
              ad-layer, by additives in the surface, and by hydrogen occlusion in the bulk of
              metallic crystallites as a result of their defective structure. Selectivity is always
              degraded by side reactions such as reactant polymerisation and isomerisation
              in alkyne hydrogenations. Adsorption of chiral modifiers, particularly
              naturally occurring alkaloids, onto the achiral surfaces of supported metals,
              induces enantioselectivity; the mechanism of pyruvate ester hydrogenation
              over cinchona-modified Pt is discussed in detail. Platinum crystals cut so as to
              expose chiral surfaces adsorb L- and D-sugars differently, this molecular

recognition leading to selectivity differences in oxidation. Selectivity is thus determined by the total environment of the catalytically active sites.

1.   INTRODUCTION

Catalysis provides the pre-eminent technology whereby rates of chemical reactions are accelerated sufficiently to make possible the large scale production of organic chemicals in industry. However, the desired product of a chemical reaction is often one of several intermediates formed along a reaction pathway and not the thermodynamically stable product. Appropriate control is therefore required in order to achieve desired product formation, and selectivity (the extent to which the desired product is formed in a multi-product process) is a prime property of a catalyst. Carbon monoxide hydrogenation provides a well-known example: methane (the thermodynamically stable product) is formed over Ni whereas methanol is produced industrially over Cu/ZnO and higher oxygenates over Ru.

For catalysed reactions to be candidates for industrial exploitation, the catalysts involved must possess both sufficient activity and appropriate selectivity. Whereas it may be possible to compensate for a shortfall in desired specific activity by appropriate plant design, any shortfall in selectivity necessarily means reduced efficiency in the use of raw materials and the need either to identify markets for the by-products or to dispose of them safely as waste.

This chapter reviews, by reference to selected examples, the electronic, structural, and mechanistic factors that influence selectivity in catalytic hydrogenations of simple unsaturated organic compounds catalysed by transition metals in a wide variety of forms.

A trivial but important matter should be emphasised at the outset. Normally, the selectivity patterns to be considered are those observed for catalysts functioning in their steady states. When metal catalysts are freshly prepared their surfaces are highly reactive towards unsaturated organic compounds, so that initial reaction may show a reactivity and selectivity very different from that achieved later when the catalyst is in the steady state. For example, ethyne hydrogenation over freshly prepared Pd/alumina may initially give rapid ethane formation but reaction soon changes to highly selective ethene formation. It is well known from experiments involving the adsorption of ^{14}C-labelled ethyne and ethene that these compounds are irreversibly adsorbed on catalytically active metals giving adsorbed hydrocarbonaceous species which are a permanent component of the active catalyst [1]. Such hydrocarbonaceous species may assist H-atom transfer between the metallic surface and adsorbed reactants and intermediates and as such are to be considered as an essential component of a healthy catalyst. Surface science has confirmed the existence of such species and has added to our knowledge of their structure [2]. In the examples that follow, unless otherwise stated, the information given refers to the behaviour of catalysts acting in a reproducible steady state.

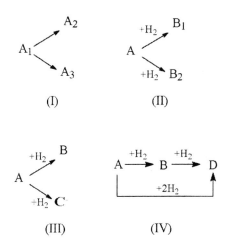

Figure 1. Some selective reaction networks

Figure 1 shows some hypothetical reaction schemes in which selectivity to any one of the products may be required, where a component A or B can exist as isomers these are represented as A_1, A_2 or B_1, B_2, B_3 etc. Thus, Scheme I might represent the isomerisation of an alk-1-ene to the *cis-* and *trans*-isomers of the corresponding alk-2-ene, Scheme II the hydrogenation of a prochiral ketone to optical isomers of the corresponding alcohol, Scheme III the hydrogenation of a compound such as acrolein containing two different functional groups, and Scheme IV the sequential hydrogenation of a di-unsaturated compound with also the possibility of direct conversion of reactant to the final product. Schemes representing particular cases are presented as they arise.

2.   SELECTIVITY AND THE ELECTRONIC NATURE OF THE CATALYST SURFACE

2.1   Hydrogenation of buta-1,3-diene over unpromoted metals

The hydrogenation of buta-1,3-diene, $CH_2=CH-CH=CH_2$, provides an instructive point of departure. This molecule is simply composed of two identical unsaturated functions, and its hydrogenation provides a mixture of the three n-butenes and butane. Before the mechanism of this reaction was investigated in detail it was commonly supposed that but-1-ene was the initial product, that the but-2-enes were formed as secondary products by isomerisation, and that any butane formed early in reaction occurred by butene re-adsorption and hydrogenation. However, D-tracer studies of the reaction over alumina-supported Fe, Co, Ni, Cu [3], and Pd [4] in a static reactor indicated a very different state of affairs.

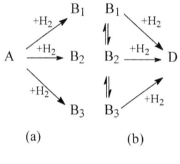

(a)   (b)

Scheme (V)

These metals were highly selective for butene formation (yields 97 to 100%), the deuterium distributions in each butene were closely similar, and butene composition was strictly independent of deuterium pressure. Since butene isomerisation would have led to D-atom redistribution in a secondary product, and since the yield of secondary product should have been favoured by increasing deuterium pressure, it became clear that all three butenes were *initial products* of reaction. Thus, highly selective butadiene hydrogenation proceeds according to Scheme Va: the three isomeric n-butenes being formed directly by 1:2- and 1:4-addition in the first step.

As strongly adsorbed butadiene is removed to the point where it ceases to dominate the active surface, butene is re-adsorbed and hydrogenated to butane in a second step (Scheme Vb). On alkene re-adsorption there is competition between butene hydrogenation and isomerisation, so that the composition moves towards thermodynamic equilibrium (*trans*-but-2-ene > *cis*-but-2-ene > but-1-ene) as the butene is removed. Such isomerisation occurs only in the second stage of the process.

Thus, the first requirement for the attainment of selectivity in a multi-step system is that reactant adsorption should be sufficiently strong to exclude re-adsorption of the required product (butenes in this case) and thereby to prevent subsequent reaction.

Table 1. Butene distributions obtained from butadiene hydrogenation over various supported and unsupported metal catalysts and the effects of electronegative additives

Catalyst	Temp /°C	Conv /%	Additive	Butenes /%			Ref
				b-1	trans-b-2	cis-b-2	
10% Cu/alumina	60	20	none	87	6	7	3
5% Pd/alumina	0	50	none	65	33	2	6
10% Co/alumina	100	10	none	82	11	7	11
			S	25	62	13	11
			P	25	69	6	11
			As	15	70	15	11
			Se	17	64	19	11
Ni powder	100	10	none	62	25	13	11
			Br	16	75	8	11
			Cl	22	66	12	11
Ni film	0	20	S, θ = 0.00	66	25	9	9
	90	20	S, θ = 0.70	32	57	11	9
		15	S, θ = 1.00	23	68	12	9
Mn film	20	20	S, θ = 0.00	86	10	4	9
			S, θ = 0.12	78	15	7	9
			S, θ = 0.32	71	21	8	9
V film	0	10	none	50	28	22	10
$VH_{0.7}$ film	0	10	H	32	60	8	10
Zr film	0	5	none	40	45	15	10
$ZrH_{2.0}$ film	0	10	H	45	52	3	10

Some representative product compositions from butadiene hydrogenation are contained in Table 1. Very variable behaviour is observed and the challenge is to extract from this data a measure of understanding.

The mechanism for reaction over alumina-supported Co and Ni proposed on the basis of D-tracer studies is one in which adsorbed alkadiene and adsorbed hydrogen are in dynamic equilibrium with σ- and σ-π-bonded half-hydrogenated states (processes (1) and (2), Figure 2) thereby providing the wide distribution of D-atoms observed in the butenes. In subsequent rate determining steps, H-atom addition to the σ-adsorbed butenyl gives solely but-1-ene whereas addition to σ-π-adsorbed butenyl occurs *via* a π-allylic transition state. In process (2), butadiene molecules adsorbed in the *transoid*-conformation provide but-1-ene and *trans*-but-2-ene by 1:2- and 1:4-addition respectively, whereas those adsorbed in the *cisoid*-conformation provide but-1-ene and *cis*-but-2-ene. If the required selectivity is for the alk-1-ene, then the selected catalyst should be one with an electronic structure that does not readily facilitate π-allyl formation in the transition state. Cu, with a nearly filled *d*-shell meets this criterion and does indeed provide a high selectivity for but-1-ene formation from butadiene (Table 1, entry 1). Mn behaves similarly (entry 14), which conforms to Dowden's postulate regarding the stability of the half-filled *d*-shell in catalysis [5]. Table 1 shows that the *trans:cis* ratios in but-2-ene formed over Co, Ni, Cu, Mn, V and Zr were typically 1.0 to 3.0 (entries 1,3,8,11,14,17,19), *i.e.* there was little selectivity within the but-2-ene yield. From this it was adduced that interconversion of the *transoid* and *cisoid* forms of adsorbed butadiene and/or the σ,π-adsorbed half-hydrogenated states was facile, as shown in process (2), Figure 2.

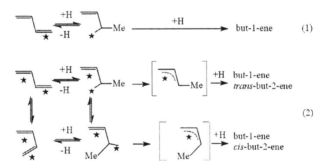

Figure 2. Simplified mechanism for buta-1,3-diene hydrogenation over metals other than Pd (full mechanism given in [3])

By contrast, Pd showed a considerable but-1-ene yield, like Ni, but a marked selectivity within the but-2-ene yield for the *trans*-isomer (Table 1, entry 2 [6]). Values of the *trans:cis* ratio as high as 20 have been recorded. Over this metal, process (1) of Figure 2 occurs but the but-2-ene-forming process is clearly different. Butadiene molecules in the gas phase undergo restricted rotation about the central carbon-carbon bond and, at any instant, far more molecules approximate in their geometry to the *transoid*-conformation than to the *cisoid* conformation. The *transoid:cisoid* ratio has been determined experimentally to be about 10-20:1 at room temperature [7]. If this conformational preference in the gas phase was frozen on adsorption, then subsequent 1:4-addition would provide a high value of the *trans:cis* ratio in the but-2-ene. Now, palladium is well known to be prolific in forming π-allylic compounds with unsaturated hydrocarbons, and the observed selectivity within the but-2-ene yield is rationalised if it is supposed that butadiene molecules adsorb directly as two conformationally distinct π-allylic states (Figure 3) which, because of their mode of bonding, do not interconvert. Hydrogenation to products then proceeds by process (3), again by sequential H-atom addition but now *via* truly π-allylic half-hydrogenated states, giving but-2-ene in which the *trans:cis* ratio reflects directly the *transoid:cisoid* ratio in the adsorbed butadiene, this in turn being closely related to that exhibited by the butadiene molecules in the gas phase. High *trans:cis* ratios in but-2-ene such as that shown in Table 1 are thereby interpreted. The *trans:cis* ratio decreases with increasing temperature as expected [8].

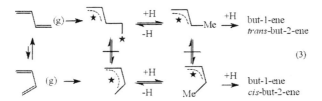

Figure 3. Mechanism of but-2-ene formation in buta-1,3-diene hydrogenation over Pd

So, reaction over Co, Ni, and Cu involves processes (1) and (2) whereas that over Pd involves processes (1) and (3), and the question arises: of the transition metals, how many follow Co, Ni, and Cu, and how many follow Pd? Butadiene hydrogenation has been studied over evaporated films of the majority of the transition elements [9,10]. Where comparisons can be made metal films and supported metals are found to give closely similar product compositions under comparable conditions. Thus, the mechanisms shown in Figures 2 and 3 are deemed to apply to the evaporated transition metals.

The most significant finding of the wide study of butadiene hydrogenaton over evaporated films was that but-1-ene yields increased as the Pauling electronegativity of the metals increased. The correlation applies to all three transition series, that for the second and third row metals is shown in Figure 4. These correlations show that the but-1-ene yield is determined by the electronic structure of the metal and that, as the *d*-band is filled, so the ability to achieve the π-allylic transition states of process (2) (Figure 2) is reduced. As mentioned above, the values of the *trans:cis* ratio in the but-2-ene were mostly in the range 1.0 to 3.0. No pure metals showed the high values typical of Pd, from which it was concluded that no other transition metals provided sites at which butadiene and the half-hydrogenated states were chemisorbed as truly π-allylic intermediates (as in Figure 3).

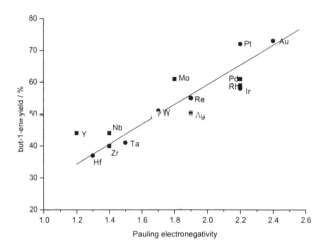

Figure 4. Buta-1,3-diene hydrogenation over pure metal films of the second and third transition series. Variation of the but-1-ene yield with the Pauling electronegativities of the elements [10]. Squares = second row elements; circles = third row elements [Value for Ir was measured over a wire and is taken from [12]].

2.2   Hydrogenation of buta-1,3-diene over metals treated with selectivity promoters

The selectivities of the transition metals for butadiene hydrogenation shown in Table 1 for V, Mn, Co, Ni, Cu, and Pd were obtained over surfaces which, as far as was known, were clean except for the presence of hydrocarbonaceous residues. We now discuss reactions at metal surfaces containing a second component which enhances 1:4-addition by promoting π-allylic intermediates and which therefore facilitates *trans*-but-2-ene formation. In the 1970s, alumina-supported Co and Ni catalysts were prepared which showed unexpectedly low but-1-ene yields and high values of the *trans:cis* ratio in but-2-ene (Table 1, entry 4), and it was eventually discovered that these catalysts were contaminated by sulphur which had migrated from the impure alumina support during catalyst preparation [11]. Once this effect was recognised, it was quickly established that the electronegative elements adjacent to sulphur in the Periodic Table, P, As, Se, Br, Cl, had a similar effect on the selectivity of Co or Ni although the effect of the halogens was transient (Table 1 entries 4-7,9,10) [11,12]. These observations rationalised hitherto anomalous reports such as that of Nozaki that nickel phosphate reduced at 650°C was a good 1:4-addition catalyst [13].

The phenomenon was investigated quantitatively by exposing evaporated films of Cr, Mn, Fe, Co, Ni, Mo, Pd, W, Re and Pt to progressive doses of hydrogen sulphide which adsorbed dissociatively at 293 K depositing sulphur and liberating molecular hydrogen [9]. Butadiene hydrogenation selectivity was studied as a function of sulphur coverage and reaction over all metals moved progressively from predominantly 1:2- addition to predominantly 1:4-addition, there being an associated increase in the *trans:cis* ratio in the but-2-ene (partial data for Ni, Table 1, entries 11-13). The effect of the adsorbed sulphur was to move the mechanism away from processes (1) and (2) shown in Figure 2 and towards process (3) of Figure 3, process (1) becoming poisoned. The origin of the effect was proposed to be a polarisation of metal atoms adjacent to adsorbed sulphur, giving $M^{\delta+}$ sites, which stabilised the half-hydrogenated states as π-allylic intermediates, and thereby enhanced 1:4-addition [9].

A clear dichotomy of surface chemistry is now evident. Pure transition metals except Pd provide 1:2-addition by process (1) and 1:2- and 1:4-addition by process (2) giving low values of the *trans:cis* ratio in the but-2-ene. Sulphur-modified transition metals facilitate 1:2- and 1:4-addition by process (3) giving high values of the *trans:cis* ratio in the but-2-ene, the low but-1-ene yields indicating that process (1) is poisoned. The transition to the second mode of behaviour is related to the ability of the sulphur-modified surfaces to adsorb the intermediates by π-allylic bonding. With the large volume of data available for pure and sulphided metal films it is interesting to see whether there is evidence for a gradual movement between these two extremes; if there is, then we should expect there to be a correlation between the but-1-ene yield and the *trans:cis* ratio in the but-2-ene. Figure 5 shows the available data for pure metals of Groups 3 to 11 of the Periodic Table and for fully sulphided metals of Groups 6 to 10. The points scatter about a smooth curve. The implications for the pure metals is interesting. Whereas the but-1-ene yield is determined for each metal by its electronegativity, the highest values of the *trans:cis* ratio (as observed for Zr, Rh, Hf for example) may indicate either that sites for processes (2) and (3) co-exist over these metals, or that the chemisorption bonds are a hybrid of those represented in processes (2) and (3).

Polarised metal atoms, $M^{\delta+}$, are also generated at the surfaces of early transition metals when they form saline hydrides. The points shown in Figure 5 for Sc, Y, I, Zr, Hf, V, Nb and Ta are for evaporated films which, after preparation, were used for butadiene hydrogenation without pre-exposure to hydrogen. There was no evidence that these films took up hydrogen to form hydrides, and the catalytic selectivities are those afforded by the unpromoted metals. However, pre-equilibration of films with hydrogen at room temperature before use in butadiene hydrogenation resulted in the absorption of hydrogen, volumetric measurements indicating the formation of the bulk phases $ScH_{1.1}$, $YH_{1.1}$, $TiH_{1.2}$, $ZrH_{2.0}$, $HfH_{\geq 1}$, $VH_{0.7}$, $NbH_{1.3}$, and $TaH_{0.9}$. (These metals had not, in all cases, absorbed hydrogen to saturation).

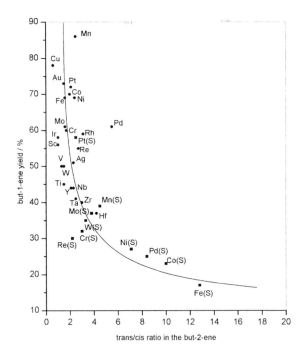

Figure 5. Buta-1,3-diene hydrogenation over pure films of the metals of Groups 3 to 11 of the Periodic Table and over sulphided films of the metals of Groups 6 to 10. Interdependence of the but-1-ene yield and the *trans:cis* ratio in the but-2-ene. Circles = metals, squares = surface-sulphided metals.

The effect of hydridation took one of three forms. Over V and Ta the but-1-ene yield was reduced and the *trans:cis* ratio increased (Table 1, entries 17 and 18), *i.e.* the effect resembled that of S-adsorption on the later transition metals. However, over Zr, Nb and Hf the but-1-ene yield increased slightly and the trans:cis ratio rose as high as 18:1 (entries 19 and 20). The simple interpretation of this last result is that process (2) has been replaced by process (3) and that process (1) has been retained, *i.e.* it has not been poisoned.

The effect of hydridation of Sc, Y, and Ti was to move the product composition slightly but not significantly in the direction shown by Zr, Nb and Hf. Further studies of the surface states of these hydrided metals are required in order to interpret the differences between these three sub-groups of behaviour. Nevertheless, this demonstates that pure Pd shares with Zr, Nb and Hf hydrides the capacity to catalyse butadiene hydrogenation by the tandem operation of processes (1) and (3).

In conclusion, attention should be drawn to five important but disparate points. First, the selectivities of *alumina-supported* Re, Fe, Ru, Os, Co, Rh, Ir, Ni, Pt, Cu, and Au generally fall on the curve given by Figure 5 (not shown), as do the sulphur-modified variants insofar as they have been studied. Thus, microcrystalline metal in supported catalysts behaves in this reaction much as the larger metal crystallites present in the evaporated films. Second, for metals over which processes (1) and (2) or (1) and (3) occur together, 1:4-addition shows a slightly higher activation energy than 1:2-addition, leading to a decrease in the selectivity for but-1-ene as temperature rises. Thus for a given metal, its position tends to move down the Figure 5 curve slightly as temperature is raised. This need not imply a concomitant favouring of π-allylic bonding over σ,π-bonding. Third, although sulphur is commonly regarded as a strong poison, modest incorporations of sulphur sufficient to modify selectivity substantially may not greatly deactivate the metal [11]. Fourth, no metal provides *cis*-but-2-ene as the major product of butadiene hydrogenation, although the Group 11 metals may give [*cis*-but-2-ene] > [*trans*-but-2-ene] but with but-1-ene in excess. The product composition *cis*-but-2-ene 70%, *trans*-but-2-ene 10%, but-1-ene 20% has been observed over MoS_2 freshly prepared from MoS_3 [15]. The behaviour was transient, and probably occurred when the material was in an over-sulphided state, causing a large fraction of the adsorption sites to be isolated. The reaction pathway may then have been confined largely to the processes shown in the lower half of Figure 3. If isolated sites are the prequirement, then the way forward might be to poison a Group 11 metal catalyst with sulphur. These metals are sometimes active by virtue of isolated sites at crystallite edges and corners; poisoning with S to remove process (1) might leave a metal surface having isolated sites capable of providing *cis*-but-2-ene as the major product. Finally, reactant streams may, on occasion, contain sulphur compounds as contaminants, and the foregoing presentation indicates that catalyst selectivities will thereby be compromised. However, the butadiene hydrogenation products can themselves be used as a sensitive test for catalyst cleanliness. The effects of small extents of contamination are measureable (see data for Mn films, Table 1, entries 14-16) and product compositions are highly reproducible. Thus, coverages of sulphur of the order of 10% can be recognised with certainty, and the detection limit would be well below 5% for well calibrated systems. The same is true for contaminations by Se, P, or As.

To summarise, in butadiene hydrogenation over pure metals the selectivity for but-1-ene formation is governed by the electronegativity of the element used as catalyst. Electronegative additives that generate positively polarised adsorption sites favour *trans*-isomer formation by 1:4-addition.

2.3   Hydrogenation of α,β-unsaturated aldehydes over unpromoted metals

More complex selectivity effects become evident in bifunctional compounds where the two functions differ. Thus, new challenges emerge on passing from buta-1,3-diene, $CH_2=CH-CH=CH_2$, to acrolein (propenal) $CH_2=CH-CH=O$, and its variously-substituted derivatives:

$CH_3CH=CH-CH=O$,	$(CH_3)_2C=CH-CH=O$
crotonaldehyde (butenal)	methyl crotonaldehyde
$CH_2=C(CH_3)-CH=O$	Ph-CH=CH-CH=O
methacrolein	cinnamaldehyde

Where a reactant contains both a carbon-carbon double bond and a carbon-oxygen double bond, thermodynamics favours the reduction of the former. For example, for crotonaldehyde hydrogenation, the free energy changes for conversion to butyraldehyde (C_3H_7CHO), crotyl alcohol ($CH_3CH=CHCH_2OH$), and *n*-butanol at 293 K are about −71, -31, and -105 kJ mol^{-1} respectively. Thus, crotonaldehyde hydrogenation over unpromoted metals of Groups 8 to 10 and copper normally gives saturated butyraldehyde as initial product, often accompanied by butanol as either a primary or secondary product.

However, the industrial demand is for α,β-unsaturated alcohols for use as intermediates in the production of certain flavours, perfumes and pharmaceuticals (*e.g.* Vitamin A) [16]. Therefore, the challenge is to design catalysts that conform to Scheme III of Figure 1 and give only saturated aldehyde and unsaturated alcohol as products with as high a selectivity as possible in favour of the latter.

Historically, a range of supported metal catalysts has been examined for selectivity to crotyl alcohol in the gas phase hydrogenation of crotonaldehyde at atmospheric pressure and modest temperatures. As expected, Ni and Cu give only butyraldehyde and butanol as initial products but, remarkably, Ni-Cu alloy provides a selectivity of 54% in favour of crotyl alcohol [17]. Platinum gives butyraldehyde selectively at 433 K, but importantly, the addition of iron generates a weak selectivity for crotyl alcohol [18]. Iridium has been little studied but its behaviour follows that of platinum [19]. In the liquid phase, Pd on many supports at 293 K gives only butanol as initial product [20] but, by contrast, Ru, Re and Os at the same temperature showed substantial selectivities to crotyl alcohol, Os/ZnO giving a reported initial yield of 97% [21]. These favourable results may be related, in part, to the use of the liquid phase because a re-examination of variously supported Ru catalysts for acrolein hydrogenation in the vapour phase has shown that, although modest selectivities to allyl alcohol are achieved initially, selectivity decays rapidly with time on stream [19]. The influence of solvent is referred to again later.

2.4 Hydrogenation of α,β-unsaturated aldehydes over metals treated with selectivity promoters

The finding in 1972 [18] that the addition of iron to platinum generated crotyl alcohol selectivity was significant and was followed in 1986 by a report that Cu/chromia behaved similarly [22]. This paved the way for more detailed studies, and promotion of the active phase has become the key feature of the most successful catalysts developed so far.

Three strategies can be envisaged which might, in principle, achieve the desired selectivity towards unsaturated alcohols: (i) selection of the most appropriate metal as catalyst, (ii) reduction in the rate of hydrogenation of the carbon-carbon double bond either by substitution, or by suitably poisoning the catalyst, and (iii) electronic modification of the surface so that the reactant undergoes enhanced activation at the carbon-oxygen double bond on adsorption. As indicated above, the alkene hydrogenation activities of the late transition metals are so high by comparison with their activities for carbonyl hydrogenation that the desired selectivity for unsaturated alcohols has not yet been obtained by the use of any unpromoted mono-metallic catalyst. Useful selectivities have been achieved by suitable substitution of the reactant, but the most promising advances have been achieved by use of metallic catalysts containing either (a) an electropositive additive or (b) an electronegative additive which, by charge conservation, can generate positive centres in the catalytically active phase. The generation of appropriate electropositive sites initiates bond formation between the aldehydic-O atom and the positive centre, causing electron depletion at the adacent C- atom and thereby its activation towards H-atom addition. These approaches will be exemplified by reference to the work of Poltarzewski,Vannice, Ponec, Hutchings, Rochester and their respective research groups on the hydrogenation of simple aliphatic unsaturated aldehydes catalysed by platinum and by copper [23-33]. The literature also contains reports of cobalt and nickel catalysed reactions and of the selective hydrogenation of cinnamaldehyde to cinnamyl alcohol [19, 23, 34-37] which are illustrative of the same principles but are not discussed here.

Table 2. Activities and selectivities observed in the vapour phase hydrogenation of some a,b-unsaturated aldehydes over Pt/silica and some metal-promoted Pt/silicas [26]

Reactant	Catalyst	Products/%				Activity[a]
		Saturated aldehyde	Unsaturated alcohol	saturated alcohol	Other products[b]	/μmol s^{-1} (g.cat)$^{-1}$
acrolein	Pt	93	2	2	3	1.4
crotonaldehyde	Pt	50	13	34	3	0.6
3-methyl crotonaldehyde	Pt	17	21	55	7	2.0
methacrolein	Pt	88	2	3	7	0.1
methyl vinyl ketone	Pt	97	trace	3	trace	57.8
acrolein	Pt/Sn[c]	67	28	3	2	1.0
crotonaldehyde	Pt/Sn[c]	50	30	19	1	5.7
3-methyl crotonaldehyde	Pt/Sn[c]	8	78	15	0	12.7
methacrolein	Pt/Sn[c]	62	26	11	1	0.9
methyl vinyl ketone	Pt/Sn[c]	96	trace	4	trace	77.2
3-methyl crotonaldehyde	Pt/Fe[c]	5	81	14	trace	7.6

[a] Pressure = 1 bar, 353 K, H$_2$: reactant = 30:1, low conversion; [b]hydrocarbons
[c] Pt:Sn and Pt:Fe = 4:1, Sn or Fe occupied a large fraction of the Pt surface

The effect on the selectivity and activity of Pt/silica at 353 K of successive methyl substitution in acrolein is shown in Table 2 [26]. The oxygenate yield from acrolein is almost entirely the saturated aldehyde. Methyl substitution at the α-carbon (methacrolein) or at the carbonyl group (methyl vinyl ketone) does not greatly perturb product selectivity, although overall reaction rates are reduced in the former reaction and dramatically enhanced in the latter. By contrast, successive substitution at the β-carbon of acrolein (crotonaldehyde and 3-methyl-crotonaldehyde) dramatically reduces the saturated aldehyde yield, raises the desired unsaturated alcohol yield to 13% (for crotonaldehyde) and 21% (for 3-methyl crotonaldehyde) but unfortunately also increases the butanol yield substantially. By contrast, the overall reaction rate is not greatly affected by substitution at the β-carbon atom.

The same group also examined the effects of incorporating cations from successive groups in the Periodic Table, Na, K, La, Ti, V, Cr, Mn, Fe, Cd, Ga, Ge, Sn, Pb, and Bi, as promoters of Pt/silica [25,26]. The most effective were Fe and Sn which raised the selectivity for unsaturated alcohol in methyl crotonaldehyde hydrogenation from 21% to about 80% with an accompanying substantial rate enhancement (Table 2).

It was inferred from these results that α,β-unsaturated aldehydes adsorb by the interaction of both unsaturated functions with the Pt surface. Substitution as in methyl vinyl ketone reduces the strength of adsorption at the carbonyl group so that the reactivity is enhanced to resemble that of a typical alkene. Substitution at the carbon-carbon double bond, particularly at the terminal carbon atom, sterically hinders the saturation of this function, thereby enhancing carbonyl hydrogenation. However, the high alkene hydrogenation activity of platinum again intervenes, causing the conversion of a major proportion of desired unsaturated alcohol to saturated alcohol and thereby reducing the desired selectivity. The addition of CO to the feed stream over Pt-Sn/silica reduced alkene hydrogenation activity somewhat and provided a further minor enhancement of the desired selectivity [25,27]]. Reactant adsorption on Pt promoted by cationic Fe or Sn clearly occurred with enhanced activation of the carbonyl group probably, as suggested above, by direct interaction of the O-atom of the reactant with the promoter cation. The promoter may also cause a diminution in alkene hydrogenation activity. Thus, an optimal combination of reactant structure and promoter provides high selectivity combined with enhanced reaction rate for more molecularly complex unsaturated aldehydes.

Comparable enhancement of selectivtity and activity had been reported earlier by Vannice and Sen for crotonaldehyde hydrogenation over Pt/titania at 318 K using a 22.7:1 ratio of H_2:aldehyde [24]. Crotyl alcohol was formed over Pt/titania in both a low temperature reduced state and a high temperature reduced (SMSI) state, the best yield being 37% over the latter at 318 K at low conversion. The specific activity of this catalyst was about 1.3 μmol s^{-1} (g.cat)$^{-1}$, *i.e.* of the same order as that recorded in Table 2 for Pt/Sn and Pt/Fe. By contrast, Pt/silica, Pt/alumina and Pt powder all gave solely butyraldehyde under the same conditions. These authors made a detailed study of the kinetics of reaction and examined also the hydrogenation of the intermediates, thereby establishing the main features of the reaction network including the importance of crotyl alcohol isomerisation to butyraldehyde. The promoting effect of titania with respect to the selectivity was attributed to the presence of Ti^{3+} or Ti^{2+} ions or oxygen vacancies that coordinated the oxygen atom of the carbonyl group thereby activating the adjacent C atom towards H-atom addition. Although the overall activities of selective Pt/titania catalysts did not differ greatly from those of unselective Pt/silica or Pt/alumina, the turnover frequencies of the former (typically 2 molecules s^{-1} $(Pt_s)^{-1}$) were greater than the latter by a factor of 50. The high activity of the less numerous selective Pt-Ti$^{\delta+}$ sites was attributed to the ready availability of dissociated hydrogen at the cognate platinum surface.

Table 3. Selectivities observed in the vapour phase hydrogenation of crotonaldehyde over various unpromoted Cu, Pd, and Cu-Pd catalysts and over the same catalysts promoted by thiophene [30,32]

Catalyst[a]	Promoter[b]	Temp	TOS[c]	Conv[d]	Products/%			
		/K	/min	/%	butyr-aldehyde	crotyl alcohol	butanol	other products[e]
Cu/alumina	-	353	5	17	39	9	51	1
Cu/alumina	-	353	5	8	59	10	27	5
Cu/alumina	-	353	60-180	5	89	10	0	1
S-Cu/alumina	Th(2)	353	5	19	9	32	48	11
S-Cu/alumina	Th(2)	353	15	10	15	40	27	18
S-Cu/alumina	Th(2)	353	30	5	20	58	16	6
S-Cu/alumina	Th(2)	353	180	3	25	37	0	38
Cu/alumina	-	373	60	16	87	9	0	4
S-Cu/alumina	Th(1)	373	60	4	42	38	0	20
S-Cu/alumina	Th(2)	373	60	4	23	56	0	21
Cu/silica	-	383	40	59	15	64	20	1
3:1Cu-Pd/silica	-	383	40	35	10	64	25	1
Pd/silica[f]	-	383	40	42	27	0	73[g]	
S-Cu/silica	Th(1)	383	40	37	22	70	7	1
S-3:1Cu-Pd/silica	Th(1)	383	40	39	29	0	65	6
S-Pd/silica[f]	Th(1)	383	40	45	20	0	80[g]	

[a] reaction conditions: pressure = 1 bar, H$_2$: crotonaldehyde = 14:1
[b] Th = thiophene; dosed onto catalyst at 483 K; dosage/µlitres shown in parenthesis
[c] TOS = time on stream   [d] Conv = conversion
[e] aldol condensation products over Cu; but-1-ene and butane over Cu-Pd
[f] 1:1 Cu-Pd/silica and 1:3 Cu-Pd/silica behaved similarly
[g] value includes the 'other products'

Enhanced selectivity has also been achieved by carrying out the reaction in solution rather than in the gas phase. The hydrogenation of acrolein over Pt/nylon-66 occurs in ethanolic solution at 318 K giving a selectivity to allyl alcohol of 19%, which was raised to over 60% by the addition of 18% Sn [23]. Again, the authors propose that Sn cations activate adsorption by the carbonyl group. There is no evidence that nylon-66 exerts a support effect.

However, the authors do not consider the effects that might arise from the dissociative adsorption of the solvent ethanol at the Pt surface to give both hydrocarbonaceous and oxygenated adsorbates [38] that might poison alkene hydrogenation.

The limitations of Pt catalysts for unsaturated alcohol formation are clearly related to their high alkene hydrogenation activities; it is perhaps not surprising, therefore, that research has also focused on a metal having a lower alkene hydrogenation activity, namely copper. Moreover, the ability of sulphur to modify the catalytic selectivity of metals by variation of the mode of H-addition (discussed above in respect of butadiene hydrogenation) has provided a means of enhancing the desired selectivity in this class of reactions also. A joint study has been undertaken by the research groups of Hutchings and Rochester [28-33] and some important features are illustrated in Table 3.

First, it should be emphasised that reactions over copper catalysts, whether sulphided or not, show product distributions that vary with time on stream. Table 3 shows that initial activity of Cu/alumina at 353 K for butanol formation is high, but that this declines to zero over the period of an hour and is replaced by activity for butyraldehyde formation. Selectivity for crotyl achohol is constant at about 10% [30]. This zero butanol yield at modest conversions does not imply that activity for full saturation has been lost because butanol formation resumes after the removal of crotonaldehyde.

The same Cu/alumina, sulphided by thiophene, shows a substantially enhanced selectivity for crotyl alcohol which passes through a maximum at 58% after 30 minutes time on stream (Table 3) [30]. Again, initial butanol formation is substantial at first but declines to zero, and is replaced by the formation not only of butyraldehyde but also of higher molecular weight products formed by aldol condensation. Acrolein hydrogenation over the same Cu/alumina is cleaner giving 100% butyraldehyde, and the sulphided catalyst 54% butyraldehyde and 46% crotyl achohol after 30 minutes time on stream. The advantageous effect of sulphur on copper for this reaction is thus clearly demonstrated.

The types of copper site available on these catalysts were determined by examining, by FTIR spectroscopy, the adsorption of CO in the absence and presence of thiophene [29]. From the shifts in band frequencies it was adduced that four types of adsorption site were present at the catalyst surface: (i) Cu^0, (ii) Cu^+ in a matrix of Cu^+, (iii) Cu^+ in a matrix of Cu^{2+}, and (iv) Cu^{2+}. Accompanying studies of the hydrogenation of crotyl alcohol and of butyraldehyde showed that crotyl alcohol isomerisation to butyraldehyde and its dehydrogenation back to reactant each occurred to a measurable degree. The reaction pathway shown in Scheme VI was proposed, curve fitting providing the values of the pseudo first order rate coefficients shown.

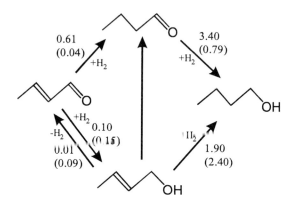

Scheme VI:
Reaction network and rate coefficients for crotonaldehyde hydrogenation over Cu/alumina at 373 K. Rate coefficients in parenthesis are for reaction over sulphided Cu/alumina.

These reveal that the rates of formation and removal of butyraldehyde are reduced by the presence of sulphur, but that the rate of crotyl alcohol hydrogenation to butanol is *increased*. Thus the enhanced selectivity for crotyl alcohol is attributable in part to a poisoning of the pathway *via* butyraldehyde and in part to an enhancement of its rate of formation. Performance would have been further improved had the presence of sulphur not facilitated the conversion of alcohol to alkane.

From the known copper surface areas of the catalysts and the amounts of sulphur retained after thiophene treatment (thiophene partially decomposed at the treatment temperature of 483 K) the ratio of sulphur atoms to copper atoms in the active surface was calculated to be about 0.002:1.000. This ratio was too low for the outcomes of S-treatments to be ascribed solely to a steric effect. Rather, the effects were attributed to the newly created Cu^+-S sites, these being responsible for activation of the carbonyl group on reactant adsorption, and the selectivity variation was therefore electronic in origin. The extent of promotion of crotyl alcohol selectivity increased with increasing dopant concentration as expected (Table 3, entries 8-10).

Support effects in the Cu-catalysed reaction appear unimportant, the selectivities described above for Cu/alumina and sulphided-Cu/alumina being fairly closely duplicated by Cu/magnesia, Cu-ZnO/alumina, and their sulphided analogues [30].

Thiophene was the most effective of a range of S-compounds studied as selectivity promoters for Cu/alumina, but its performance was closely emulated by dimethylsulphoxide, thiophane, and dimethyl sulphide. Sulphur dioxide and carbon disulphide were rather less effective [31]. Attempts to enhance selectivity by treating Cu/alumina with pyrrole, furan, and cyclopentadiene met with little success [33a]. FTIR studies of the effects of these compounds on the spectra of adsorbed-CO revealed that pyrrole adsorbed at both Cu^0 and Cu^+ sites and enhanced the population of the latter by adsorbing as the pyrrolate ion, thereby inducing a small improvement in crotyl alcohol selectivity. By contrast, cyclopentadiene acted as an indiscriminate poison, and furan was insufficiently strongly adsorbed to be effective. Gold suported on ZnO is also promoted by sulphur for selective cinnamaldehyde hydrogenation [33b].

Finally, an interesting attempt was made by Hutchings, Rochester and co-workers to create Cu^+ sites at the surfaces of silica-supported Pd-Cu alloys and to investigate the effect on crotonaldehyde hydrogenation of treating these materials with thiophene [32]. FTIR spectroscopy of CO adsorption revealed that, as the mol % Pd diminished from 100% to 25%, so the Pd ensembles at the surface were reduced in size by the presence of the Cu, as expected. Four types of sites were identified from the CO-spectra: type I Pd sites (at which CO adsorbed as the bridged structure), type II Pd sites (at which CO adsorbed as the linear structure), Cu^+ sites, and Cu^0 sites. 25 mol% Pd-Cu/silica behaved much as Cu/silica giving a substantial selectivity towards crotyl alcohol (Table 3). However, the Pd-rich catalysts favoured carbon-carbon double bond hydrogenation and crotyl alcohol did not appear in the product. The order of resistance of these sites to modification by thiophene was:

$$Pd(\text{type I}) > Pd(\text{type II}) > Cu^+ > Cu^0$$

Consequently the desired Cu^+ sites, although created, were preferentially poisoned, and this left the Pd sites active for the hydrogenation of the carbon-carbon double bond in crotonaldehyde. Hence all the sulphided Pd-Cu alloys showed a zero selectivity for crotyl alcohol formation.

In view of this information the 2:1 Ni-Cu/alumina catalyst, which showed an increase in selectivity over Ni/alumina and Cu/alumina [17], would merit re-examination.

An attempt to enhance allyl alcohol selectivity in acrolein hydrogenation over Pt by treatment with thiophene was only marginally successful [25]. This is not surprising since platinum, of the metals in Group 10, has the weakest metal-sulphur bond, and indeed, platinum was the only late transition metal for which, in butadiene hydrogenation, sulphur adsorption did not induce a substantial change from 1:2-addition to 1:4-addition [9].

The reactions over supported metal catalysts described above occur at the range of crystal faces normally present in microcrystalline metal. Delbecq and Sautet have made a theoretical study of the adsorption geometries of acrolein, crotonaldehyde, and 3-methyl crotonaldehyde at the low index faces of Pt and Pd [39]. Depending on the face selected, so the adsorption geometries were found to be different, di-σ forms being preferred on Pt(111), planar-η_4 forms on Pt(110) and π_{CC} forms on Pt(110). The matrix of binding energies calculated provide insights into the possible origins of variations in experimentally observed activities and selectivities. Futhermore, this data should motivate experimentalists to devise methods for the preparation of face-specific supported catalysts, with a view to achieving yet more highly selective and predictable behaviour.

Hydrogenations of α,β-unsaturated aldehydes at Pt single crystal surfaces have given results which do not differ greatly from those obtained over supported Pt catalysts but which may disturb some of the interpretations advanced above. The hydrogenation of crotonaldehyde and of 3-methyl crotonaldehyde (0.005 Pa) with hydrogen (3 Pa) over Pt(111) at 330 K gave selectivities to the unsaturated alcohol of 10% and 56% respectively [40], which compares with values of 13% and 21 % obtained over Pt/silica at higher pressures (Table 2). A later study gave a confirmatory value for 3-methyl crotonaldehyde of 60% (reactant pressure 0.02 Torr, hydrogen pressure 400 Torr, temperature = 353 K) [41]. Selectivities were a little higher for the surface $Pt_{80}Fe_{20}(111)$, being 13% for crotonaldehyde and 70% for 3-methyl crotonaldehyde [40] although the turnover frequencies for the alloy surface were higher by a factor of 5. Iron present as $Fe^{\delta+}$ on Pt, proposed for supported catalysts as providing carbonyl activation by an electronic effect, do not feature at this alloy surface, and so the enhanced selectivity might have to be construed in terms of changes in the geometric or electronic structure of Pt atoms in the surface layer. The lower the electronic density on Pt atoms the greater is the extent of activation of carbon-carbon double bond [42], and Pt atoms of low electron density will exist at steps, edges, corners and kinks in polycrystalline metal, but not in the plane surface of Pt(111). The possibility must be countenanced, therefore, that high selectivity is provided by low index planes of Pt and that the promoting effect of additive metals might be to block Pt sites of low electron density at edges and kinks, thereby reducing the number of sites having the highest activity for carbon-carbon double bond saturation. If this were the case, one would expect selectivity to increase as supported Pt catalysts were progressively sintered – a simple experiment that appears not to have been performed. Although the single crystal experiments are carried out under different pressure conditions, and are sometimes thought to be difficult to compare on that count, it is interesting that the turnover frequency for crotonaldehyde hydrogenation measured over Pt(111) at 330 K and 0.02 Torr hydrogen pressure is 0.05 molecules s^{-1} Pt_s^{-1} [40], the same as that recorded by Vannice and Sen for Pt/silica at 319 K and 730 Torr [24].

To summarise, the hydrogenation of α,β-unsaturated aldehydes to the industrially valuable unsaturated alcohols follows Scheme VI. Selectivity for unsaturated alcohol is achieved by increasing the rate of its formation by creating positively polarised sites at the surface that preferentially activate the carbonyl group of the reactant. It would be inviting, now, to re-investigate Os/alumina and Os/zinc oxide that were so effective in the liquid phase [21] and to determine their performance in gas phase reactions. Osmium catalysts are extraordinarily difficult to free of oxygen because the Os-O bond is so strong, and there is a possibility that residual oxygen creates $Os^{\delta+}$ centres that are effective in carbonyl activation. This would establish that high selectivity can indeed be achieved without an added promoter. Finally, given the toxic properties of osmium, which present problems for industrial application, similar chemistry might be shown by the preceding element in the Periodic Table, rhenium, which also forms a very strong metal-oxygen bond.

3. SELECTIVITY AND DEFECTIVE METALLIC STRUCTURE

Butadiene hydrogenation over alumina-supported Ru, Rh, Os, Ir, and Pt differs from that over Fe, Co, Ni, Cu, Pd, and Au in giving butane as a substantial primary product. Typical initial butane yields from reactions in a static reactor at 298 K have been recorded as Ru, 18%, Rh, 21%, Os, 57%, Ir, 75%, Pt 42% [6]. The capacity of these metals to provide alkane formation directly from the alkadiene in a step additional to those shown in Scheme V is evident in the hydrogenation of all simple alkynes and alkadienes over these noble metals and has been widely reported [43-46]. This reactivity is troublesome in that alkane is not normally a desired product and the required selectivity to alkene is thereby degraded. Taken together with the low butane yields given by Fe, Co, and Ni under the same conditions (<5%), it is evident that the selectivity, S, for alkene formation (defined as S = [alkene]/{[alkene] + [alkane]}) decreases on passing down each of the Groups 8, 9, and 10, *i.e.* $S_{Fe} > S_{Ru} > S_{Os}$, $S_{Co} > S_{Rh} > S_{Ir}$, and $S_{Ni} \sim S_{Pd} > S_{Pt}$.

Attempts were made to rationalise this behaviour in terms of the relative strengths of adsorption of reactant and alkene on the various metals (which influence the lifetime of adsorbed alkene) and the relative activities of the metals for alkene hydrogenation [43-45]. However, this purely mechanistic explanation lacked independent support.

Serendipity played a role in an eventual alternative interpretation. In an unrelated project, the present author had cause to measure by H_2/D_2 exchange the concentrations of hydrogen remaining 'occluded' in well-pumped metal powders freshly prepared by reductions of metal salt precursors [47]. For those metals of Groups 8 to 10 that dissolve hydrogen endothermically (*i.e.* for all metals except Pd) the extent of hydrogen occlusion, x in MH_x, decreased in the order:

$IrH_{0.10-0.46} > OsH_{0.07-0.11} > RuH_{0.04} > RhH_{0.02-0.03} > PtH_{0.006} > CoH_{0.004} \sim NiH_{0.001}$

The Group 11 metals occluded no measurable amount of hydrogen. These concentrations of hydrogen exceeded the true solubility by many orders of magnitude, and these metals differ from the early transition metals in not forming saline hydrides. Nevertheless, this 'occluded' hydrogen was chemically bound to the metal, as witnessed by the isotope exchange, and was not removable by pumping.

The correspondence between the hydrogen occlusion sequence and the selectivity sequences raised the question as to whether there was a causal connection between the two.

When precursor salts are reduced the metal atoms so produced migrate to form the crystalline metallic lattice. Where metal atom mobility is high well ordered metal crystallites will result, but where metal atom mobility is very low the metallic particles may be defective or cavitated. During catalyst reduction all metal atoms at the moment of their formation are in a surface and Huttig's rule says that self-diffusion in surface regions of metallic specimens becomes significant at about one third of the Kelvin melting point. Thus it is to be expected that, at one extreme, catalysts containing osmium or iridium particles (m.pts. 3320 and 2720 K respectively) prepared by reduction, typically at 473 K, will be defective, whereas at the other extreme cobalt and nickel particles (m.pts. 1495 and 1445 K respectively) prepared typically at 723 K will be well annealed. On the basis of this model, and the assumption that the rate of cavity creation decreases exponentially with increasing temperature, it was predicted that the extent of hydrogen occlusion, x, should vary with reduction temperature, T_R, according to the equation: $\ln x = E(T_H - T_R)/RT_HT_R$, where T_H is the Huttig temperature. This equation was satisfactorily obeyed (Figure 6).

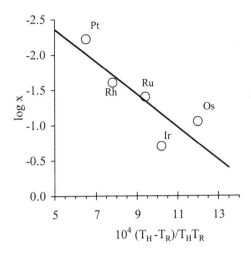

Figure 6. Function test for the cavity model of hydrogen occlusion

Thus, in this model, the active surface is conceived as containing (a) regions at which the metallic structure is well ordered, hydrogen occlusion does not occur, and selectivity for alkene formation from alkadiene or alkyne is high and (b) regions at which the metallic structure is disordered, hydrogen is occluded, and alkane formation is dominant [47].

The possibility that this alkane formation was related to a physical property of the metallic active phases was attractive because it suggested that the widely reported selectivity pattern had arisen only because investigators had, by and large, used common catalyst preparation conditions. By the same token, much improved selectivities should be attainable by use of exceptional preparation conditions. The crucial test was to prepare iridium powders at a range of temperatures, some above the Huttig temperature and some below, and to measure their extents of hydrogen occlusion and selectivities in butadiene hydrogenation. Iridium powders $IrH_{0.46}$ and $IrH_{0.12}$ prepared below the Huttig temperature gave butane yields of 80% and 40% respectively at 300 K, whereas $IrH_{0.03}$ and $IrH_{0.02}$ prepared above the Huttig temperature gave values of 5% and 3% under the same conditions [9]. It is clear therefore that annealed Ir behaves much as annealed cobalt or nickel and that high alkane yields are attributable to the defective nature of the metallic active phase as conventionally prepared.

The concept that cavitated or dislocated regions in catalyst metal particles favoured alkane formation was open to a further test, because supported metal catalysts can be prepared containing exceedingly small metal particles and, as particle size diminishes, so it becomes physically impossible for individual particles to contain cavities although structural crystallinity may still be poor. Accordingly, a series of Ir/silica catalysts was prepared by reduction *below* the Huttig temperature but loading was varied so that 20% and 10% Ir/silicas contained Ir particles in the range 1 to 5 nm in diameter, easily visible by HRTEM, whereas 0.3%, 0.1%, and 0.01% Ir/silicas contained no particles visible by HRTEM (detection limit, 0.7 nm) The firrt two catalysts behaved 'normally' giving alkane yields of 70% at 340 K whereas the lightly loaded catalysts gave substantially lower yields of 50%, 38%, and 25% respectively. The butane yield over 0.01% Ir/silica, though not small, was nevertheless the lowest that has yet been reported for a supported iridium catalysts prepared in the conventional fashion.

Thus, where the selective hydrogenation of an alkadiene or alkyne to alkene is an important target, special care should be taken when reducing the catalyst, to minimise the creation of structurally defective regions in the metallic active phase.

4.   SELECTIVITY AND MOLECULAR CONGESTION AT THE ACTIVE SITE

Having demonstrated that selectivity is influenced by the electronic state of the active surface and by the defect state of the active phase, we turn now to consider some molecular interaction effects that can arise in the adsorbed layer and how they can influence selectivity.

The isomerisation of an alk-1-ene to *cis-* and *trans-*alk-2-ene is an intriguing class of formally simple reactions which may be selective for the formation of either product (Scheme I, Figure 1). In principle, such reactions may occur either by H-atom addition (to give σ-alkyl intermediates) followed by the abstraction of a different H-atom, or by H-atom abstraction (to give π-allylic intermediates) followed by H-addition to a different carbon atom. These processes can be distinguished kinetically and by D-tracer studies, and consideration here is restricted to reactions proceeding by the addition/abstraction mechanism involving alkyl intermediates.

But-1-ene hydrogenation over supported platinum metals is accompanied by isomerisation the but-2-ene product normally showing a *trans:cis* ratio of 1.5-2.0:1. Such ratios have been interpreted in terms of the relative energies of the conformations of the adsorbed 2-butyl group at a uniform metal surface and the energy profiles associated with the H-atom abstraction that result in their conversion to the *trans*- and *cis*-products [48]. Thus, selectivity is under simple kinetic control where the intermediate is free to adopt its minimum energy conformation at the active site.

However, markedly different behaviour may be observed when the catalyst is a metal cluster carbonyl. $Ru_6C(CO)_{17}$/silica catalyses butene isomerisation after mild thermal treatment which causes the desorption of a small fraction of the carbonyl ligands but probably leaves the nuclearity of the cluster intact. This catalyst favours the *cis*-product giving 70% *cis*-but-2-ene at 253 K and 61% at 293 K [9]. Decomposition of the cluster at 358 K gives a polycrystalline Ru/silica catalyst which, under the same conditions, gives 42% *cis*-product, *i.e.* selectivity favours the *trans*-product and this is to be compared with 37% *cis*-product provided by Ru/alumina prepared conventionally from $RuCl_3$ [49]. Thus, use of a carbonyl cluster as catalyst precursor provides an atypical selectivity.

$$CHD=CDPr + X \underset{*}{\overset{}{\rightleftharpoons}} \underset{*}{CHDCDXPr} \longrightarrow CHD=CXPr + \underset{*}{D} \qquad (1) \qquad p$$

$$CHD=CDPr + X \rightleftharpoons \underset{*}{CHDXCDPr} \longrightarrow CHX=CDPr + \underset{*}{D}$$

$$Or \qquad (2) \qquad s$$

$$CDX=CDPr + \underset{*}{H}$$

$$\underset{*}{CHDXCDPr} \longrightarrow CHDXCD=CHEt + \underset{*}{H} \qquad (3) \qquad i$$

X = H or D; Pr = propyl; Et = ethyl

Figure 7. The three processes that account for hydrogen isotope exchange and isomerisation in 1,2-dideuteropent-1-ene.

Selectivity in favour of *cis*-alkene is obtained wherever there is substantial molecular congestion at the active site. This was shown for a series of nickel catalysts (one homogeneous and three heterogeneous), for all of which it was demonstrated that the isomerisation of the labelled pent-1-ene $DHC=CDC_3H_7$ occurred by the addition/abstraction mechanism involving pentyl intermediates [50]. These catalysts were (i) $NiH[P(OEt)_3]_4$ in solution in benzene, (ii) $[Na^+]_4[Ni \text{ phthalocynanine}]^{4-}$ supported on quartz wool, (iii) Ni/alumina, and (iv) evaporated Ni film. The first two catalysts contain single metal atom sites, whereas the last two contain multiple metal atom sites. The nickel atom sites in (i) are congested in three dimensions by the bulky triethylphosphite ligands (which are retained in solution under the conditions used). In (ii) the nickel atom sites are congested in two dimensions by the phthalocyanine ring. In (iii) and (iv) the nickel atom sites have nickel atom neighbours; sites located at edges and corners of the metal crystallites, most common in (iii), are the least congested.

Table 4. Product compositions and reaction parameters for $CHD=CDC_3H_7$ isomerisation catalysed by various nickel catalysts [50]

	$NiH[P(OEt)_3]_4{}^+$	$(Na^+)_4(NiPc^{4-})$	Ni/alumina	Ni film
	single metal atom sites	single metal atom sites	multiple metal atom sites[a]	multiple metal atom sites[a]
	3-D site congestion	2-D site congestion	2-D site congestion	2-D site congestion
	reaction in homogeneous solution	reaction at the gas-solid interface	reaction at the gas-solid interface	reaction at the gas-solid interface
Products /%				
cis-pent-2-ene	70	42	36	32
trans-pent-2-ene	30	58	64	68
Parameters[b]				
p	0.10	0.17	0.10	0.35
s	0.82	0.25	0.17	0.05
i	0.08	0.58	0.73	0.60

[a]metal atoms have other metal atoms as neighbours
[b]parameters defined in Figure 7

Isomerisation and isotope exchange in 1,2-dideuteropent-1-ene involves the three steps shown in Figure 7, of which the first two involve exchange at C_2 and C_1 and the third is the isomerisation step.

The probabilities of these steps occurring, designated p, s, and i respectively, are obtained by fitting the experimental and computed D-distributions in the products. Table 4 shows the product compositions (selectivities) obtained for these catalysts and the calculated values of p, s, and i. Considering the parameters first, for the homogeneous catalyst and Ni/alumina the relative chances of forming the 1-pentyl and 2-pentyl intermediates, $p:(i + s)$, are the same, but the chances of achieving isomerisation, i, are low for the former (8%) and high for the latter (73%).

By contrast, for the two catalysts having all or most Ni atom sites in a planar environment (Ni phthalocyanine and Ni film), the chance of achieving isomerisation is virtually the same (58 and 60%). For the catalysts having single metal atom sites, isomerisation requires a critical rotation of the secondary pentyl group as shown in Figure 8.

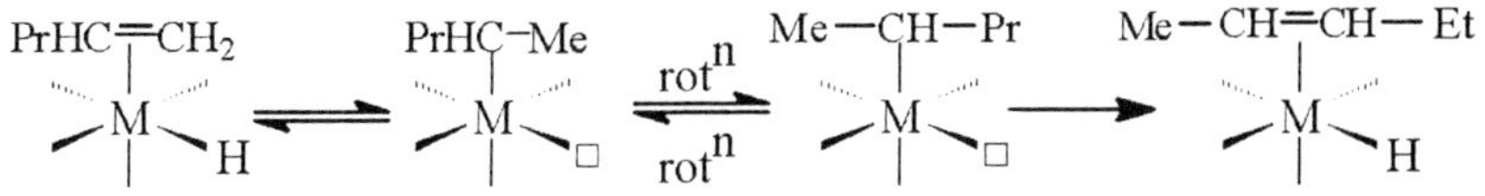

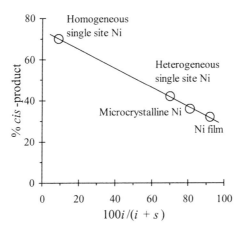

Figure 8. The critical rotation required to effect pent-1-ene isomerisation at a single metal atom site ($\square$ = ligand vacancy)

The value of the reaction parameter i signifies that this rotation is difficult in the three-dimensionally congested environment of $NiH[P(OEt)_3]_4^+$ but is much more easily achieved at the two-dimensionally congested site $NiPc^{4-}$. There is an uncertainty as to whether pent-2-ene is adsorbed at catalysts (iii) and (iv) as a π-complex to a single metal atom or by a di-σ-complex to two adjacent metal atoms, and in consequence the interpretation of the parameter i for these catalysts is to some degree ambiguous. Nevertheless, the chance of isomerisation over conventional Ni/alumina is the highest recorded and this is consistent with this catalyst having the least congested sites.

Figure 9. Graphical representation of the relationship between the four catalyst types presented in Table 4: variation of the selectivity for *cis*-alkene formation with the chance of the 2-pentyl intermediate being converted to pent-2-ene, $100i/(i + s)$

Clearly, molecular congestion at active sites influences the behaviour of the reaction intermediate. But how is it that molecular congestion promotes *cis*-selectivity, whereas a relative absence of molecular congestion provides *trans*-selectivity? Where congestion is high the 2-pentyl group adopts the conformation of lowest molar volume and in this conformation the H-atom directed towards the adsorption site is the one which, when lost, gives *cis*-pent-2-ene as product. However, where molecular congestion is less significant, the 2-pentyl group adopts a conformation closer to the minimum energy conformation for which H-atom loss to the site favours *trans*-pent-2-ene formation.

The incremental variation in the behaviour of these four diverse types of catalyst is further demonstrated in Figure 9 which shows a consistent correlation of selectivity for the *cis*-product with the chance of 2-pentyl being converted to pent-2-ene.

This model was considered to apply also to butene isomerisation over activated $Ru_6C(CO)_{17}$/silica cited above, and it was further corroborated by studies of dideuteropent-1-ene isomerisation catalysed by two homogenous Ru catalysts in benzene solution at 353 K [51]. In these reactions selectivity changed from *cis*-alkene to *trans*-alkene as the catalyst complex was decongested by solvation, bulky triphenyl phosphine ligands being exchanged for smaller ligands (solvent molecules). Thus, for $RuHCl(PPh_3)_3$, selectivity changed smoothly from 57% in favour of the *cis*-pent-2-ene to 77% in favour of the *trans*-pent-2-ene with increasing dilution whereas for $RuHCl(CO)(PPh_3)_3$ the values changed from 63% in favour of the *cis*-product to 91% in favour of the *trans*-product. Again the distribution of deuterium in the product confirmed that the mechanism of isomerisation *via* alkyl intermediates had remained the same at all dilutions, confirming that the selectivity changes were steric in origin.

Hydrogenations are normally conducted under conditions of high surface coverage of adsorbates and there may be many instances where steric control of selectivity is operative but not suspected.

5.   SPECIAL FACTORS AFFECTING SELECTIVITY IN ALKYNE HYDROGENATIONS

Hydrogenations of simple alkynes have features in common with the hydrogenations of the isomeric alka-1,3-dienes. One such shared feature is the pattern of alkane yields over the metals of Groups 8-10 (discussed above), from which it is concluded that disordered regions of the active phase again catalyse unwanted alkane formation thereby degrading alkene selectivity. However, alkyne hydrogenations show some unique complexities that bear upon alkene selectivity, which are discussed here.

5.1 Higher molecular weight products

The dominant process in alkyne hydrogenation is the stepwise addition of two hydrogen atoms across the carbon-carbon triple bond. This is demonstrated most clearly in but-2-yne hydrogenation catalysed by Cu/alumina [52] and by Pd/alumina [53] each of which is completely selective for the formation of *cis*-but-2-ene. The reaction with deuterium over Pd gives $CH_3CD=CDCH_3$ as the sole product [54] by the process shown in Figure 10. However, Pd-catalysed ethyne hydrogenation at ambient temperature shows two complications.

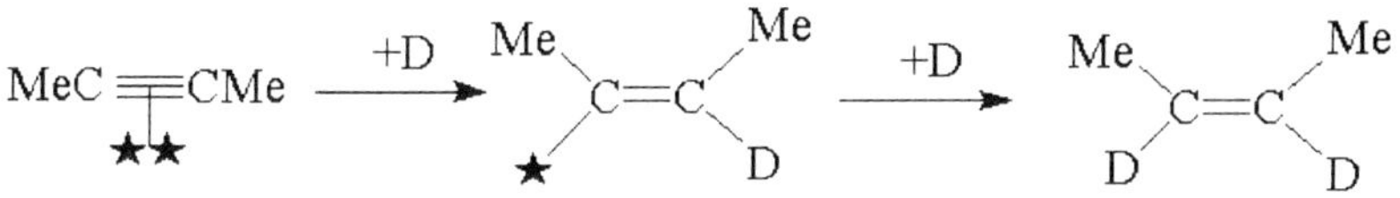

Figure 10. Stereospecific addition of deuterium to adsorbed but-2-yne

First, a substantial fraction of the reactant is dimerised to C_4-product (butadiene and butene) together with smaller yields of higher hydrocarbons; one third may be so consumed [55]. Second, the reaction with deuterium does not give 100% *cis*-$C_2H_2D_2$ as would be expected by analogy with Figure 10, but a distribution of isotopically distnguishable ethenes of which the following is typical: C_2H_4 3%, C_2H_3D 20%, $C_2H_2D_2$ 70%, C_2HD_3 6%, C_2D_4 1%. Of the $C_2H_2D_2$, 85% had the *cis*-configuration, 14% the *trans*-configuration and 1% was *asymmetric* $H_2C=CD_2$ [56].

Sheridan and Bond have suggested that the normal form of the half-hydrogenated state, adsorbed-vinyl, might exist transiently in a free radical form that initiates the formation of higher hydrocarbons by a straight chain reaction [57,58]. One may speculate as to whether a free radical could exist as a distinct adsorbed intermediate so close to the metal surface or whether the free electron would become paired with one from a neighbouring surface site to give (in Kemball's notation) an α,α,β-tri-adsorbed species. Deuterium addition to either the free radical form or to the tri-adsorbed form of vinyl would provide equal quantities of *cis*- and *trans*-$C_2H_2D_2$. On this basis, the palladium catalysed reaction proceeds as shown in Figure 11, with about half of the ethyne being hydrogenated directly to *cis*-$C_2H_2D_2$, the other half being converted to the radical species of which a majority is converted to C_4-

product and a minority is hydrogenated to an equimolar mixture of *cis-* and *trans*-$C_2H_2D_2$. Over the other platinum group metals, the C_4-yields are lower and the *trans*-$C_2H_2D_2$ yields higher, which suggests that multiply-bonded vinyl spends a smaller fraction of time in the free radical form and a larger fraction as the α,α,β-species. The higher yields of *trans*-$C_2H_2D_2$ are accompanied by higher yields of the asymmetric variant. Alumina-supported Ru and Os provide the extreme cases in which the fraction of ethyne converted to C_4-products is only 8%, *trans*-$C_2H_2D_2$ is the major ethene formed, and the *asym*-variant appears in yields of up to 16% [59].

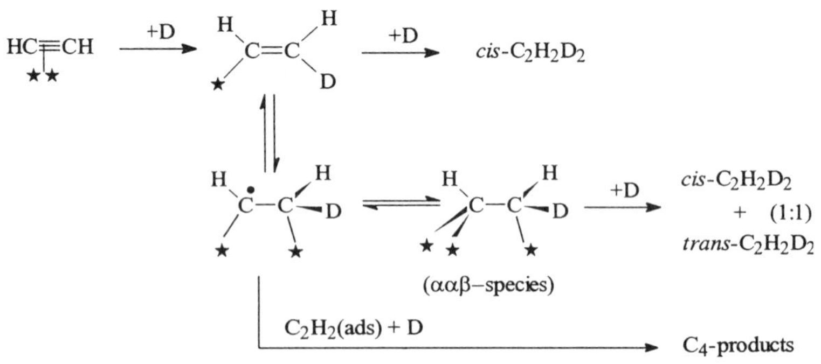

Figure 11. Ethyne/D_2 reaction: processes contributing to the formation of *cis-* and *trans*-$C_2H_2D_2$ and to C_4-products.

A steady state scheme has been presented which interprets the various product distributions observed [56]. This indicates that the pool of adsorbed 'hydrogen' contains both H and D and that ethyne exchange occurs producing both adsorbed-C_2HD and adsorbed-C_2D_2. *Asymm*-$C_2H_2D_2$ is then formed by processes such as:

$$HC\equiv CD(ads) + H \rightarrow H_2C=CD(ads)$$
$$H_2C=CD(ads) + D \rightarrow H_2C=CD_2(ads)$$

Direct evidence for the existence of radicals in ethyne and ethene hydrogenations has recently been sought by Roberts and Rowlands using EPR spectroscopy [60,61]. A solution in dichloromethane of the spin trap compound 2,4,6 tri-tert-butyl nitroso-benzene was added to a reactor in which ethyne hydrogenation over Pd/alumina was taking place at room temperature [60]. A radical-adduct was detected but the radical was assigned an allylic not a vinylic structure, and the spectrum was attributed to that of a C_4-intermediate extracted from the butene-forming process. Adsorbed-vinyl was, perhaps, too reactive to be detected by this technique at this temperature.

A Pd/alumina catalyst is used industrially for the selective removal of ethyne from ethene streams to levels of <5 ppm before the ethene is transferred to polymerisation plants [62]. Ethyne polymerisation again occurs and, because this process is operated in the presence of CO, the polymerisation product, known as 'green oil', includes oxygenates. These side reactions deactivate the catalyst.

In the hydrogenation of di-substituted alkynes, the alkyl substituents protect the half-hydrogenated states against dimerisation and selectivity is thereby enhanced.

5.2   Slow loss of alkene selectivity as reaction progresses

In alkyne and alkadiene hydrogenations under excess hydrogen over late transition metals other than Pd, the alkane yield may increase as reaction progresses even though the reactant is present at a sufficient partial pressure to dominate the surface. For this situation it is important to determine whether alkene readsorption is occurring, or whether the increased alkane is formed directly during one residence of the reactant on the surface due to an increasing hydrogen coverage as the instantaneous value of the alkyne/hydrogen ratio increases. The extent of product readsorption has been assessed by Webb and co-workers in isotopic tracer experiments involving the hydrogenation of $^{12}C_2H_2/^{14}C_2H_4$ mixtures and measurement of the yields of $^{12}C_2H_6$ and $^{14}C_2H_6$. One such series of experiments over the standard reference catalyst EUROPT-1, 6.3% Pt/silica, at 295 K showed that re-adsorption and hydrogenation of ethene accounted for only 1% of the 75% ethane yield at conversions up to 50% [63]. Such studies provided evidence that the Pt surface contains three distinct types of site which facilitate (i) ethyne conversion to ethene, (ii) ethyne conversion directly to ethane, and (iii) ethene conversion to ethane. Type (i) sites were considered dominant. This firmly establishes that this reaction is described by Scheme IV of Figure 1 in which the direct conversion of A to C is important and the conversion of B to C is unimportant. Direct conversion of ethyne to ethane at type (ii) sites was conceived as involving adsorbed-vinylidene and other multiply-bonded intermediates but direct evidence for the bonding of adsorbed intermediates is not provided by this technique. It is tempting to suggest that these sites exist at the surfaces of the disordered regions of metal crystallites discussed earlier [47] but further investigations are required to establish any such connection.

5.3   Sudden loss of alkene selectivity as reaction progresses.

Over Pd the effect of conversion on selectivity, in reactions in a static reactor under an excesss of hydrogen, can take an altogether more dramatic turn. In the initial stage of reaction Pd is highly selective for alkene formation giving a C_2- and C_4-yields that are almost entirely ethene and butene, but suddenly reaction may convert to extremely fast non-selective ethane and butane formation [55]. The effect is reproducible, is unique to palladium, and occurs also in propadiene and buta-1,2-dione hydrogenation under similar conditions [64]. The origin of the effect was unclear until it was encountered in buta-1,2-diene hydrogenation; in this reaction, the initial products are but-1-ene and *cis*-but-2-ene, but when sudden selectivity breakdown occurs the butane yield is accompanied by the formation of *trans*-but-2-ene. A D-tracer study revealed that the butane and *trans*-but-2-ene were highly exchanged, the exchange pattern being consistent with these products having been formed *via* adsorbed-alkylidene species. Translated back into the context of ethyne hydrogenation, this means that ethane formation after the sudden breakdown of selective reaction occurs by the process shown in Figure 12. The suddenness and reproducibility of the selectivity breakdown is so remarkable that it suggests that a surface phase change takes place.

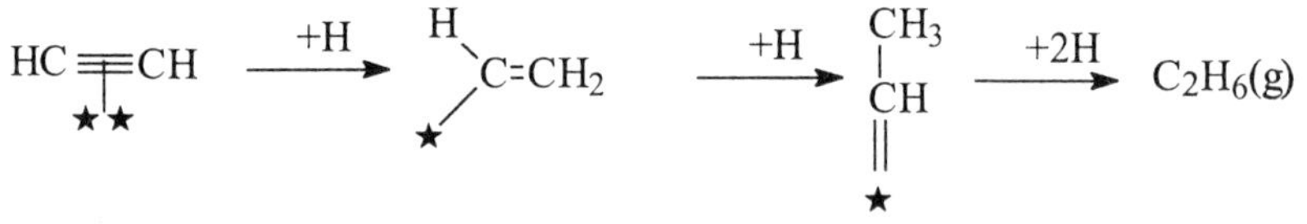

Figure 12. Proposed mechanism for ethane formation in the non-selective hydrogenation of ethyne over Pd

It is as though a sudden switch occurs in the reactivity of adsorbed-vinyl as the instantaneous hydrogen coverage increases beyond a certain point, giving adsorbed-C_2H_4 in a form (adsorbed ethylidene) that cannot desorb and which is rapidly hydrogenated to ethane.

5.4   Isomerisation of the reactant

Of the various relevant isomers of C_4H_6, buta-1,3-diene is the most thermodynamically stable. Thus, in the hydrogenations of the higher energy isomers, it is to be expected that some reactant isomerisation might occur to the detriment of reaction selectivity. In buta-1,2-diene hydrogenation over Ni powder at 348 K, 3% of the product was buta-1,3-diene [64]. The overall effect of this isomerisation (which D-tracing showed to be largely an intramolecular H-shift) was to enhance the very small yield of *trans*-but-2-ene that was formed directly from buta-1,2-diene. By contrast, over Pd/alumina buta-1,2-diene was isomerised to but-2-yne during hydrogenation [54]. The reverse isomerisation occurs in but-2-yne hydrogenation over Group 8-10 metals; it is less evident because the buta-1,2-diene seldom desorbs in analytical quantities but this isomerisation forms the basis of another interesting selectivity pattern. The almost completely selective conversion of but-2-yne to *cis*-but-2-ene over Pd has been noted. Slightly lower selectivities to *cis*-alkene are given by Ni and Pt which provide small yields of but-1-ene (1 to 5%); the metals of Group 9 give more but-1-ene (5 to 10%), and the metals of Group 8 more again (15 to 25%) [52,65]. D-tracer studies showed that the *cis*-but-2-ene contained, on average, about 2 D-atoms per molecule, as expected, but that the but-1-ene contained 3 D-atoms per molecule. This was consistent with the isomerisation of but-2-yne to buta-1,2-diene of overall composition C_4H_5D (the H-atom transfer not being intramolecular in this case) and subsequent addition of two D-atoms to give but-1-ene of overall composition $C_4H_5D_3$. The locations of deuterium in the product expected from this mechanism were confirmed by NMR. Thus the potentially high selectivity for *cis*-alkene formation in but-2-yne hydrogenation is diminished by reactant isomerisation, the effect being most pronounced at Group 8; Scheme VII describes this situation.

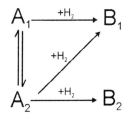

Scheme VII

Selectivity in but-2-yne hydrogenation over Groups 8-10 metals except Pd is further diminished by small yields (~5%) of *trans*-but-2-ene; in tracer experiments its D-content is intermediate between that of the *cis*-but-2-ene and that of the but-1-ene. This product arises partly by direct hydrogenation of the free-radical form of the half-hydrogenated state (giving $C_4H_6D_2$) and partly from buta-1,2-diene hydrogenation (giving $C_4H_6D_3$) [65]. These processes occur in addition to those shown in Scheme VII.

5.5   A Remaining Puzzle.

Despite the intensive studies that have been undertaken there is still a need for clarification of the behaviour of adsorbed-alkene formed in alkyne and alkadiene hydrogenations. The general pattern of butane formation in the hydrogenation of alkynes and alkadienes over the Group 8-10 metals is similar, but D- and ^{14}C-tracer evidence for the participation of alkylidene intermediates exists only for reactions of molecules having at least one *sp*-carbon atom (*i.e.* alkynes and alka-1,2-dienes). There is a problem here. Either alkylidene intermediates participate in buta-1,3-diene hydrogenation in the direct butane-forming step but have gone undetected, or the participation of alkylidene species is restricted to the conditions of sudden selectivity breakdown, and some other property of the adsorbed intermediates is crucial in determining the general selectivity pattern. Evidence was presented above that the alkane forming step is associated with the presence of disorder in the metallic active phase. Such disorder may be associated with an increased concentration of surface defects such as kinks and steps in the crystal surface. The possibility exists that supported metals contain sites for both strong and weak adsorption of alkene; at the former the chances of hydrogenation would be high and of desorption low, and at the latter desorption would occur in preference to hydrogenation. If, for the sake of argument, adsorbed alkene desorbed from weakly binding terrace sites but was further hydrogenated at more strongly binding kink sites, then this could be subjected to experimental enquiry by promoting catalysts with adsorbates which are known (from surface science studies) to adsorb selectively at kink sites. Thus the tools are available to enable the role of surface morphology in selective catalysis to be further investigated, and this approach may well provide the next step forward in our understanding of these formally simple but molecularly complex processes.

6.   ENANTIOSELECTIVITY

Hydrogenations which give optically active products constitute a special class of selective reactions, and the creation of catalyst surfaces which guide reactions so as to provide one enantiomer in substantial excess over the other is a particular challenge. Legislative requirements that chiral compounds for use as pharmaceuticals and agrochemicals should be pure enantiomers and not racemates has stimulated much research in this area, and industrial targets have spurred the development of both heterogeneous and homogeneous catalysts for this purpose.

6.1   The historical path

The small polycrystalline metal particles that constitute the active phase in supported metal catalysts possess no overall chiral quality, and hence the hydrogenation of a pro-chiral compound gives racemic product. The surfaces of these small metal particles consist of low index terraces, steps, and kinks [66], of which terraces and steps have no inherent chiral quality. Kinks can exist as mirror image structures, but in randomly grown microcrystals the enantiomorphs are present in equal numbers and so the metal surface as a whole has no overall chiral quality. The generation of adsorption sites that provide an asymmetric environment for an adsorbate can be achieved in principle by either of two routes. First, use of natural chiral solids as supports might cause metal particles grown at its surface to contain an excess of one type of kink site over another. Alternatively, the adsorption of a chiral auxilliary onto a metal surface which might cause a prochiral molecule adsorbing at an adjacent site to be exposed to a chiral environment.

In 1932 Schwab and Rudolph explored the first alternative when they supported the Group 10 metals on cleaved quartz; these catalysts showed enantiomer discriminations of up to 10% in the dehydration of secondary butanol [67]. This was followed by studies of metals supported on silk (Akabori,1956), polysaccharides (Balandin,1959) and cellulose (Harada, 1970) [68,69]. Although Pd/silk gave values of the enantiomeric excess (ee = $\{|[R] - [S]|\}/\{[R] + [S]\}$) as high as 66% these materials were not sufficiently well characterised to establish beyond doubt that they were functioning solely as heterogeneous catalysts. This methodology has not, so far, provided a positive way forward..

A second approach originated in various studies undertaken by Erlenmeyer in the 1920s. First, he reported an enantiomeric excess of 50% in the bromination of cinnamic acid catalysed by ZnO modified by fructose [70] and later a positive result was obtained in the bromination of the hydrocinchonine salt of cinnamic acid in homogeneous solution [71]. The latter observation prompted Lipkin and Stewart to investigate whether a hydrocinchonine salt would modify a Pt surface so as to induce enantioselective hydrogenation, and in 1939 they reported the enantioselecive hydrogenation of hydrocinchonine β-methylcinnamate to β-phenylbutyric acid catalysed by Adams Pt [72]. This chemistry was rediscovered by Orito, Imai, and Niwa in 1978 [73-75] and has formed the foundation for the extensive studies of noble metals modified by cinchona alkaloids that have appeared over the last decade and in which values of the enantiomeric excess of up to 98% have been reported in α-ketoester hydrogenations over cinchona-modified Pt (Table 5A) [76,77]. Almost coincident with Lipkin and Stewart's report was that of Nakamura, who showed in 1940 that modification of metals by chiral acids provides an enantioselective outcome in β-diketone hydrogenation [78], and this was taken up first by Izumi, who developed the Raney Ni/(tartaric acid)/NaBr catalyst for methylacetoacetate hydrogenation [79], and subsequently by Harada, Tai, and Sugimura [80,81] who have refined the system to the point where an enantiomeric excess of 98% has been reported in the hydrogenation of 3-cyclopropyl-3-oxopropanoate [87,83].

Both the (chiral organic acid)/(base metal) systems and the alkaloid/(noble metal) systems have been reviewed [79-81,84-88]. The former represents catalysts of great complexity in that performance is strongly dependent on catalyst preparation variables, catalyst modification variables, reactant variables, and reaction condition variables [81]. In the hands of experts these catalysts are very effective, but the less-expert (the author addresses himself) must be warned that much practical experience is necessary if optimum results are to be obtained with consistency. Three sources of difficulty in this system are (i) that Raney Ni when modified with tartaric acid is both physically and chemically an exceedingly complex material, (ii) that when modified it is a very delicate material, and (iii) that adsorption of the chiral auxillary is corrosive. By comparison, the alkaloid/(noble metal) systems are much more robust and better defined. Conventional, well characterised supported metal catalysts can be used, alkaloid adsorption is not corrosive and can be investigated by a variety of instrumental techniques, and reproducibility from laboratory to laboratory is good. However, the underlying principles are common to both systems and they will be illustrated here with reference to alkaloid/(noble metal) catalysts.

6.2   The Orito reaction

Pt catalyses the hydrogenation of α-ketoesters to α-hydroxy esters in a variety of protic and aprotic solvents at room temperature; elevated hydrogen pressure accelerates the rate but is not a requirement for measurable reaction. Addition of a cinchona alkaloid such as cinchonidine or quinine (Figure 13) as a *modifier* to the hydrogenation of a pyruvate ester, say, MeCOCOOEt, results in the formation of the corresponding lactate, MeC*H(OH)COOEt, in which the R-enantiomer predominates (Table 5A) and so the product is optically active.

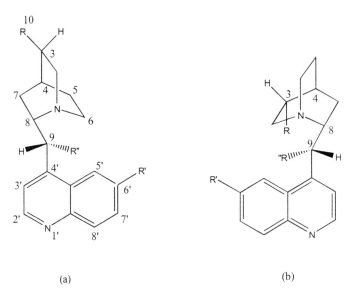

Figure 13. Structures of simple cinchona alkaloids.

By contrast, addition of the near enantiomers cinchonine or quinidine results in preferential formation of S-ethyl lactate; the product is again optically active with a rotation in the opposite sense. Values of the enantiomeric excess in the region of 70% can be obtained using ethanol as solvent but with no other optimisation [89], and any conventional supported Pt catalyst having a mean metal particle size of 1.5 nm or above is effective. Improvements can be achieved by hydrogenation of the vinyl group at C_3 and by substituting –OH for –OMe at C_9 and by optimising the solvent (Table 5, entries 3 and 5).

Three remarkable features of the reaction are (i) that enantioselective reaction shows a 10 to 50-fold rate enhancement over racemic reaction (*i.e.* reaction in the absence of alkaloid) [90], (ii) minute quantities of alkaloid suffice, considerable enantioselectivity being obtained when the Pt surface is only partially covered with alkaloid [91], and (iii) enantioselectivity is lost at temperatures above about 45°C [90].

Table 5. Values of the enantiomeric excess (ee) observed in hydrogenations at ambient temperature and elevated pressures over Pt and Pd catalysts modified by natural alkaloids and simple derivatives

A: pyruvate ester hydrogenations catalysed by cinchona-modified Pt;

B: other hydrogenations catalysed by cinchona-modified Pt or Pd

C: various hydrogenations catalysed by Pt or Pd modified by other alkaloids

	Catalyst	Modifier[a]	Reactant[b]	Solvent	ee /%	Ref
A	Pt/silica	CD	MePy	ethanol	66-74(R)	89
	Pt/alumina	CN	EtPy	ethanol	56(S)	107
	Pt/silica	HCD	MePy	ethanol	80(R)	90
	Pt colloid[c]	CD	MePy	ethanol	98(R)	77
	Pt/alumina	(MeO)HCD	EtPy	acetic acid	98(R)	76
	Pt/silica	epiquinidine	MePy	ethanol	0	65,103
	Pt/alumina[d]	Bz(CD)Cl	EtPy	ethanol	0	107
B	Pt/silica	CD	B-2,3-dione	ethanol	80(R)[e]	87
	Pt/alumina	CD	PDM-acetal	acetic acid	96(R)	130,131
	Pt/alumina	CD	PhCOCF$_3$	CH$_2$Cl$_2$	56(R)	132
	Pd/alumina	CD	Tiglic acid	toluene	45(S)	133
	Pd/carbon	CD	MePyrazine-C	ethanol	24(S)	87,138
	Pd/carbon	CN	MePyrazine-C	ethanol	14(R)	87,138
C	Pt/silica	brucine	MePy	ethanol	20(S)	103,123
	Pt/silica	codeine	MePy	ethanol	5(S)	123
	Pt/alumina	DHV	EtPy	methanol	30(S)	126
	Pd/alumina	ephedrine-A	PA-oxime	ethanol	26(S)	137
	Pd/carbon	ephedrine-B	MePyrazine-C	ethanol	17(S)	87,138

[a]Modifiers: CD = cinchonidine; CN = cinchonine; HCD = 10,11-dihydrocinchonidine; (MeO)HCD = HCD having methoxy substitutent at C$_9$; Bz(CD)Cl = benylcinchonidinium chloride; DHV = dihydrovinpocetine; ephedrine-A = 1R,2S-ephedrine; ephedrine-B = 1R,2R-ephedrine; ephedrine = PhCH(OH)CH(Me)NHMe.

[b]Reactants: MePy = methyl pyruvate; EtPy = ethyl pyruvate; B-2,3-dione = butane-2,3-dione; PDM-acetal = pyruvaldehyde dimethyl acetal; MePyrazine-C = methyl pyrazine-2-carboxylate; PA-oxime = pyruvic acid oxime.

[c]polyvinylpyrolidone-stabilised Pt colloid of mean particle size 1.4 nm.

[d]2%(R) observed over Pt/silica in MePy hydrogenation [65].

[e]at high conversion, see [87,127]

6.3 Alkaloid adsorption

The adsorption of 10,11-dihydrocinchonidine on Pt(111) has been studied by XP and NEXAFS spectroscopies and by LEED. XPS and LEED showed that high coverage is achieved, that adsorption is not ordered, and that the expected stoichiometry was retained up to 275°C [92]. The angular dependence of the N *K*-edge spectra demonstrated that the quinoline ring was oriented parallel to the Pt surface at room temperature but inclined at about 60° at 50°C [93]. Exchange between this alkaloid and D_2 over Pt/silica at room temperature showed that H-atoms at all positions in the quinuclidine ring system were exchangeable for D-atoms and that the OH-group at C_9 rapidly became OD; however, there was no exchange in the quinuclidine ring system [94]. Thus studies using both single crystal Pt surfaces and supported Pt microcrystallites confirm that these alkaloids adsorb at room temperature by interaction of the π-electron system of the quinoline moiety with that of the Pt surface. Isotherms for the adsorption of cinchonidine from ethanol onto Pt/silica at room temperature show that adsorption occurs on both the metallic active phase and on the support [89]. Adsorption is strong but can be largely though not completely reversed by assiduous washing [95].

6.4 The alkaloid-reactant interaction

Studies of the manner in which enantiomeric excess increases with increasing loading of cinchonidine, and the manner in which enantioselectivity changes as R-directing alkaloids are mixed with S-directing cinchona alkaloids, together showed that the phenomenon of enantioselectivity was a result of a 1:1-interaction of alkaloid molecules with reactant molecules in the adsorbed state [96]. [This superseded an earlier proposal that the enantioselective site resulted from a templating of the surface by the alkaloid in which two or three modifier molecules interacted with each pyruvate molecule [85,89]]. Molecules of α-ketoester may adsorb on a planar metal surface by either of their enantiofaces, as shown in Figure 14.

Those adsorbed as shown in A give R-product on hydrogenation whereas those adsorbed as B give S-product. Provided no kinetic factor operates, *i.e.* if $k_2^A = k_2^B$, (for a discussion, see [97]) then the observed enantiomeric excess is thermodynamically determined and is given by $(\theta_A - \theta_B)/(\theta_A + \theta_B)$. Let us suppose that the adsorption of alkaloid molecules generates enantioselective sites in their vicinity, and that α-ketoester molecules adsorb at these sites, some as enantioface A with a free energy change $-\Delta G^\circ_{ads,A}$ and some as enantioface B with a free energy change $-\Delta G^\circ_{ads,B}$, then, provided adsorption equilibrium is achieved, the relative surface coverages, θ_A/θ_B, will be a direct function of $\exp(-\delta\Delta G^\circ_{ads}/RT)$ where $\delta\Delta G^\circ_{ads}$ is the difference in the free energies of adsorption of the ester as enantiomers A and B. Thus a chiral environment that distinguishes only modestly between A and B in energy terms will provide a high (or low) value of θ_A/θ_B (because of the exponential quality of the expression) and hence a substantial enantiomeric excess.

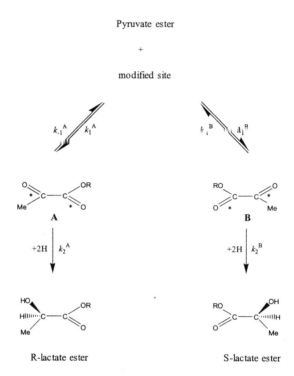

Figure 14. Adsorption of pyruvate ester by each of its enantiofaces and the consequent formation of R- and S-lactate ester by hydrogenation

Where the outcome of reaction is thermodynamically controlled in this way, enantioselectivity is a direct result of *selective enantioface adsorption.* Racemic reaction will, of course, occur at other adjacent Pt metal atom sites that have no chiral quality because they are not appropriately disposed with respect to the adsorbed modifier molecules. Whereas this can be neglected, to a first approximation, because the reaction rate at enantioselective sites is 10 to 50 times that at sites that give racemic product, it is the case that such racemic reaction always degrades the observed enantioselectivity. [N.B. It cannot be neglected in those classes of reaction, such as unsaturated acid hydrogenation (see below), where there is no rate enhancement at enantioselective sites].

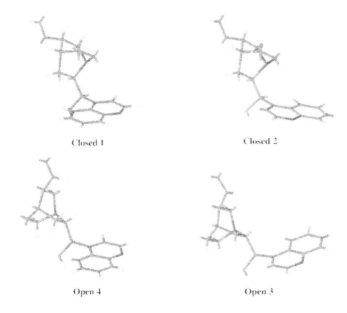

Closed 1

Closed 2

Open 4

Open 3

Figure 15. The four lowest energy confirmations of cinchoinidine

The enantioselective site has been envisaged by (i) computing the minimum energy conformations of the cinchona alkaloids, (ii) modelling the 1:1-interaction of alkaloid with pyruvate ester, and (iii) by making the assumption that all relative energies remain valid for the alkaloid-pyruvate complexes in the adsorbed state [96]. The alkaloids undergo rotation about the carbon-carbon single bonds at $C_{4'}$-C_9, C_9-C_8, and C_3-C_{10}, and about C_9-O (Figure 13), the first two rotations dominating the total energy. Minimum energy conformations are identified from maps of potential energy which respond to the variation of each torsion angle. Such energy minimisation calculations for cinchonidine reveal the presence of four conformations of lowest energy [96,98-102] which are shown in Figure 15. These are described as 'open' or 'closed' depending on whether the lone pair on the quinuclidine-N points away from, or towards, the quinoline ring. For present purposes, the conformation 'open-3' is of crucial importance because, in this state, the lone pair points into the cleft formed by the quinoline and quinuclidine ring systems. Docking of pyruvate ester with cinchonidine in the open-3 conformation can be computed for these entities in vacuo for any desired relative positions, and can be envisaged on a Pt(111) surface as shown in Figure 16.

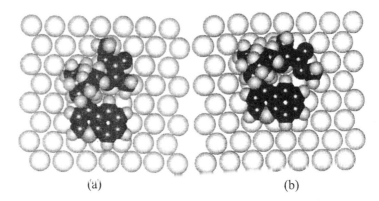

(a)   (b)

Figure 16. Methyl pyruvate adsorbed by each of its enantiofaces at the enantioselective site adjacent to cinchonidine adsorbed in the open-3 configuration on Pt(111). Upon hydrogenation (a) gives R-lactate and (b) gives S-lactate

[The calculations do not take the surface into account. Assumptions as to the mode of adsorption are made, *i.e.* that alkaloid is adsorbed to Pt atoms located beneath the centres of the two aromatic rings and that pyruvate is adsorbed to Pt atoms located beneath the two carbonyl groups [96]]. In Figure 16a, methyl pyruvate is adsorbed as enantioface A of Figure 14 and gives R-lactate ester on hydrogenation, whereas it is adsorbed as enantioface B in Figure 16b and gives S-lactate on hydrogenation. Calculations show that the interaction energy for the situation in Figure 16b exceeds that in Figure 16a [96], and it is therefore expected that the yield of R-lactate will exceed that of S-lactate. That is, the observed sense of the enantioselectivity is correctly predicted. Corresponding calculations for cinchonine give the reverse result of this expectation, the same procedures and arguments leading to an expectation in agreement with experiment that enantioselectivity should favour S-lactate ester. No conditions for selective enantioface adsorption of pyruvate ester are evident for 1:1-interactions with cinchonidine or cinchonine in open-4 or closed-1 or closed-2 conformations.

Epiquinidine, a cinchona-type alkaloid in which both C_8 and C_9 have the R-configuration, provides no minimum energy conformation in which the lone pair on the quinuilclidine-N points into a cleft, and no 1:1-interactions with pyruvate ester exist which appear to lead to selective enantioface adsorption [103]. This alkaloid is, therefore, not expected to be an effective modifier of Pt in pyruvate hydrogenation, and in fact gives racemic lactate as product (Table 5, entry 6) with very little rate enhancement [85,103].

Because enantioselectivity and rate enhancement appear to go hand in hand, it should be possible to interpret the latter by reference to the 1:1-interactions exemplified in Figure 16. The Figure shows that the α-keto group that undergoes reduction is in close proximity to the quinuclidine-N atom. Now, it is known that racemic pyruvate hydrogenation is accelerated in the presence of achiral N-bases [104,105] and that the extent of such acceleration is related to the pK_a of the base. However, the 10 to 50-fold rate enhancement that accompanies pyruvate hydrogenation substantially exceeds that attributable to the basicity of the cinchona alkaloids. The author's group has offered an interpetation based on H-bonding effects [85,105], whereas others have preferred to clothe their interpretation in the formalism of ligand acceleration of reaction rates [106]. D-tracer studies show that pyruvate ester hydrogenation occurs by the simple addition of two D-atoms to the carbon-oxygen double bond [90], and this is conceived (as in carbon-carbon double bond saturation) as occurring by consecutive addition, the first addition being reversible and the second being the rate-determining step. Molecular mechanics calculations show that the half-hydrogenated state formed by the addition of a H-atom to the O-atom of the carbonyl group leads to an energetically favourable state in which that H-atom is involved in H-bonding to the quinuclidine-N atom of the adsorbed modifier. Such H-bonding would stabilise the half-hydrogenated state, increasing its lifetime and hence its concentration, and an increase in the concentration of the organic species involved in the rate-determining step would result in an increase in reaction rate.

It follows from these arguments that quaternisation of the quinuclidine-N in any cinchona alkaloid should have a catastrophic effect on enantioselective reaction. First, the presence of a fourth substituent on the quinuclidine-N would prevent the 1:1-interaction shown in Figure 16, and second, the mechanism for rate enhancement would be invalid because H-bond formation could not occur. In practice, enantioselectivity is zero (Table 5, entry 7), and rate-enhancement is slight [85, 107].

This model therefore provides a self-consistent interpretation of the three principal observations (i) the sense of the enantioselectivity in relation to alkaloid structure, (ii) the rate enhancement, and (iii) the effect of quaternisation on enantioselectivity and reaction rate.

It should be noted that Margitfalvi and co-workers have advanced an alternative 'chemical shielding' model in which it is proposed that alkaloid-reactant complexes are formed in solution and are in adsorption-desorption equilibrium with the Pt surface which supplies H-atoms for hydrogenation [108,109]. One might expect, if this mechanism was important, that the other platinum group metals which adsorb hydrogen efficiently would be equally efficient for enantioselective pyruvate hydrogenation; this is not the case.

6.5   Solvent effects

There is a substantial solvent effect in enantioselective pyruvate ester hydrogenation, the enantiomeric excess being higher in solvents of lower dielectric constant, ε, although acidic solvents are important exceptions. The origin of this effect lies in the influence of solvent on alkaloid conformation. Since only the open-3 conformation participates in the creation of the enantioselective sites, conditions in which its concentration is reduced will result in a reduced enantiomeric excess because Pt atoms adjacent to alkaloid adsorbed in the other low energy states will constitute sites for production of racemic product. Molecules of cinchonidine in the solid state exhibit the open-3 conformation [110] but, on dissolution, they populate the four low energy state conformations shown in Figure 15. Each low energy state has a different dipole moment and the relative concentrations of the four states is influenced by the dielectric constant of the reaction medium. Burgi and Baiker have calculated (by use of a reaction field model in combination with Hartree Fock and density functional calculations [98]) the electronic energies, Gibbs free energies, and dipole moments for the four lowest energy conformations of cinchonidine relative to that of open-3 in media of dielectric constant 2.0, 4.8, 20.7, and 78.5. They have shown, first, that open-4 is the highest energy conformation by such a margin that it can be neglected, and second, that the energies of closed-1 and closed-2 approach that of open-3 as dielectric constant increases. This is confirmed by ^{1}H NMR spectroscopy where, after assignment of the spectrum of an alkaloid in solution, the conformational balance can be deduced from inter-ring NOEs and quantification of certain JJ coupling constants. Such experiments show that the fraction of the open-3 configuration is 70% in toluene ($\varepsilon = 2.34$), 62% in tetrahydrofuran ($\varepsilon = 7.6$), 40% in acetone ($\varepsilon = 20.7$) and 30% in water ($\varepsilon = 78.5$) [98] and in hydrogenations the enantiomeric excess declines with increasing dielectric constant of the reaction medium [98]. Exceptional behaviour is shown by solutions in acetic acid where the alkaloid molecules are protonated at the quinuclidine-N atom and the concentration of open-3 is almost 100%, and by solutions in ethanol for which the value is unexpectedly high at 77%. The latter result has been attributed to H-bonding between the alkaloid and solvent, but any such effect would be compounded if small quantities of acetic acid were generated at the Pt surfaces by ethanol oxidation in the presence of dissolved air.

It is thus clear that the highest enantioselectivities in α-ketoester hydrogenation are to be obtained in solvents of low dielectric constant, or those that contain free organic acid.

6.6 Importance of adventitious oxygen

Experiments in which catalysts were modified in a separate stage before use in a reactor (as in Orito's original study) have shown that, under strictly anaerobic conditions, alkaloid-modified catalysts are poisoned and that little or no hydrogenation, enantioselective or racemic, is achieved [111]. However, where modification is carried out in the presence of oxygen, or air, or nitrous oxide (aerobic modification) enantioselectivity is achieved normally. This suggests that co-adsorption of alkaloid and oxygen is required to generate the surface conditions appropriate for enantioselective reaction, adsorbed-O being removed as water during the early stages of reaction. Possibly, in the absence of a competitor adsorbate the alkaloid achieves such high coverage that it poisons the surface. These proposals were tested by examining competitor adsobates of a very different chemical type. It was found that catalyst modification by co-adsorption of cinchonidine with propyne or buta-1,3-diene was effective (anaerobic modification), giving active enantioselective catalysts and reactions in which the C_3- or C_4- co-adsorbate was stripped from the surface as hydrogenated products in the early stages of pyruvate ester hydrogenation. Thus, it is most probable that, when investigators normally charge solvent, reactant, and alkaloid into a reactor in a single operation, and subsequently pressurise with hydrogen, they adventitiously include sufficient dissolved air to achieve aerobic modification, and such reactions are normally conducted in the presence of traces of water.

6.7 Synthetic modifiers

The cinchona alkaloids have complex structures and it is natural to enquire whether or not comparable enantioselectivties could be obtained by use of simpler synthetic compounds as modifier. From the foregoing paragraphs it is evident that the cinchonas function because they contain an aromatic 'anchor' to provide strong adsorption and a basic-N atom to interact with the reactant, these functions being separated by two carbon atoms at least one of which is chiral. Baiker, Pfaltz and co-workers have synthesised and tested a variety of compounds which conform to these requirements [88,112-115] of which those shown in Figure 17 are typical.

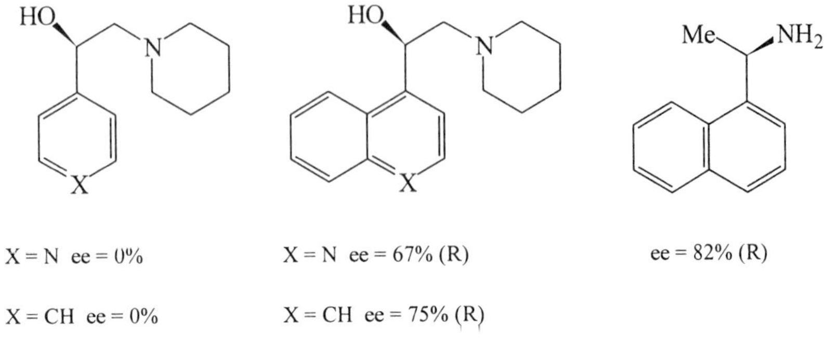

X = N ee = 0%   X = N ee = 67% (R)   ee = 82% (R)

X = CH ee = 0%   X = CH ee = 75% (R)

Figure 17. Performance of synthetic modifiers in the enantioselective hydrogenation of ethyl pyruvate

Clearly, the naphthalene and quinoline ring systems provide a sufficient anchor whereas benzene and pyridine ring systems do not. This is consistent with the observation that 5',6',7',8',10,11-hexahydrocinchonodine was less effective than 10,11-dihydro-cinchonidine [116] and that use of histidine and ephedrine as modifiers provides values of the enatiomeric excess of only a few percent [96]. Anthracene provides an even more effective anchor, but the triptycene analogue was ineffective (ee = 0%), showing that planarity in the aromatic anchor is an additional requirement [117]. The compound 1-(1-naphthyl)ethylamine also directed reaction enantioselectively, giving an enantiomeric excess of up to 82%, notwithstanding the fact that only one carbon atom separates the aromatic system and the basic-N atom. In this case the compound added to the reaction was not the active modifier; it underwent alkylation by ethyl pyruvate and subsequent reduction to give the true modifier NpCHMeNHCMeCOOEt (N-[1'-(1-naphthyl)ethyl]-2-aminopropionic acid ethyl ester), which was the effective agent [118-120]. In this species, only the chiral carbon atom adjacent to the naphthalene anchor is effective in determining the sense of the observed enantioselectivity. Alkylation of naphthylethylamine with other ketonic compounds gave a variety of other modifiers. The authors demonstrated that their modelling procedures, used originally to describe the crucial cinchona-pyruvate interaction, apply *mutatis mutandis* to the modelling of all these synthetic modifiers, in each case the sense of the observed enantioselectivity being correctly predicted [88].

6.8   The wider picture

In the 1980s the Orito reaction appeared almost enzymic in its specificity; any departure from Pt as catalyst, or from cinchona alkaloids as modifiers, or from α-ketoesters as reactants, appeared to result in an almost total loss of enantioselectivity. These apparent restrictions have now been overcome to a considerable degree. Thus, Ir and Ru have been rendered enantioselective for pyruvate hydrogenation but only when prepared under very specific conditions [111,121]; Pd is active but catalyses the reaction by a more complex mechanism directing the reaction in the opposite sense to that stated above for Pt [122]. Members of the strychnos, morphine, and vinca families (Figure 18) of alkaloids modifiy Pt, rendering it enantioselective for α-ketoester hydrogenation (Table 5C), but they are less effective than the cinchona alkaloids [87,123-126].

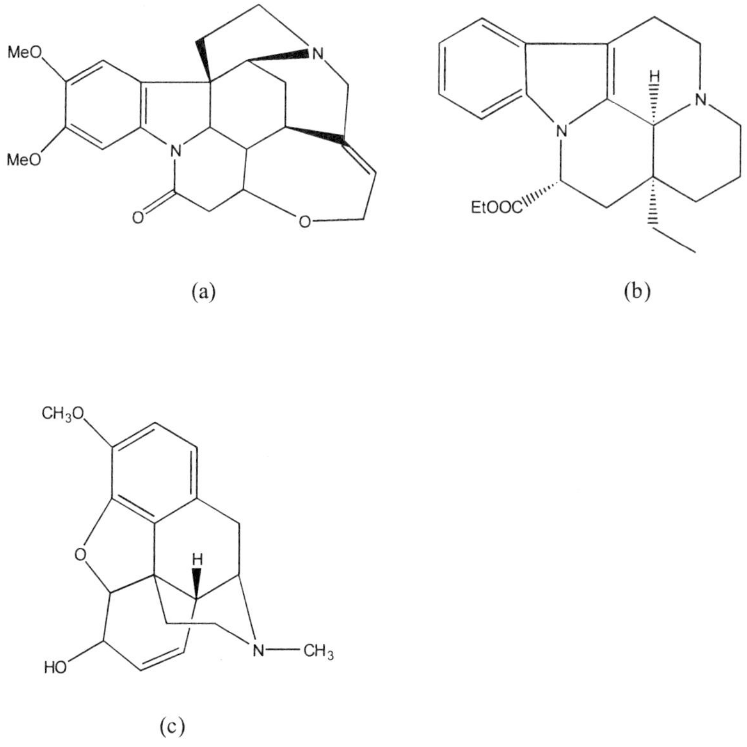

(a)   (b)

(c)

Figure 18. Structures of (a) brucine (a strychnos alkaloid), (b) dihydrovinpocetine (a vinca alkaloid), and (c) codeine (a morphine alkaloid)

They each contain an aromatic ring to provide for adsorption to the active surface but, by comparison with the cinchonas, they are rigid molecules and cannot respond conformationally to the requirements of the 1:1 modifier-reactant interaction. And finally, as Table 5B indicates, cinchona-modified

Pt or Pd catalyses the enantioselective hydrogenation of (i) a variety of diketones such as butane-2,3-dione [127] and ketopantolactone [97,128,129] and activated monoketones such as pyruvaldehyde dimethyl acetal MeCOCH(OMe)$_2$ [130,131] and PhCOCF$_3$ [132]), (ii) α,β-unsaturated acids such as tiglic acid ((E)-2-methyl-2-butenoic acid) and phenylcinnamic acids by saturation of the >C=C< bond [122, 133-135], (iii) oximes such as pyruvic acid oxime by saturation of the >C=N– bond [136,137] and (iv) heterocyclic compounds such as methyl pyrazine-2-carboxylate by the stereocontrolled saturation of the heterocyclic ring system [87,138].

The Pd-catalysed reactions of the α,β-unsaturated acids, just mentioned, merit further brief comment. The se molecules each form a H-bond to the quinuclidine-N atom of an adsorbed alkaloid molecule to form a precursor state. For this reason, the esters cannot be enantioselectively reduced, and once again quaternised alkaloids are ineffective as modifiers. Molecules in the precursor state then undergo selective enantioface adsorption as either the monomer [122] or the dimer [133] and thereafter are hydrogenated to product, cinchonidine providing the S-product in excess and cinchonine the R-product in excess, as predicted by the model [87]. However, since the carbon-carbon double bond undergoing hydrogenation is distant from the quinuclidine-N atom there is no rate enhancement. Thus, reaction at the enantioselective sites has no advantage over that at the sites that give racemic product, and consequently the values of the enantiomeric excess are more modest than those observed in α-ketoester hydrogenation, the best values being in the range 35 to 70%.

6.9   Inherently chiral metal surfaces

Since kinks confer local chirality on metal surface structure (see above), it is of considerable importance to determine whether single crystal surfaces having high concentrations of kink sites catalyse enantioselective reaction in the absence of an organic chiral directing agent. For example, the (321), (531) and (643) planes of face-centred cubic crystals have the required kinked structures and, depending on the angle at which the crystal is cut, so the kinks exposed are enantiomorphic (*i.e.* (321)R and (321)S etc). The Cahn-Ingold-Prelog system of nomenclature can be applied defining substitutent priority in terms of atomic surface density, *i.e.* (111) > (100) > (110). Thus, when the component (111), (100), (110) planes occur in a clockwise rotational sequence as viewed from the fluid phase, the kink possesses the R-configuration, and when they occur in an anticlockwise rotational sequence the kink has the S-configuration as shown for Pt(321) in Figure 19 [141].

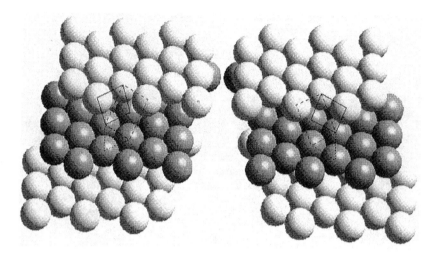

Figure 19. The configurations of the R-kink site on Pt(321)R (left) and the S-kink site on Pt(321)S (right).

Attard and co-workers have prepared a number of chiral electrode surfaces and have reported molecular recognition in the adsorption and electro-oxidation of D- and L- glucose [139-141]. Figure 20 shows the two pairs of asymmetry-controlled responses at Pt(321) - those for D-glucose oxidation at the R-surface and for L-glucose oxidation at the S-surface are identical, and those for L-glucose oxidation at the R-surface and for D-glucose oxidation at the S-surface are different from the first pair but again identical to each other. This behaviour is directly ascribable to the inherent left- or right-handedness of the kink sites at the surfaces. Surfaces with higher kink densities gave stronger discrimination. Pt(332) and Pt(211) which are stepped surfaces having no kink features showed no discrimination. This work has been seminal in establishing cyclic voltammetry as a technique both for characterisation of chiral surfaces and for detecting molecular recognition at chiral surfaces, and demonstrating chiral catalysis.

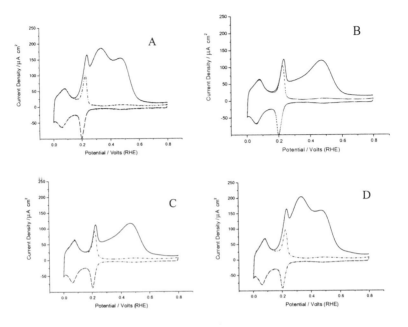

Figure 20. Cyclic voltammograms of glucose electro-oxidation on Pt(321). (A) L-glucose on Pt(321)R; (B) D- glucose on Pt(321)R; (C) L-glucose on Pt(321)S; (D) D-glucose on Pt(321)S. The pairs of asymmetry-controlled responses are (A) and (D), and (B) and (C). [Conditions: electrolyte = 0.005M glucose in 0.05M H_2SO_4; sweep rate = 50 mV s^{-1}]

Moreover, cyclic voltammetry is applicable to the characterisation of polycrystalline metal catalysts supported on conducting solids such as graphite [142] and is capable of providing information as to whether alkaloids adsorb preferentially on one type of site rather than another at polycrystalline metal surfaces [143]. The next challenge is to create stable high area (supported) chiral metal surfaces that sustain enantioselective reaction *per se*.

7.   IN CONCLUSION

This account has shown that selectivity in metal-catalysed hydrogenations is influenced by the electronic characteristics of the catalyst surface and by all aspects of the environment of the active sites. Careful studies of reaction mechanism are essential and, even in this age of direct spectroscopic interrogation of the surface, isotope tracer techniques remain extremely valuable as a means of identifying the important molecular pathways to products. The environment of active sites has many aspects, involving conditions above the site, in the plane of the site, and below the site. And the situation becomes the more demanding when the requirement is for sites having a chiral environment. The small metal crystallites that typify the active phases of supported metals have detailed morphologies involving terraces, steps, and kinks, which have not yet been systematically exploited as an aid to improving catalyst selectivity. To do so is the next major objective of catalyst design; to achieve it will require wide participation of researchers from the catalysis and surface science community.

8. ACKNOWLEDGEMENTS

The author wishes to acknowledge his gratitutde to Professors Geoffrey Bond and Dennis Dowden who introduced him to the delights of selective hydrogenation, to Dr. Richard Moyes with whom he shared over forty years of research and to Dr. Arthur Ibbotson who initiated his work in enantioselective hydrogenation. Acknowledgement is also due to his many colleagues and students at the University of Hull, and now at Cardiff University, whose arguments have provided so much stimulation. Gratitude is also due to the many funding bodies that have supported the author's research, in particular EPSRC and its predecessors, and many industrial companies of which Johnson Matthey and ICI have been to the fore in their support of selective hydrogenation.

9. REFERENCES

1   K. Campbell and S.J. Thomson, *Prog. Surf. Membrane Sci.,* 9 (1975) 163.
2   G.A. Somorjai *in* 'Introduction to Surface Chemistry and Catalysis' Wiley, New York, (1994) p. 507.
3   J.J. Phillipson, P.B. Wells and G.R. Wilson, *J. Chem. Soc. (A),* (1969) 1351.
4   A.J. Bates, Z.K. Leszczynski, J.J. Phillipson, P.B. Wells and G.R. Wilson, *J. Chem. Soc. (A)*, (1970) 2435.
5   D.A. Dowden, *Bull. Soc. Chim. Belg.,* 67 (1958) 439.
6   G. C. Bond, G. Webb, P.B. Wells and J.M. Winterbottom, *J.Chem. Soc.,* (1965) p.3218.
7   L.B. Smith and J.L. Massingil, *J. Am. Chem. Soc.,* 83 (1961) 4301.
8   P.B. Wells and A.J. Bates, *J. Chem.* Soc. *(A),* (1968) 3064.
9   A.G. Burden, J. Grant, J. Martos, R.B. Moyes and P.B. Wells, *Discuss. Faraday Soc.,* 72 (1981) 95.
10   R.B. Moyes, P.B. Wells, J. Grant and N.Y. Salman, *Appl. Catal.,* in press.
11   M. George, R.B. Moyes, D. Ramanarao and P.B. Wells, *J. Catal.,* 52 (1978) 486.

12   J. Grant, R.B. Moyes and P.B. Wells, *J. Catal.,* 51 (1978) 355.
13   F. Nozaki and R. Adachi, *J. Catal.,* 40 (1975) 166.
14   W. Palczewska *in* 'Hydrogen Effects in Catalysis' (eds. Z. Paal and P.G. Menon) Marcel Dekker, New York and Basel, (1988, P373.
15   M. Eyre, R.C. Hoodless, R.B. Moyes and P.B.Wells unpublished work.
16   K.Bauer and D. Garbe *in* "Ullmann's Encyclopaedia of Industrial Chemistry", 3rd Edition, VCH, New York, (1988) Vol. A11, p. 141.
17   H. Noller and W.M. Lin, *J. Catal.,* 85 (1984) 25.
18   J. Simonik and P. Beranek, *Coll. Czech Chem. Commun.,* 37 (1972) 353; *J.Catal.,* 24 (1972) 348.
19   B. Coq, F. Figueras, P. Geneste, C. Moreau, P. Moreau and M. Warawdekar, *J. Mol. Catal.,* 78 (1993) 211.
20   D.V. Sokolskii, A.K. Zharmagambetova and N.V. Anisilmova, *React. Kin. Catal. Lett.,* 30 (1986) 101.
21   D.V. Sokolskii, A.K. Zharmagambetova, N.V. Anisilmova and A. Ualikhanova, *Dokl. Akad. Nauk. SSSR* 273 (1983) 151.
22   R. Hubault, M. Daage, and J.P. Bonnelle, *Appl. Catal.,* 22 (1986) 231.
23   Z. Poltarzewski, S. Galvagno, R. Pietropaolo and P. Staiti, *J. Catal.,* 102 (1986) 190.
24   M.A. Vannice and B. Sen, *J. Catal.,* 115 (1989) 65.
25   T.B.L.W. Marinelli, J.H. Vleeming and V. Ponec, Proc. 10th International Congress of Catalysis, B, (1993) 1211.
26   T.B.L.W. Marinelli, S. Nabuurs and V. Ponec, *J. Catal.,* 151 (1995) 431.
27   T.B.L.W. Marinelli and V. Ponec, *J. Catal.,* 156 (1995) 51.
28   G.J. Hutchings, F. King, I.P. Okoye and C.H. Rochester, *Appl. Catal. A: General,* 83 (1992) L7.
29   M.B. Padley, C.H. Rochester, G.J. Hutchings, and F.King, *J. Catal.,* 148 (1994) 438.
30   G.J. Hutchings, F. King, I.P. Okoye, M.B. Padley and C.H. Rochester, *J. Catal.,* 148 (1994) 453.
31   G.J. Hutchings, F.King, I.P. Okoye, M.B. Padley and C.H. Rochester, *J. Catal.,* 148 (1994) 464.
32   S.S. Ashour, J.E. Bailie, C.H. Rochester, J. Thomson and G.J. Hutchings, *J. Mol. Catal. A: Chemical,* 123 (1997) 65.
33   (a) J.E. Bailie, G.J. Hutchings, H.A. Abdullah, J.A. Anderson and C.H. Rochester, *Phys. Chem. Chem. Phys.,* 2 (2000) 283. (b) J.E. Bailie and G.J. Hutchings, *J. Chem. Soc. Chem. Commun.,* (1999) 2151.
34   Y. Nitta, K. Ueno and T Imanaka, *Appl. Catal.,* 56 (1989) 9.
35   Y. Nitta, Y. Hiramatsu and T.Imanaka, *J. Catal.,* 126 (1990) 235.
36   A.N. Patil, M.A. Banares, X. Lei, T.P.Fehlner and E.E.Wolf, *J. Catal.,* 159 (1996) 458.
37   J.E. Bailie, C.H. Rochester and G.J. Hutchings, *J. Chem. Soc. Faraday Trans.,* 93 (1997) 4389.
38   M.K. Rajumon, M.W. Roberts, F. Wang and P.B. Wells, *J. Chem. Soc. Faraday Trans. I,* 94 (1998) 3699.
39   F. Delbecq and P. Sautet, *J. Catal.,* 152 (1995) 217.
40   P. Beccat, J.C. Bertolini, Y. Gauthier, J. Massardier and P. Ruiz, *J. Catal.,* 126 (1990) 451.
41   T. Birchem, C.M. Pradier, Y. Berthier and G. Cordier, *J. Catal.,* 146 (1994) 503.
42   D.G. Blackmond, R. Oukaci, B. Blanc and P. Gallezot, *J. Catal.,* 131 (1991) 401.
43   G.C. Bond, G. Webb, P.B. Wells and J.M. Winterbottom, *J. Catal.,* 1 (1962) 74.
44   G.C. Bond and P.B. Wells, *Advan. Catal.,* 15 (1964) 91.

45 P.B. Wells, *Plat. Met. Rev.*, 7 (1963) 18. See references therin.
46 P.B. Wells, *Chem. and Ind.*, (1964) p. 1742. See references therin.
47 P.B. Wells, *J. Catal.*, 52 (1978) 498.
48 G.C. Bond, J.J. Phillipson, P.B. Wells and J.M. Winterbottom, *Trans. Faraday Soc.*, 60 (1964) 1847.
49 G. C. Bond, G. Webb and P.B. Wells, *Trans. Faraday Soc.*, 64 (1968) 3077.
50 D. McMunn, R.B. Moyes and P.B. Wells, *J. Catal.*, 52 (1978) 472.
51 D. Bingham, D.E. Webster and P.B. Wells, *J. Chem. Soc. Dalton Trans.* (1974) 1519.
52 J.J. Phillipson, P.B. Wells and D.W. Gray, *Proc. 3rd Intern. Congr. Catal.*, 2 (1964) 1250.
53 W.M. Hamilton and R.L. Burwell Jr., *Proc. 2nd Intern. Congr. Catal.*, (1961) 987.
54 E.E. Meyer and R.L. Burwell Jr., *J. Amer. Chem. Soc.*, 83 (1963) 2881.
55 G.C. Bond and P.B. Wells, *J. Catal.*, 5 (1966) 65.
56 G.C. Bond and P.B. Wells, *J. Catal.*, 6 (1966) 397.
57 J. Sheridan, *J. Chem. Soc.*, (1945) p. 133.
58 G.C. Bond, *J. Chem. Soc.*, (1958) p. 4288.
59 G.C. Bond, G. Webb and P.B. Wells, *J. Catal.*, 12 (1968) 157.
60 H.A. Edwards, PhD thesis, Cardiff University (1996)
61 A.F. Carley, H.A. Edwards, B. Mile, M.W. Roberts, C.C. Rowlands, F.E. Hancock and S.D. Jackson, *J. Chem. Soc. Faraday Trans.*, 90 (1994) 3341.
62 V. Ponec and G.C. Bond *in* 'Catalysis by Metals and Alloys' *Stud. Surf. Sci. Catal.*, 95 (1995) p. 497.
63 E.A. Arafa and G. Webb, *Catal. Today,* 17 (1993) 411.
64 R.G. Oliver and P.B. Wells, *J. Catal.*, 47 (1977) 364.
65 G. Webb and P.B. Wells, *Trans. Faraday Soc.*, 61 (1965) 1232.
66 R. van Hardeveld and F. Hartog, *Surf. Sci.*, 15 (1969) 189.
67 G.M. Schwab and L. Rudolph, *Naturwiss.*, 20 (1932) 363.
68 S. Akabori, S. Sakurai, Y. Izumi, and Y. Fujii, *Nature*, 178 (1956) 323.
69 (a) A.A. Balandin, E.I. Klabunovski, and Y.I. Petrov, *Dokl. Akad. Nauk. SSSR*, 127 (1959) 557; (b) K. Harada and T. Yoshika, *Naturwiss.* 57 (1970) 306.
70 E. Erlenmeyer and H. Erlenmeyer, *Biochem. Zeitschr.* 233 (1922) 52.
71 H. Erlenmeyer, *Helv. Chim. Acta,* 13 (1930) 731.
72 D. Lipkin and T.D. Stewart, *J. Am. Chem. Soc.*, 61 (1939) 3295, 3297.
73 Y. Orito, S. Imai and S. Niwa, Collected papers of the 43rd Catalyst Forum, Japan, (1978) p.30.
74 Y. Orito, S. Imai and S. Niwa, *Nippon Kagaku Kaishi* (1979) p. 1118.
75 Y. Orito, S. Imai and S. Niwa, *Nippon Kagaku Kaishi* (1980) p. 670.
76 H-U. Blaser, H.P. Jalett and J. Wiehl, *J. Mol. Catal.*, 68 (1991) 215.
77 X. Zuo, H. Liu and M. Liu, *Tetrahedron Lett.*, 39 (1998)1941.
78 Y. Nakamura, *Bull. Chem. Soc. Japan,* 16 (1941) 367.
79 Y. Izumi, *Adv. Catal.*, 32 (1983) 215.
80 A. Tai and T. Harada *in* 'Taylored Metal Catalysts' (Ed. Y. Iwasawa) Reidel, Dordrecht, 1986, p.265.
81 A. Tai and T. Sugimura *in* 'Chiral Catalyst Immobilisation and Recycling' (Eds. D.E. de Vos, I.F.J. Vankelecom and P.A. Jacobs) Wiley-VCH, Weinheim, 2000, p.173.
82 S. Nakagawa, T. Sugimura and A. Tai, *Chem. Lett.*, (1997) p. 859.
83 S. Nakagawa, T. Sugimura and A. Tai, *Chem. Lett.*, (1998) p. 1257.
84 H-U. Blaser and M. Muller, *Stud. Surf. Sci. Catal.*, 59 (1991) 73.
85 G. Webb and P.B. Wells, *Catal. Today,* 12 (1992) 319.

86   A. Baiker and H-U. Blaser *in* 'Handbook of Heterogeneous Catalysis' Vol. 5 (Eds. G. Ertl, H. Knozinger and J. Wetkamp) Wiley-VCH, Weinheim, 1997, p.2422.

87   P.B. Wells and R.P.K. Wells *in* 'Chiral Catalyst Immobilisation and Recycling' (Eds. D.E. de Vos, I.F.J. Vankelecom and P.A. Jacobs) Wiley-VCH, Weinheim, 2000, p.123.

88   A. Baiker *in* 'Chiral Catalyst Immobilisation and Recycling' (Eds. D.E. de Vos, I.F.J. Vankelecom and P.A. Jacobs) Wiley-VCH, Weinheim, 2000, p.155.

89   I.M. Sutherland, A. Ibbotson, R.B. Moyes and P.B. Wells, *J. Catal.,* 125 (1990) 77.

90   P.A. Meheux, A. Ibbotson and P.B. Wells, *J. Catal.,* 128 (1991) 387.

91   G. Bond, K.E. Simons, A. Ibbotson, P.B. Wells and D A. Whan, *Catal. Today,* 12 (1992) 421.

92   A.F. Carley, M.K.Rajumon, M.W. Roberts and P.B. Wells, *J. Chem. Soc. Faraday Trans.,* 91 (1995) 2167.

93   T. Evans, A.P. Woodhead, A, Gutierrez Sous, O. Thornton, P.J. Hall, A.A. Davis, N.A. Young, P.B. Wells, R.J. Oldman, O. Plashkevych, O. Vatras, H. Agren and V. Carravetta, *Surf. Sci.,* (1999) L691.

94   G. Bond and P.B. Wells, *J. Catal.,* 150 (1994) 329.

95   S.R. Watson, PhD thesis, University of Hull, 1995.

96   K.E. Simons, P.A. Meheux, S.P. Griffiths, I.M. Sutherland, P. Johnston, P.B. Wells, A.F. Carley, M.K. Rajumon, M.W. Roberts and A. Ibbotson, *Rec. Trav. Chim. Pays-Bas,* 113 (1994) 465.

97   P.B. Wells and A.J. Wilkinson, *Topics in Catal.,* 5 (1999) 39.

98   T. Burgi and A. Baiker, *J. Am. Chem. Soc.,* 120 (1998) 12920.

99   G.D.H. Dijsktra, R.M. Kellogg and H. Weinberg, *Rec. Trav. Chim. Pay-Bas,* 108 (1989) 105.

100   G.D.H. Dijkstra, R.M. Kellogg,, H. Weinberg, J.S. Svendsen, I. Marko and B. Sharpless, *J. Am. Chem. Soc.,* 111 (1989) 8070.

101   G.D.H. Dijkstra, R.M. Kellogg and H. Weinberg, *J. Org. Chem.,* 55 (1990) 6121.

102   F.I. Carrol, R. Abraham, K. Gaetano, S.W. Mascarella, R.A. Wohl, J. Lind and K. Petzoldt, *J. Chem. Soc. Perkin Trans. I,* 1 (1991) 3107.

103   P.B. Wells, K.E. Simons, J.A. Slipszenko, S.P. Griffiths and D.F. Ewing, *J. Mol. Catal. A: Chemical,* 146 (1999) 159.

104   H-U. Blaser, H.R. Jalett, D.M. Monti, J.F. Reiber and J.T. Wehrli, *Stud. Surf. Sci. Catal.,* 41 (1988) 153.

105   G. Bond, P.A. Meheux, A. Ibbotson and P.B. Wells, *Catal. Today,* 10 (1991) 371.

106   M. Garland and H-U. Blaser, *J. Am. Chem. Soc.,* 112 (1990) 7048.

107   H-U. Blaser, H.P. Jalett, D.M. Monti, A. Baiker and J.T. Wehrli, *Stud. Surf. Sci. Catal.,* 67 (1991) 147.

108   J.L. Margitfalvi, M. Hegedus and E. Tfirst, *Stud. Surf. Sci. Catal.,* 101 (1996) 241.

109   J.L. Margitfalvi, E. Talas, E. Tfirst, C.V. Kumar and A. Gergely, *Appl. Catal.,* 191 (2000) 177.

110   B. Oleksyn, *Acta Cryst.,* B38 (1982) 1832.

111   S.P. Griffiths, P. Johnston and P.B. Wells, *Appl. Catal. A: General,* 191 (2000) 193.

112   A. Baiker and T. Heinz, *Topics in Catal.,* 4 (1999) ??

113   G. Wang, T. Heinz, A. Pfaltz, B. Minder, T. Mallat and A. Baiker, *J. Chem. Soc. Chem. Commun.,* (1994) 2047.

114   K.E. Simons, G. Wang, T. Heinz, A. Pfaltz and A. Baiker, *Tetrahedron Asymm.,* 6 (1995) 505.

115   B. Minder, T. Mallat, A. Baiker, G. Wang, T. Heinz and A. Pfaltz, *J. Catal.,* 154 (1995) 371.

116 J.L. Margitfalvi, P. Marti, A. Baiker, L. Botz and O. Sticher, *Catal. Lett.,* 6 (1990) 281.

117 M. Schurch, T. Heinz, R.Aeschimann, T. Mallat, A. Pfaltz and A. Baiker, *J. Catal.,* 173 (1998) 187.

118 B. Minder, M. Schurch, T. Mallat and A. Baiker, *Catal. Lett.,* 31 (1995) 143.

119 T. Heinz, G. Wang, A. Pfaltz, B. Minder, M. Schurch, T. Mallat and A. Baiker, *J. Chem. Soc. Chem. Commun.,* (1995) 1421.

120 B. Minder, M. Schurch, T. Mallat, A. Baiker, G. Wang, T. Heinz and A. Pfaltz, *J. Catal.,* 160 (1996) 261.

121 K.E. Simons, A. Ibbotson, P. Johnston, H. Plum and P.B. Wells, *J. Catal.,* 150 (1994) 321.

122 T.J. Hall, P. Johnston, W.A.H. Vermeer, S.R. Watson and P.B. Wells, *Stud. Surf. Sci. Catal.,* 101 (1996) 221.

123 S.P. Griffiths, P. Johnston, W.A.H. Vermeer and P.B. Wells, *J. Chem. Soc. Chem. Commun.,* (1994) 2431.

124 P.B. Wells, K.E. Simons, J.A. Slipszenko, S.P. Griffiths and D.F. Ewing, *J. Mol. Catal. A: Chemical,* 146 (1999) 159.

125 A. Tungler, T. Mathe, T. Tarnai, K. Fodor, G. Toth, J. Kajtar, I. Kolossvary, B. Herenyi and P.A. Sheldon, *Tetrahedron Asymm.,* 5 (1994) 1171.

126 A. Tungler, T. Tarnai, T. Mathe, J. Petro and R.A. Sheldon, *in* 'Chiral Reactions in Heterogeneous Catalysis' (eds. G. Jannes and V. Dubois) Plenum, New York (1995) p. 121.

127 J.A. Slipszenko, S.P. Griffiths, P. Johnston, K.E. Simons, W.A.H. Vermeer and P.B. Wells, *J. Catal.,* 179 (1998) 267.

128 M. Schurch, O. Schwalm, T. Mallat, J. Weber and A. Baiker, *J. Catal.,* 169 (1997) 275.

129 M. Schurch, N. Kunzle, T. Mallat and A. Baiker, *J. Catal.,* 176 (1998) 569.

130 B. Torok, K. Felfoldi, K. Balazik and M.Bartok, *J. Chem. Soc. Chem. Commun.,* (1999) 1725.

131 M. Studer, S. Burkhardt and H-U. Blaser, *J. Chem. Soc. Chem. Commun.,* (1999) 1727.

132 T. Mallat, M. Bodmer and A. Baiker, *Catal. Lett.,* 44 (1997) 95.

133 K. Borszeky, T. Burgi,, Z. Zhaohui, T. Mallat and A. Baiker, *J. Catal.,* 187 (1999) 160.

134 Y. Nitta, K. Kobiro and Y. Okamoto, *Proc. 70th Ann. Mtg.Chem. Soc. Japan,* 1 (1996) 573.

135 Y. Nitta and A. Shibata, *Chem. Lett.,* (1998) 161.

136 T. Yoshida and K. Harada, *Bull. Chem. Soc. Japan,* 44 (1971) 1062.

137 K. Borszeky, T. Mallat, R. Aeschiman, W.B. Schweizer and A. Baiker, *J. Catal.,* 161 (1996) 451.

138 J.R. Carroll, PhD thesis, University of Hull, (2000).

139 G.A. Attard, A. Ahmadi, J. Feliu, A. Rhodes, E. Herrero, S. Blais and G. Jerkiewicz, *J. Phys. Chem.,* 103 (1999) 1381.

140 A. Ahmadi, G.A. Attard, J. Feliu and A. Rhodes, *Langmuir,* 15 (1999) 2420.

141 G.A. Attard, *J. Phys. Chem,* B, 105 (2001) 3158.

142 J. Petro, T. Mallat, E. Polyanski and T. Mathe *in* 'Hydrogen Effects in Catalysis' (eds. Z. Paal and P.G. Menon) Marcel Dekker, New York and Basel, (1988) p. 225.

143 G.A. Attard, J.E. Gillies, C.A. Harris, D. Jenkins, P. Johnston, M. Price, D. Watson, and P.B. Wells, *Applied Catal.,* in press.

Appendix 1

M.W. ROBERTS' PUBLICATIONS

1. **A method of surface analysis and its application to reduced nickel powder.**
    M.W. Roberts and K.W. Sykes,
    Proceedings of the Royal Society, A, 242, 534, (1957).
2. **Nickel powder with adsorptive properties approaching those of evaporated nickel films.**
    M.W. Roberts and K.W. Sykes,
    Transactions of The Faraday Society, 54, 548, (1958).
3. **Kinetics of nitridation of calcium.**
    M.W. Roberts and F.C. Tompkins,
    Proceedings of the Royal Society, A, 251, 369, (1959).
4. **Diskussionsbeitrage.**
    M.W. Roberts, R. Suhrmann, G. Wedler, F.C. Tompkins and W.J. Moore,
    Bd. 63, No 7, (1959).
5. **Stability of evaporated films.**
    M.W. Roberts,
    Nature, 182, 1151, (1958).
6. **La Nitruration du calcium et autres reactions ternissant les metaux.**
    M.W. Roberts and F.C. Tompkins,
    Journal de Chimie Physique, 48, 562, (1960).
7. **High vacuum techniques.**
    M.W. Roberts,
    J. Royal Inst. Chemistry, August 1960, p.275.
8. **Heats of chemisorption of simple diatomic molecules on metals.**
    M.W. Roberts,
    Nature, 188, 1020, (1960).
9. **The interaction of krypton, oxygen and hydrogen with iron films.**
    M.W. Roberts,
    Transactions of The Faraday Society, 56, 128, (1960).
10. **Direct observation in the electron microscope of oxide layers on aluminium.**
    K. Thomas, and M.W. Roberts,
    Journal of Applied Physics, 32, 70, (1961).

352 *Appendix 1*

11. **Mechanism of the oxidation of iron films at temperatures from -195°C to 120°C.**
    M.W. Roberts,
    Transactions of The Faraday Society, 57, 99, (1961).
12. **Interaction of hydrogen sulphide with nickel, tungsten and silver films.**
    J.M. Saleh, C. Kemball, and M.W. Roberts
    Transactions of The Faraday Society, 57, 1771, (1961).
13. **Factors which may influence the initial reaction of gases with metals.**
    M.W. Roberts,
    First International Congress on Metallic Corrosion, Butterworths, 84, (1961).
14. **Metal oxidation.**
    M.W. Roberts,
    Quarterly Reviews, 16, No. 1, 71 (1962).
15. **The interaction of methyl mercaptan with nickel and tungsten films.**
    J.M. Saleh, M.W. Roberts, and C. Kemball,
    Transactions of The Faraday Society, 58, 1642, (1962).
16. **Adsorption of carbon dioxide on nickel and the influence of adsorbed species on subsequent hydrogen chemisorption.**
    C.M. Quinn and M.W. Roberts,
    Transactions of The Faraday Society, 58, 569, (1962).
17. **Surface potential measurements during the oxidation and subsequent reduction of nickel and iron films.**
    C.M. Quinn and M.W. Roberts,
    Proc. Chem. Soc. p.246, July 1962.
18. **Hydrogenation of ethylene on sintered nickel films.**
    E. Crawford, M.W. Roberts and C. Kemball,
    Transactions of The Faraday Society, 58, 1761, (1962).
19. **Probability of gas adsorption on metal films. Part 1 - Carbon monoxide and nitrogen on molybdenum.**
    M.W. Roberts,
    Transactions of The Faraday Society, 59, 698, (1963).
20. **High-temperature oxidation of zirconium ribbons.**
    C.M. Quinn and M.W. Roberts,
    Transactions of The Faraday Society, 59, 985, (1963).
21. **Interaction of carbon dioxide at low pressures with tantalum and zirconium.**
    C.M. Quinn and M.W. Roberts,
    2nd International Congress on Metallic Corrosion, New York, (1963).
22. **Adsorption of gases by molybdenum films at low pressures.**
    M.W. Roberts,
    Conference organised by the Institute of Physics on 'Sorption properties of vacuum-deposited metal films', The University of Liverpool, April 1963.
23. **Field-emission studies of the interaction of hydrogen sulfide and sulfur with tungsten.**
    J.M. Saleh, M.W. Roberts and C. Kemball,
    Journal of Catalysis, 2, 189, (1963).

24. Nature of thin oxide films on metals as revealed by work function
    measurements.
    C.M. Quinn and M.W. Roberts,
    Nature, 200, 648, (1963).
25. Chemisorption and displacement processes on molybdenum films.
    J.G. Little, C.M. Quinn and M.W. Roberts,
    Journal of Catalysis, 3, 57, (1964).
26. Surface-potential measurements of nitrogen on tungsten.
    C.M. Quinn and M.W. Roberts,
    Journal of Chemical Physics, 40, 237, (1964).
27. Chemisorption of oxygen and subsequent processes on metal films: Work
    function measurements.
    C.M. Quinn and M.W. Roberts,
    Transactions of The Faraday Society, 60, 899, (1964).
28. Mechanism of the sulphidation of lead and oxidized lead films.
    J.M. Saleh, B.R. Wells and M.W. Roberts,
    Transactions of The Faraday Society, 60, 1865, (1964).
29. A lead hydride of high stability.
    B.R. Wells and M.W. Roberts,
    Proceedings of the Chemical Society, 173, (1964).
30. Sticking probability of gases on metal films determined by the flow
    method. Nitrogen on molybdenum.
    C.S. McKee and M.W. Roberts
    Chemical Communications, 4, 59, (1965).
31. Photoelectric investigation of the nickel + oxygen system.
    C.M. Quinn and M.W. Roberts,
    Transactions of The Faraday Society, 61, 1775, (1965).
32. Physical adsorption of gases on pyrex glass. Evidence for superactivity.
    J.R.H. Ross and M.W. Roberts,
    Journal of Catalysis, 4, 620, (1965).
33. Nature and reactivity of nickel and oxidized nickel surface.
    M.W. Roberts and B.R. Wells,
    Discussions of The Faraday Society, 41, 162, (1966).
34. Kinetics of the dissociation of hydrogen sulphide by iron films.
    M.W. Roberts and J.R.H. Ross,
    Transactions of The Faraday Society, 62, 2301, (1966).
35. Chemisorption and incorporation of oxygen by nickel films.
    M.W. Roberts and B.R. Wells,
    Transactions of The Faraday Society, 62, 1608, (1966).
36. Synthesis of ammonia and related processes on reduced molybdenum
    dioxide.
    M.R. Hillis, C Kemball and M.W. Roberts,
    Transactions of The Faraday Society, 62, 3570, (1966).
37. Modern views on Adsorption.
    *M.W. Roberts, "Encyclopaedic Dictionary of Physics", Supplementary Volume
    2, Pergamon Press (1966).*

38. **Probability of gas adsorption on metal films, Part 2. Nitrogen on tungsten.**
    C.S. McKee and M.W. Roberts,
    Transactions of The Faraday Society, 63, 1418, (1967).
39. **Surface rearrangement involving chemisorbed oxygen: The aluminium-oxygen system.**
    M.W. Roberts and B.R. Wells,
    Surface Science, 8, 453, (1967).
40. **Chemisorption of oxygen by aluminium.**
    M.W. Roberts and B.R. Wells,
    Surface Science, 15, 325, (1969).
41. **Chemisorption and displacement reactions on tungsten, molybdenum, and tantalum films.**
    M.W. Roberts and I. Whalley
    Transactions of The Faraday Society, 65, 1377, (1969).
42. **The interaction of hydrogen sulphide and hydrogen with palladium and tantalum films.**
    M.W. Roberts and J.R.H. Ross,
    "Reactivity of Solids,", John Wiley & Sons, Inc. 411, (1969).
43. **The nature of some metal, oxide and reduced oxide surfaces.**
    M.W. Roberts, Comptes Rendus des Journees d'etudes sur les solides finement divises, Saclay, September 1967, 207, (1969), Ed. J. Ehretsmann, Direction de la Documentation Francaise.
44. **The incorporation of chemisorbed species.**
    M.W. Roberts,
    "Recent Progress in Surface Science", Academic Press, 3, 1, (1970).
45. **Low Energy Electron Diffraction (LEED) and Auger Electron Spectroscopy (AES).**
    C.S. McKee and M.W. Roberts,
    Chemistry in Britain, 6, 106, (1970).
46. **Interaction of atomic hydrogen with evaporated lead films.**
    M.W. Roberts and N.J. Young,
    Transactions of The Faraday Society, 66, 2636, (1970).
47. **Rate of hydrogen dissociation at a hot tungsten surface.**
    J.R. Anderson, I.M. Ritchie and M.W. Roberts,
    Nature, 227, 704, (1970).
48. **The interaction of tetramethylsilane with an electron-emitting tungsten filament.**
    M.W. Roberts and J.R.H. Ross,
    Chemical Communications, 1170, (1970).
49. **Thermal transpiration data for xenon and nitrous oxide at high temperature.**
    J.G. Hardy and M.W. Roberts,
    Journal of The Chemical Society, 1683 (1971).
50. **The adsorption of carbon monoxide on Cu(001). LEED and Auger emission studies.**
    R.W. Joyner, C.S. McKee and M.W. Roberts,
    Surface Science, 26, 303, (1971).

51. The interaction of hydrogen sulphide with Cu(001).
   R.W. Joyner, C.S. McKee and M.W. Roberts,
   Surface Science, 27, 279, (1971).

52. Mechanism of the catalytic decomposition of methanol on gold filaments.
   J.G. Hardy and M.W. Roberts,
   Chemical Communications, 494, (1971).

53. Chemisorption and decomposition of tetramethylsilane over tungsten and iron surfaces.
   M.W. Roberts and J.R.H. Ross,
   Journal of The Chemical Society, Faraday Transactions I, 68, 221, (1972).

54. The surface chemistry of manganese.
   R.I. Bickley, M.W. Roberts and W.C. Storey,
   Journal of The Chemical Society, (A), 2774, (1971).

55. Adsorption of neopentane on tungsten and palladium films.
   J.R.H. Ross, M.W. Roberts and C. Kemball,
   Journal of The Chemical Society, Faraday Transactions I, 68, 914, (1972).

56. Contact angle studies of some low energy polymer surfaces.
   W.J. Murphy, M.W. Roberts and J.R.H. Ross,
   Journal of The Chemical Society, Faraday Transactions I, 68, 1190 (1972).

57. Mechanism of formation and some surface characteristics of thin polymer films formed on metal surfaces by electron bombardment.
   S. Frost, W.J. Murphy, M.W. Roberts, J.R.H. Ross and J.H. Wood,
   Faraday Special Discussions of The Chemical Society, 2, 198, (1972).

58. Surface studies by photoemission.
   M.W. Roberts,
   "Surface and defect properties of solids". Specialist Periodical Report,
   The Chemical Society, 1, 144, (1972).

59. Chemisorption, decomposition, and oxidation of methanol over gold and nickel filaments.
   M.W. Roberts and T.I. Stewart,
   Proceedings of the Conference "Chemisorption and Catalysis",
   Ed. Peter Hepple organised by the Institute of Petroleum p.16 (1972).

60. Evidence for surface activation in the photolysis of adsorbed lead tetraethyl.
   D.L. Perry and M.W. Roberts,
   J.C.S. Chem. Comm., 147, (1972).

61. Some observations on the surface sensitivity of photoelectron spectroscopy.
   C.R. Brundle and M.W. Roberts,
   Proceedings of The Royal Society London, A 331, 383, (1972).

62. Surface sensitivity of ESCA for sub-monolayer quantities of mercury adsorbed on a gold substrate.
   C.R. Brundle and M.W. Roberts,
   Chemical Physics Letters, 18, 380, (1973).

63. Auger electron spectroscopy studies of clean polycrystalline gold and of the adsorption of mercury on gold.
   R.W. Joyner and M.W. Roberts,
   Journal of The Chemical Society, Faraday Transactions I, 69, 1242, (1973).

64. **Surface sensitivity of HeI Photoelectron Spectroscopy (UPS) for H₂O**
   Adsorbed on gold.
   C.R. Brundle and M.W. Roberts,
   Surface Science, 38, 234, (1973).

65. **ESCA studies of chemisorption on metals: carbon monoxide on**
   molybdenum and tungsten films.
   S.J. Atkinson, C.R. Brundle and M.W. Roberts,
   Journal of Electron Spectroscopy and Related Phenomena, 2, 105, (1973).

66. **Models for an adsorbed layer and their evaluation by comparison of LEED**
   and optical diffraction patterns: the system W(112)-O₂.
   C.S. McKee, D.L. Perry and M.W. Roberts,
   Surface Science, 39, 176, (1973).

67. **A study of the preparation of atomically clean tungsten surfaces by Auger**
   electron spectroscopy.
   R.W. Joyner, J. Rickman and M.W. Roberts,
   Surface Science, 39, 445, (1973).

68. **Low temperature adsorption of CO on polycrystalline molybdenum**
   studied by x-ray and vacuum uv photoelectron spectroscopy.
   S.J. Atkinson, C.R. Brundle and M.W. Roberts,
   Chemical Physics Letters, 24, 175, (1974).

69. **An ultra high vacuum electron spectrometer for surface studies.**
   C.R. Brundle, D. Latham, M.W. Roberts and K. Yates,
   Journal of Electron Spectroscopy and Related Phenomena, 3, 241, (1974).

70. **Evidence for the nature of CO adsorbed on nickel from electron**
   spectroscopy.
   R.W. Joyner and M.W. Roberts,
   Journal of The Chemical Society, Faraday Transactions I, 70, 1819, (1974).

71. **Chemisorption of nitrogen on tungsten studied by Auger electron**
   spectroscopy.
   R.W. Joyner, J. Rickman and M.W. Roberts,
   Journal of The Chemical Society, Faraday Transactions I, 70, 1825, (1974).

72. **Reference levels in photoelectron spectroscopy**
   A.F. Carley, R.W. Joyner and M.W. Roberts,
   Chemical Physics Letters, 27, 580, (1974).

73. **Oxygen 1s binding energies in oxygen chemisorption on metals.**
   R.W. Joyner and M.W. Roberts,
   Chemical Physics Letters, 28, 246, (1974).

74. **Oxygen (1s) binding energies in carbon monoxide adsorption on metals.**
   R.W. Joyner and M.W. Roberts,
   Chemical Physics Letters, 29, 447, (1974).

75. **Ultra-violet and x-ray photoelectron spectroscopy (UPS and XPS) of CO,**
   CO₂ and H₂O on molybdenum and gold films.
   S.J. Atkinson, C.R. Brundle and M.W. Roberts,
   Faraday Discussions of The Chemical Society, 58, 62, (1974).

76. Carbon monoxide adsorption on iron in the temperature range 85 to 350K as revealed by x-ray and vacuum ultraviolet (He(11)) photoelectron spectroscopy.
    K. Kishi and M.W. Roberts,
    Journal of The Chemical Society, Faraday Transactions I, 71, 1715, (1975).

77. Mechanism of the interaction of hydrogen sulphide with adsorbed oxygen on lead studied by x-ray induced photoelectron spectroscopy.
    K. Kishi and M.W. Roberts,
    Journal of The Chemical Society, Faraday Transactions I, 71, 1721, (1975).

78. Auger electron spectroscopy and its applications in surface chemistry.
    R.W. Joyner and M.W. Roberts,
    "Surface and Defect Properties of Solids", Specialist Periodical Report, The Chemical Society IV, 68, (1975).

79. Interaction of oxygen with Cu(100) studied by low energy electron diffraction (LEED) and x-ray photoelectron spectroscopy (XPS).
    M.J. Braithwaite, R.W. Joyner and M.W. Roberts,
    Faraday Discussions of The Chemical Society, 60, 89, (1975).

80. Development of stepped surface regions on polycrystalline gold. Low energy electron diffraction and Auger studies.
    S.A. Isa, R.W. Joyner and M.W. Roberts,
    Journal of The Chemical Society, Faraday Transactions I, 72, 540, (1976).

81. The application of electron spectroscopy in the study of molecular processes at solid surfaces.
    M.W. Roberts,
    Proceedings of the 6th Czechoslovak Conference on Electronics and Vacuum Physics, Bratislava 4, 45, (1976).

82. Low temperature oxygen and activated nitrogen faceting of Ni(210) Surfaces.
    R.E. Kirby, C.S. McKee, and M.W. Roberts,
    Surface Science, 55, 725, (1976).

83. Photoelectron spectroscopic investigation of the adsorption and catalytic decomposition of formic acid by copper, nickel, and gold.
    R.W. Joyner and M.W. Roberts,
    Proceedings of The Royal Society London, A350, 107, (1976).

84. The adsorption of nitric oxide by iron surfaces studied by photoelectron spectroscopy.
    K. Kishi and M.W. Roberts,
    Proceedings of The Royal Society London, A352, 289, (1976).

85. Photoelectron spectroscopy and surface chemistry.
    M.W. Roberts,
    "Photoelectron Emission", Proceedings of the Daresbury Study Week-end, DL/SRF/R8, (1976).

86. Adsorption of nitrogen and ammonia by polycrystalline iron surfaces in the temperature range 80-290K studied by electron spectroscopy.
    K. Kishi and M.W. Roberts,
    Surface Science, 62, 252, (1977).

87. **Defect surface structures studied by LEED.**
    C.S. McKee, M.W. Roberts and M.L. Williams,
    Advances in Colloid and Interface Science, 8, 29, (1977).

88. **Adsorption of carbon monoxide on copper (100) at 295K, characterized by photoelectron spectroscopy.**
    S.A. Isa, R.W. Joyner and M.W. Roberts,
    J.C.S. Chem. Comm., 377, (1977).

89. **Electron spectroscopic study of nitrogen species adsorbed on copper.**
    M.H. Matloob and M.W. Roberts,
    Journal of The Chemical Society, Faraday Transactions I, 73, 1393, (1977).

90. **The mechanism of the oxidation and passivation of iron by water vapour - an electron spectroscopic study.**
    M.W. Roberts and P R Wood,
    Journal of Electron Spectroscopy and Related Phenomena, 11, 431, (1977).

91. **Interaction of cobalt with oxygen, water vapour, and carbon monoxide. X-ray and ultraviolet photoemission studies.**
    R.B. Moyes and M.W. Roberts,
    Journal of Catalysis, 49, 216, (1977).

92. **Low energy electron diffraction and electron spectroscopic studies of the oxidation and sulphidation of Pb(100) and Pb(110) surfaces.**
    R.W. Joyner, K. Kishi and M.W. Roberts
    Proceedings of The Royal Society London, A358, 223, (1977).

93. **Adsorption of hydrazine on iron studied by x-ray photoelectron spectroscopy.**
    M.H. Matloob and M.W. Roberts,
    Journal of Chemical Research (S), 336, (1977).

94. **Adsorption of carbon monoxide on copper (100) studied by photoelectron spectroscopy and low energy electron diffraction.**
    S.A. Isa, R.W. Joyner and M.W. Roberts,
    Journal of The Chemical Society, Faraday Transactions I, 74, 546, (1978).

95. **New Perspectives in Surface Chemistry and Catalysis - Tilden Lecture**
    M.W. Roberts,
    Chemical Society Reviews, 6, 373, (1977).

96. **The nature of catalytic sites on solid surfaces.**
    M.W. Roberts,
    British Association Meeting, University of Aston (1977).

97. **Electron spectroscopic study of nitric oxide adsorbed on copper**
    M.H. Matloob and M.W. Roberts
    Physica Scripta, 16, 420, (1977).

98. **Adsorption of nitric oxide on cu(100) surfaces: an electron spectroscopic study.**
    D.W. Johnson, M.H. Matloob and M.W. Roberts,
    J.C.S. Chem. Comm., 40, (1978).

99. **The adsorption of oxygen on Cu(210).**
    C.S. McKee, L.V. Renny and M.W. Roberts,
    Surface Science, 75, 92, (1978).

100. Chemistry of the Metal-Gas Interface.
M.W. Roberts and C.S. McKee, pp. 594
Oxford University Press, (1978).
Russian Translation, Moscow, (1982).
101. An x-ray photoelectron spectroscopic study of the interaction of oxygen and nitric oxide with aluminium.
A.F. Carley and M.W. Roberts,
Proceedings of The Royal Society London, A.363, 403, (1978).
102. Contact angle studies of polymer surfaces.
K.M. Byrne, M.W. Roberts and J.R.H. Ross,
Adhesion 1, 19, (1977).
103. The effect of reduction and temperature on the electronic core levels of tungsten and molybdenum in WO_3 and $W_xMo_{1-x}O_3$. A photoelectron spectroscopic study.
E. Salje, A.F. Carley and M.W. Roberts,
Journal of Solid State Chemistry, 29, 237, (1979).
104. A study of the interaction of nitric oxide with Cu(100) and Cu(111) surfaces using low energy electron diffraction and electron spectroscopy.
D.W. Johnson, M.H. Matloob and M.W. Roberts,
Journal of The Chemical Society, Faraday Transactions I, 75, 2143, (1979).
105. A study of the adsorption of oxygen on silver at high pressure by electron spectroscopy.
R.W. Joyner and M.W. Roberts,
Chemical Physics Letters, 60, 459, (1979).
106. Chemisorption of nitric oxide by nickel.
A.F. Carley, S. Rassias, M.W. Roberts and W. Tang-han,
Surface Science, 84, L227, (1979).
107. Surface segregation of potassium in nickel induced by oxidation.
A.F. Carley, S. Rassias and M.W. Roberts,
Journal of Chemical Research (S), 208, (1979).
108. Nitrogen chemisorption by iron.
D.W. Johnson and M.W. Roberts,
Surface Science, 87, L255, (1979).
109. A "High-Pressure" Electron Spectrometer for Surface Studies.
R.W. Joyner and M.W. Roberts,
Surface Science, 87, 501, (1979).
110. Hydroxylation and dehydroxylation at Cu(111) Surfaces.
C.T. Au, J. Breza and M.W. Roberts,
Chemical Physics Letters, 66, 340, (1979).
111. A study of the interaction of nitric oxide with nickel and oxidized nickel surfaces by x-ray photoelectron spectroscopy.
A.F. Carley, S. Rassias, M.W. Roberts and W.T. Han,
Journal of Catalysis, 60, 385, (1979).
112. The oxidation of cadmium (0001) studied by low energy electron diffraction (LEED) and Auger Electron Spectroscopy (AES).
R.W. Joyner, M.W. Roberts and G.N. Salaita,
Surface Science, 84, L505, (1979).

113. **The critical surface tension of wool**
   K.M. Byrne, M.W. Roberts and J.R.H. Ross,
   Textile Research Journal, 49, 34, (1979).

114. **Adsorption of hydrazine and ammonia on aluminium.**
   D.W. Johnson and M.W. Roberts,
   Journal of Electron Spectroscopy and Related Phenomena, 19, 185, (1980).

115. **XPS studies of surface charge on nickel oxide.**
   M.W. Roberts and R. St. C. Smart,
   Chemical Physics Letters, 69, 234, (1980).

116. **XPS studies of donor and acceptor chemisorption of NO and CO on nickel oxide surfaces.**
   M.W. Roberts and R. St.C. Smart,
   Surface Science, 100, 590-604 (1980).

117. **Photoelectron spectroscopic study of the surface of some high-performance liquid chromatography substrates.**
   M.W. Roberts, A.F. Carley and L. Moroney,
   Faraday Symposium, No. 15, 39, (1980).

118. **Photoelectron spectroscopic evidence for the activation of adsorbate bonds by chemisorbed oxygen.**
   M.W. Roberts and C.T. Au,
   Chemical Physics Letters, Vol. 74, Number 3, 472 (1980).

119. **A study of the interaction of formic acid and propionic acid with oxidised lead and copper surfaces by photoelectron spectroscopy and LEED.**
   M.W. Roberts, S.A. Isa, R.W. Joyner and M.H. Matloob,
   Applications of Surface Science, 5, 345-360, (1980).

120. **Photoelectron spectroscopy and surface chemistry.**
   M.W. Roberts,
   Advances in Catalysis, Volume 29, 55, (1980).

121. **Chemisorption of HCl and H$_2$S by Cu(111)-O surfaces.**
   M.W. Roberts, L. Moroney and S. Rassias,
   Surface Science 105, L249-L254, (1981).

122. **X-ray induced effects during the oxidation of Bi(0001).**
   M.W. Roberts, R.W. Joyner and S.P. Singh-Boparai,
   Surface Science, 104, L199-L203, (1981).

123. **Evidence from photoelectron spectroscopy for dissociative adsorption of oxygen on nickel oxide.**
   M.W. Roberts and R. St. C. Smart,
   Surface Science, 108, 271-280, (1981).

124. **Molecular events at solid surfaces.**
   M.W. Roberts,
   Pure and Applied Chemistry, 53, 2269-2281, (1981).

125. **Surface Chemistry. Photoelectron spectroscopy and surface chemistry.**
   M.W. Roberts,
   Chemistry in Britain, Volume 17, Number 11, (1981).

126. An XPS study of the influence of chemisorbed oxygen on the adsorption of ethylene and water vapour by Cu(110) and Cu(111) surfaces.
M.W. Roberts and C.T. Au,
Journal de chimie physique, 78, No. 11/12, 921-926, (1981).

127. Surface hydroxylation at a Zn(0001)-O surface.
M.W. Roberts, C.T. Au and A.R. Zhu,
Surface Science, 115, L117-L123, (1982).

128. Coordination and activation of simple molecules at metal surfaces.
M.W. Roberts and R. Mason,
Inorganica Chimica Acta, 50, 53-58, (1982).

129. A novel reaction at a Pb(110) surface.
M.W. Roberts, A.F. Carley and M.S. Hegde,
Chemical Physics Letters, Volume 90, Number 2, 108-110, (1982).

130. New approaches to surface chemistry.
M.W. Roberts,
Science Progress (Oxford), 68, 93-110, (1982).

131. The dual role of oxygen in the interaction of hydrogen chloride with a Pb(110)-O Surface.
M.W. Roberts, P.G. Blake and A.F. Carley,
Surface Science, 123, L733-L738, (1982).

132. Chemistry of the Metal-Gas Interface.
M.W. Roberts and C.S. McKee,
Russian Translation (O.U.P., Moscow, p.p. 539, (1982)).

133. Chemisorption of oxygen at Ag(110) surfaces and its role in adsorbate activation.
M.W. Roberts, Chak-tong Au, and Sunder Singh-Boparai,
J. Chem. Soc., Faraday Trans. I, 79, 1779-1791, (1983).

134. Studies of the thermal decomposition of βNiO(OH) and nickel peroxide by x-ray photoelectron spectroscopy.
M.W. Roberts, Lee M. Moroney and Roger St. C. Smart,
J. Chem. Soc., Faraday Trans. I, 79, 1769-1778, (1983).

135. The specificity of surface oxygen in the activation of adsorbed water at metal surfaces.
M.W. Roberts, A.F. Carley and S. Rassias,
Surface Science, 135, 35-51, (1983).

136. XPS studies on $WO_{2.90}$ and $WO_{2.72}$ and the influence of metallic impurities
R. Gehlig, E. Salje, A.F. Carley and M.W Roberts,
Journal of Solid State Chemistry, 49, 318-324, (1983).

137. Photoelectron spectroscopic evidence for Ni^{3+} species in chemisorption at a Ni(100) surface.
M.W. Roberts, Albert F. Carley and Stephen R. Grubb,
J.C.S. Chem. Comm, 459-460, (1984).

138. A photoelectron spectroscopic study of the adsorption and catalytic decomposition of formic acid at Zn(0001) and Zn(0001)-O surfaces.
C.T. Au and M.W. Roberts,
8th International Congress on Catalysis - Berlin 1984.

139. **Chemisorption of nitric oxide at a Zn(0001) surface and the role of water vapour in its hydrogenation.**
C.T. Au and M.W. Roberts,
Proceedings of The Royal Society London, A396, 165-181, (1984).

140. **The defect structure of nickel-oxide surfaces as revealed by photoelectron spectroscopy.**
M.W. Roberts and R. St. C. Smart,
J. Chem. Soc. Faraday Trans. I, 80, 2957-2968, (1984).

141. **The role of water vapour in the hydrogenation of nitric oxide at a Zn(0001) surface.**
C.T. Au, M.W. Roberts and A.R. Zhu,
J.C.S. Chem. Comm, 737, (1984).

142. **Photoelectron spectroscopy and the surface chemistry of Wool**
C.N. Carr, S.F. Ho, D.M. Lewis, E.D. Owen and M.W. Roberts,
J. Text. Inst., 1985, No. 6. p.419.

143. **Structure of the chloride overlayer at a magnesium surface**
C.T. Au and M.W. Roberts,
Surface Science 149, L18-L24, (1985).

144. **XPS determination of band bending in defective semiconducting oxide surfaces**
M.W. Roberts and R. St. C. Smart,
Surface Science, 151, 1-8, (1985).

145. **Defects in oxide overlayers at nickel single-crystal surfaces**
A.F. Carley, P.R. Chalker and M.W. Roberts,
Proceedings of The Royal Society London, A399, 167-179, (1985).

146. **The impact of photoelectron spectroscopy on surface chemistry and catalysis**
M.W. Roberts
Proc. Indian Natn. Sci. Acad. 51, A. No. 1, pp. 165-179, (1985).

147. **An XPS study of the interaction of NO with a magnesium surface**
R.G. Copperthwaite, A.F. Carley and M.W. Roberts,
Surface Science, 165, L1-L6, (1986).

148. **Specific role of transient O⁻(s) at Mg(0001) surfaces in activation of ammonia by dioxygen and nitrous oxide.**
C.T. Au and M.W. Roberts,
Nature, Vol. 319, No. 6050, pp. 206-208, (1986).

149. **Photoelectron spectroscopy: a strategy for the study of reactions at solid surfaces.**
C.T. Au, A.F. Carley and M.W. Roberts,
International Reviews in Physical Chemistry, Vol. 5, No. 1, 57-87, (1986).

150. **Surface reactivity as revealed by photoelectron spectroscopy**
C.T. Au, A.F. Carley and M.W. Roberts,
Royal Society Discussion Meeting London 1985, Phil. Trans. R. Soc. Lond. A318, 61-79, (1986).

151. **Reaction of carbon dioxide with the magnesium surface**
S. Campbell, P. Hollins, E. McCash and M.W. Roberts,
Journal of Electron Spectroscopy and Related Phenomena, 39, 145-153, (1986).

152. Chemisorptive replacement of surface oxygen by hydrogen halides (HCl and HBr) at Pb(110) surfaces.
P.G. Blake, A.F. Carley, V. Di Castro and M.W. Roberts,
J. Chem. Soc. Faraday Trans. I, 82, 723-737, (1986).

153. The promotion of surface-catalysed reactions by gaseous additives: the role of a surface oxygen transient.
Chak-tong Au and M. Wyn Roberts,
J. Chem. Soc., Faraday Trans. I, 83, 2047-2059, (1987) (Faraday Symp. 21).

154. The role of surface oxygen in reactions of propylene at Mg(0001) Surfaces
C.T. Au, Li Xing-chang, Tang Ji-an and M.W. Roberts,
Journal of Catalysis, 106, 538-543, (1987).

155. Activation of carbon dioxide at low temperatures at aluminium surfaces
A.F. Carley, D. Gallagher and M.W. Roberts,
Surface Science, 183, L263-L268, (1987).

156. The identification and characterization of mixed oxidation states at oxidised titanium surfaces.
A.F. Carley, P.R. Chalker, J.C. Riviere and M.W. Roberts,
J. Chem. Soc. Faraday Trans. I, 83, 351-370, (1987).

157. Oxygen induced dissociation of carbon monoxide at an sp-metal (aluminium) surface
A.F. Carley and M.W. Roberts,
J. Chem. Soc. Chem. Commun. 355 (1987).

158. An x-ray photoelectron and electron spin resonance study of wool treated with aqueous solutions of chromium and copper ions.
C.M. Carr, J.C. Evans and M.W. Roberts,
Textile Research Journal, Vol. 57, No. 2, Feb. 1987.

159. Activation of carbon dioxide and carbon monoxide at aluminium surfaces
A.F. Carley, D.E. Gallagher and M.W. Roberts,
Spectrochimica Acta., Vol. 43A, No. 12 pp. 1447-1453, (1987).

160. Electron spectroscopic studies of the chemical interaction of benzene with transient O⁻(s) on Mg(0001) surfaces.
C.T. Au, Tang Ji-an and M.W. Roberts,
Journal of Xiamen University (Natural Sci) Vol. 26 No. 2. March 1987. (In Chinese).

161. Chemistry in two dimensions
M.W. Roberts
University of Wales Science and Technology Review No. 2. 58, (1987).

162. The chemical reactivity of oxidised lead surfaces studied by XPS: the mechanism of "halogen induced" surface etching.
A.F. Carley and M.W. Roberts,
Vacuum, Vol. 38, No. 4/5 pp. 397-399 (1988).

163. Evidence from coadsorption studies for a molecular precursor state in the oxidation of Zn(0001)
A.F. Carley, M.W. Roberts and Song Yan,
J. Chem. Soc. Chem. Commun. p. 267-268, (1988).

164. **Metal oxide overlayers and oxygen induced chemical reactivity studied by photoelectron spectroscopy.**
M.W. Roberts.
Surface and Near-Surface Chemistry of Oxide Materials edited by J. Nowotny and L.-C. Dufour, Elsevier Science Publishers Chapter 5, 219-244.(1988).

165. **The reactive chemisorption of carbon dioxide at magnesium and copper surfaces at low temperature**
R.G. Copperthwaite, P.R. Davies, M.A. Morris, M.W. Roberts and R.A. Ryder, Catalysis Letters 1 11-20, (1988).

166. **Intermolecular charge-transfer and the cleavage of the dioxygen bond at metal surfaces: oxygen at Zn(0001)**
A.F. Carley, M.W. Roberts and Song Yan, Catalysis Letters 1, 265-270, (1988).

167. **The nature and reactivity of chemisorbed oxygen and oxide overlayers at metal surfaces as revealed by photoelectron spectroscopy**
M.W. Roberts, in
Structure and Reactivity of Surfaces, Elsevier p. 787-797, (1989).
Eds. C. Morterra, A. Zecchina and G. Costa.

168. **Activation of carbon dioxide leading to a chemisorbed carbamate species at a Cu(100) surface**
A.F. Carley, P.R. Davies and M.W. Roberts
J.C.S. Chem. Comm., 677 1989.

169. **Chemisorption and reaction pathways at metal surfaces: the role of surface oxygen.**
M.W. Roberts,
Chem. Soc. Rev. 18, 451-475, (1989).

170. **Computer modelling of the kinetics of the coadsorption of ammonia and dioxygen at a Mg(0001) surface**
P.G. Blake and M.W. Roberts,
Catalysis Letters 3, 399-404, (1989).

171. **The influence of pre-oxidation on the adsorption of CO at a Zn(0001) surface: characterisation of a weakly chemisorbed species by XPS and UPS.**
A.F. Carley, M.W. Roberts and Song Yan,
Applied Surface Science, 40, 289-293, (1990).

172. **Dissociative chemisorption and localized oxidation states at titanium surfaces**
A.F. Carley, J.C. Roberts and M.W. Roberts,
Surface Science Letters, 225, L39-L41, (1990).

173. **Role of oxygen transients in the chemistry of dioxygen at atomically clean metal surfaces: the Zn(0001)-dioxygen-ammonia system.**
A.F. Carley, Song Yan and M.W. Roberts,
J. Chem. Soc. Faraday Trans. 86, No. 15 2701-2710, (1990).

174. **X-ray photoelectron spectroscopic study of the high-T_c superconductor $YBa_2Cu_3O_{7-x}$: evidence for Cu^{3+} and surface oxygen excess.**
A.F. Carley, M.W. Roberts, J.S. Lees and R.J.D. Tilley
J. Chem. Soc. Faraday Trans. 86 3129-3134 (1990).

175. Hydroxylation of molecularly adsorbed water at Ag(111) and Cu(100)
    surfaces by dioxygen: photoelectron and vibrational spectroscopic studies.
    A.F. Carley, P.R. Davies, M.W. Roberts and K.K. Thomas,
    Surface Science Letters 238 L467-L472 (1990).

176. A survey of experimental techniques in surface chemical physics
    J.H. Block, A.M. Bradshaw, P.C. Gravelle, J. Haber, R.S. Hansen,
    M.W. Roberts, N. Sheppard and K. Tamaru,
    Pure and Appl. Chem. Vol. 62, No. 12 pp. 2297-2322 (1990).

177. HREELS and XPS Evidence for Facile H-transfer from NH$_3$(a) to
    chemisorbed N Adatoms at a Zn(0001) Surface
    A.F. Carley, Song Yan and M.W. Roberts
    J. Chem. Soc. Faraday Trans., 86 3827-3828 (1990).

178. Activation of carbon dioxide at bismuth, gold and copper surfaces
    V.M. Browne, A.F. Carley, R.G. Copperthwaite, P.R. Davies, E.M. Moser and
    M.W. Roberts
    Applied Surface Science, 47 375-379 (1991).

179. A perspective of surface chemistry
    M.W. Roberts, Catalysis Today, 12 501-505 (1992).

180. Mixed oxidation states of titanium at the metal-oxide interface
    M.W. Roberts and M. Tomellini, Catalysis Today, 12 443-452 (1992).

181. Evidence for the role of surface transients and precursor states in
    determining molecular pathways in surface reactions.
    M.W. Roberts, Plenary Lecture, 8th International Seminar on Electron
    Spectroscopy, Poland 1990, Applied Surface Science 52 133-140 (1991).

182. The reactive chemisorption of formic acid at Al(111) surfaces and the
    influence of surface oxidation and coadsorption with water: a combined
    XPS and HREELS investigation.
    P.R. Davies, M.W. Roberts and N. Shukla,
    Ninth Interdisciplinary Surface Science Conference, University of Southampton,
    April 1991
    Journal of Physics: Condensed Matter 3 S237-S244 (1991).

183. Kinetics of coadsorption of dioxygen and ammonia at a Zn(0001) surface:
    a theoretical model.
    A.F. Carley, M.W. Roberts and M. Tomellini,
    J. Chem. Soc. Faraday Transactions, 87 3563-3567 (1991).

184. Molecular mechanisms and catalysis at metal surfaces - the role of surface
    sensitive spectroscopies.
    M.W. Roberts
    Spectroscopy World Vol. 3. No. 6 (1991).

185. The role of a dioxygen precursor in the selective formation of imide NH(a)
    species at a Cu(110) surface.
    B. Afsin, P.R. Davies, A. Pashuski and M.W. Roberts
    Surface Science Letters 259 L724-L728 (1991).

186. Activation of carbon dioxide by ammonia at Cu(100) and Zn(0001)
    surfaces leading to the formation of a surface carbamate.
    P.R. Davies and M.W. Roberts
    J. Chem. Soc. Faraday Trans. 88 361-368 (1992).

187. **Surface structure and the instability of the formate overlayer at a Pb(110) surface.**
B. Afsin and M.W. Roberts
Catalysis Letters, 13 277-282 (1992).

188. **Catalytic cleavage of dioxygen bond at a Zn(0001)-Ba surface: the role of a dioxygen surface transient**
A.F. Carley, M.W. Roberts and Wang Fancheng
J.C.S. Chem. Comm Issue 10 738-739 (1992).

189. **Molecular events in the coadsorption of molecules at metal surfaces**
M.W. Roberts
Journal of Molecular Catalysis, Vol. 74 No. 1-3, p11-22 (1992).

190. **A new approach to the mechanism of heterogeneously catalysed reactions: the oxydehydrogenation of ammonia at a Cu(111) surface.**
A Doronin, A. Pashusky, and M.W. Roberts
Catalysis Letters 16 345-350 (1992).

191. **Co-adsorption studies of ethene with isotopically labelled water and hydrogen on copper/silica**
S.D. Jackson, A. Owens and M.W. Roberts
React. Kinet. Catal. Lett., 46, No. 2, 245-348 (1992).

192. **Reaction pathways in the oxydehydrogenation of ammonia at Cu(110) surfaces.**
B. Afsin, P.R. Davies, A. Pashusky, M.W. Roberts and D. Vincent
Surface Science 284 109-120 (1993).

193. **Chemisorption and reactions at metal surfaces**
M.W. Roberts
Special commemorative issue of Surface Science - The First Thirty Years 299/300 769-784 (1994).

194. **Electronic structure of copper particles supported on TiO$_2$, graphite, and Al$_2$O$_3$ substrates: a comparative study.**
A.F. Carley, M.K. Rajumon and M.W. Roberts
Journal of Solid State Chemistry, 106 156-163 (1993).

195. **Surface oxygen and chemical reactivity**
C.T. Au, A.F. Carley, A. Pashuski, S. Read, M.W. Roberts and A. Zeini-Isfahan
Springer Series in Surface Science, Vol. 33, Eds E. Umbach and H.-J. Freund (Adsorption on Ordered Surfaces of Ionic Solids and Thin Films) 241-253, (1994).

196. **Formation of an oxy-chloride overlayer at a Bi(0001) surface**
B. Afsin and M.W. Roberts
Spectroscopy Letters, 27, 139-146 (1994).

197. **Oxygen sites active in H-abstraction at a Cu(110)-O surface: comparison of a Monte Carlo simulation with imide formation studied by XPS and VEELS.**
A.F. Carley, P.R. Davies, M.W. Roberts and D. Vincent
Topics in Catalysis 1 35-42 (1994).

198. An EPR study of a palladium catalyst using PBN (N-benzylidene-tert-butylamine N-oxide) spin trap: direct demonstration of hydrogen spillover.
   A.F. Carley, H.A. Edwards, B. Mile, M.W. Roberts, C.C. Rowlands, S.D. Jackson and F.E. Hancock
   J.Chem. Soc. Chem. Commun. Vol. 12 1407-1408, (1994).

199. Characterization of oxygen adsorbed at Ba-modified Zn(0001) surfaces: evidence for peroxo species.
   A.F. Carley, M.K. Rajumon, M.W. Roberts and Wang Fancheng
   Solid State Communications 91, 791-794 (1994).

200. Low energy pathway for the formation of a Pt(111)-N(2x2) overlayer.
   T.S. Amorelli, A.F. Carley, M.K. Rajumon, M.W. Roberts and P.B. Wells
   Surface Science Vol. 315 L990-L994 (1994).

201. Applications of EPR to study the hydrogenation of ethene and benzene over a supported Pd catalyst: detection of free radicals on a catalyst surface
   A.F. Carley, H.A. Edwards, F.E. Hancock, S.D. Jackson, B. Mile, M.W. Roberts and
   C.C. Rowlands. J. Chem. Soc. Faraday Trans. 90 3341-3346 (1994).

202. Oxygen dimerization at a Zn(0001)-O surface
   A.F. Carley, M.K. Rajumon, M.W. Roberts and Wang Fancheng
   Journal of Solid State Chemistry 112, 214-216 (1994).

203. Activation of carbon monoxide and carbon dioxide at cesium-promoted Cu(110) and Cu(110-O) surfaces
   A.F. Carley, M.W. Roberts and A.J. Strutt
   The Journal of Physical Chemistry, American Chemical Society 98, 9175-9181 (1994).

204. The hydroxylation of Cu(111) and Zn(0001) surfaces
   A.F. Carley, P.R. Davies, M.W. Roberts, N. Shukla, Y. Song, K.K. Thomas
   Applied Surface Science 81, 265-272 (1994).

205. A model for the enantioselective hydrogenation of pyruvate catalysed by alkaloid-modified platinum
   K.E. Simons, P.A. Meheux, S.P. Griffiths, I.M. Sutherland, P. Johnston, P.B. Wells, A.F. Carley, M.K. Rajumon, M.W. Roberts and A. Ibbotson
   Recl. Trav. Chim. Pays-Bas 113, 465-474 (1994).

206. Chemical reactivity of CO and CO_2 at a Cu(110)-Cs surface
   A.F. Carley, M.W. Roberts and A.J. Strutt
   Catalysis Letters 29, 169-175 (1994).

207. The reactive chemisorption of carbon dioxide at Mg(100) surface
   Zeini-Isfahani, Asghar, Roberts, M.W., Carley, A.F. and Read, S.
   Iran J. Chem & Chem. Eng., Vol. 13, No. 1, (1994).

208. Oxygen states at a Cu(111) surface: the influence of coadsorbed ammonia
   P.R. Davies, M.W. Roberts, N. Shukla and D.J. Vincent
   Surface Science, Vol. 325, 50-56 (1995).

209. Nature of the oxygen species at Ni(110) and Ni(100) surfaces revealed by exposure to oxygen and oxygen-ammonia mixtures: evidence for the surface reactivity of O⁻ type species
    G.U. Kulkarni, C.N.R. Rao and M.W. Roberts
    The Journal of Physical Chemistry 99, 3310-3316, (1995).

210. Coadsorption of dioxygen and water on the Ni(110) surface: role of O⁻ type species in the dissociation of water
    G.U. Kulkarni, C.N.R. Rao and M.W. Roberts
    Langmuir, 11, 2572-2575, (1995).

211. Surface oxygen transients and their role in providing low energy reaction pathways
    A.F. Carley, P.R. Davies and M.W. Roberts
    Perspectives in Solid State Chemistry, Narosa, Ed. K.J. Rao, 150-159, (1995).

212. XPS AND LEED studies of 10,11-Dihydrocinchonidine adsorption at Pt(111)
    A.F. Carley, M.K. Rajumon, M.W. Roberts and P.B. Wells
    J. Chem. Soc. Faraday Trans. 91, 2167-2172, (1995).

213. The oxygen state active in the catalytic oxidation of carbon monoxide at a caesium surface: isolation of the reactive anionic $CO_2^{\delta-}$ species
    G.U. Kulkarni, S. Laruelle and M.W. Roberts,
    Chem. Commun. (1996)

214. Oxidation of methanol at copper surfaces
    A.F. Carley, A.W. Owens, M.K. Rajumon and M.W. Roberts
    Catalysis Letters, 37 79-87 (1996).

215. Evidence for the instability of surface oxygen at the Zn(0001)-O-Cu interface from core-level and X-ray induced Auger spectroscopies
    A.F. Carley and M.W. Roberts
    Topics in Catalysis 3, 91-102 (1996).

216. Facile hydrogenation of carbon dioxide at Al(111) surfaces: the role of coadsorbed water
    A.F. Carley, P.R. Davies, Eva M. Moser and M. Wyn Roberts
    Suface Science 364, L563-L567 (1996).

217. Surface chemistry of carbon dioxide
    H.-J. Freund and M.W. Roberts
    Surface Science Reports, Vol. 25, No. 8, 225-273, (1996).

218. The role of short-lived oxygen transients and precursor states in the mechanisms of surface reactions; a different view of surface catalysis
    M.W. Roberts
    Chem. Soc. Reviews Issue No. 6 437 (1996)

219. Surface oxygen and chemical specificity at copper and caesium surfaces
    A.F. Carley, A. Chambers, P.R. Davies, G.G. Mariotti, R. Kurian and M.W. Roberts
    Faraday Discuss., 105, 225 (1996).

220. Reactivity of oxygen states at caesium surfaces towards carbon monoxide and carbon dioxide
    G.U. Kulkarni, S. Laruelle and M.W. Roberts
    J. Chem. Soc. Faraday Trans. 92, 4793 (1996).

221. Coadsorption of dioxygen and carbon monoxide on a Mg(100) surface
Zeini-Isfahani, Asghar, Roberts, M.W., Carley, A.F. and Read, S.
Iran J. Chem & Chem. Eng., Vol. 15, No. 1, (1996).

222. XPS study of oxygen adsorption on supported silver: effect of particle size
V.I. Buktiyarov, A.F. Carley, L.A. Dollard and M.W. Roberts
Surface Science, 381, L605 (1997).

223. The active site in oxygenation catalysis at single crystal metal surfaces
A.F. Carley, P.R. Davies and M.W. Roberts
Current Opinion in Solid State Materials Science, 2, 525 (1997);
Editors: A.K. Cheetham, H. Inokuchi and Sir John Meurig Thomas.

224. Oxygen states present at a Ag(111) surface in the presence of ammonia; evidence for a NH_3-$O_2^{\delta-}$ complex
A.F. Carley, P.R. Davies, M.W. Roberts and S. Yan
J.C.S. Chem. Comm. 35 (1998).

225. Interaction of oxygen and carbon monoxide with CsAu surfaces
A.F. Carley, M.W. Roberts and A.K. Santra
J. Phys. Chem. B. 101, No.48, 9978 (1997)

226. Coadsorption of carbon monoxide and nitric oxide at Ag(III): evidence for a CO-NO complex.
A.F. Carley, P.R. Davies, M.W. Roberts, A.K. Santra and K.K. Thomas
Surface Science 406, No.1-3, L587 (1998).

227. An STM-XPS study of ammonia oxidation : the molecular architecture of chemisorbed imide strings at Cu(110) surfaces.
A.F. Carley, P.R. Davies and M.W. Roberts
J.C.S. Chem. Comm. 1793,(1998).

228. Chemisorption of ethanol at Pt(111) and Pt(111)-O surfaces
M. K. Rajumon, R. S. Roberts, F. Wang and P. B. Wells
J.C.S. Faraday Trans. 94, 3699 (1998).

229. Oxygen states present at a Ag(111) surface in the presence of ammonia: evidence for a NH_3-$O_2^{\delta-}$ complex
A. F. Carley, P. R. Davies, M. W. Roberts, K. K. Thomas and S. Yan
Chem.Comm., 35 (1998).

230. Selective oxidation of propene at cesium and cesium-modified Ag(100) surfaces
A. F. Carley, A. Chambers, M. W. Roberts and A. K. Santra
Israel Journal of Chemistry 38, 393 (1998).

231. Oxygen chemisorption at Cu(110) at 120 K: dimers, clusters and mono-atomic oxygen states
A. F. Carley, P. R. Davies, G. U. Kulkarni and M. W. Roberts
Catal.Lett. 58, 93 (1999).

232. Spectroscopic investigation of potassium-doped Ni(110)
A. F. Carley, S. D. Jackson, J. N. OShea and M. W. Roberts
Surf. Sci. 440, 868 (1999).

233. The reactivity of copper clusters supported on carbon studied by XPS
A. F. Carley, L. A. Dollard, P. R. Norman, C. Pottage and M. W. Roberts
J.Elec.Spec.Rel.Phen. 99, 223; 233 (1999).

234. The formation and characterisation of Ni^{3+} - an X-ray photoelectron spectroscopic investigation of potassium-doped Ni(110)-O
A. F. Carley, S. D. Jackson, J. N. O'Shea and M. W. Roberts
Surf. Sci. 440, L868 (1999).

235. Reactions of co-adsorbed carbon dioxide and dioxygen at the Mg(0001) surface at low temperatures
A. F. Carley, G. Hawkins, S. Read and M. W. Roberts
Top. Catal. 8, 243 (1999).

236. Flexibility of the Cu(110)-O structure in the presence of pyridine
A. F. Carley, P. R. Davies, R. V. Jones, G. U. Kulkarni and M. W. Roberts
J.C.S. Chem.Comm., 687 (1999).

237. Controlling oxygen states at a Cu(110) surface: the role of coadsorbed sulfur and temperature
A. F. Carley, P. R. Davies, R. V. Jones, K. R. Harikumar and M. W. Roberts
J.C.S.Chem.Comm., 185, (2000).

238. The structure of sulfur adlayers at Cu(110) surfaces: an STM and XPS study
A. F. Carley, P. R. Davies, R. V. Jones, K. R. Harikumar, G. U. Kulkarni and M. W. Roberts
Surf. Sci. 447, 39, (2000).

239. Structural aspects of chemisorption at Cu(110) revealed at the atomic level
A. F. Carley, P. R. Davies, R. V. Jones, K. R. Harikumar, G. U. Kulkarni and M. W. Roberts
Top. Catal. 11, 299 (2000).

240. Heterogenous catalysis since Berezelius: some personal reflections
M. W. Roberts
Catal. Lett. 67, I (2000).

241. Alkali metal reactions with Ni(110)-O and NiO(100) surfaces
A. F. Carley, S. D. Jackson, M. W. Roberts and J. O'Shea
Surf. Sci. 454, 141 (2000).

242. Charles Kemball, CBE
M.W Roberts
Biog. Mems. Fell. R. Soc. Lond 46 285 (2000)

243. The chemisorption of nitric oxide and the oxidation of ammonia at Cu(110) surfaces: a X-ray photoelectron spectroscopy (XPS) and scanning tunnelling microscopy (STM) study
A. F. Carley, P. R. Davies, K. R. Harikumar, R. V. Jones, M. W. Roberts and G. U. Kulkarni
Top. Catal. 14, 101 (2001).

244. Oxidation states at alkali-metal-doped Ni(110)-O surfaces
A. F. Carley, S. D. Jackson, J. N. O'Shea and M. W. Roberts
Phys.Chem.Chem.Phys. 3, 274 (2001).

245. A combined XPS and STM study of the adsorption of methyl mercaptan at a Cu(110) surface
A. F. Carley, P. R. Davies, K. R. Harikumar, R. V. Jones and M. W. Roberts
Surf. Sci. 490 L585, (2001)

246. **Surface science - Editorial overview**
   J. M. Thomas and M. W. Roberts
   Current Opinion in Solid State & Materials Science 5 65 (2001)
247. **A combined XPS, STM and TPD study of the adsorption of methyl mercaptan at a Cu(110) surface**
   A. F. Carley, P. R. Davies, K. R. Harikumar, R. V. Jones, M. W. Roberts and C. J. Welsby
   Top. Catal. (Submitted).

PUBLICATIONS FOR WHICH M.W. ROBERTS HAS BEEN AN EDITOR.

Surface and Defect Properties of Solids.[+] Specialist Periodical Report, The Chemical Society, London.
Editors: M.W. Roberts and J.M. Thomas
Volume 1 (1972)
Volume 2 (1973)
Volume 3 (1974)
Volume 4 (1975)
Volume 5 (1976)
Volume 6 (1977)
[+]The Chemical Society recommended that the title of this series be changed to: 'The Chemical Physics of Solids and their Surfaces'
Volume 7 (1978)
Volume 8 (1979)

Reactivity of Solids: Proceedings of the Seventh International Symposium on the Reactivity of Solids.
Bristol, 1972. Chapman and Hall (1972)
Editors: J.S. Anderson, M.W. Roberts and F.S. Stone

Interfacial Science: A Chemistry for the 21st Century Monograph
Edited by M.W. Roberts; International Union of Pure and Applied Chemistry, Blackwell Science (1997).

Appendix 2

M. W. Roberts' Students

J.M.	Saleh	Ph.D.	1962	J.A.	Morris	Ph.D.	1976
E.	Crawford	Ph.D.	1962	J.	Rickman	Ph.D.	1976
C.M.	Quinn	Ph.D.	1963	F.	Gobal	Ph.D.	1977
M.R.	Hillis	Ph.D.	1964	S.A.	Isa	Ph.D.	1977
J. G.	Little	M.Sc.	1964	M.A.	Lunn	Ph.D.	1977
C.	Mckee	Ph.D.	1966	A.E.	Barber	Ph.D.	1978
J.R.H.	Ross	Ph.D.	1966	A.	Kazempour	Ph.D.	1978
B.R.	Wells	Ph.D.	1966	M.H.	Matloob	Ph.D.	1978
W.C	Storey	Ph.D.	1968	W.	Tang-Han	Ph.D.	1979
N.J.	Young	Ph.D.	1968	P.C.T.	Au	Ph.D.	1980
R.W	Joyner	Ph.D.	1969	A.F.	Carley	Ph.D.	1980
T.E.	Bridden	Ph.D.	1970	P.G.	Harris	Ph.D.	1980
G.	Manor	M.Sc.	1970	S.F.	Ho	Ph.D.	1980
L.P.	Metcalfe	Ph.D.	1970	C.	Howells	Ph.D.	1980
W.R.	Murphy	Ph.D.	1970	P.	Meehan	Ph.D.	1980
D.L.	Perry	Ph.D.	1970	O.A.	Na Lamphun	Ph.D.	1980
T.I.	Stewart	Ph.D.	1970	L.M.	Moroney	Ph.D.	1981
J.G.	Hardy	Ph.D.	1971	S.	Rassias-Soulis	Ph.D.	1981
W.J.	Murphy	Ph.D.	1971	C.M.	Carr	Ph.D.	1983
J.H.	Wood	Ph.D.	1971	D.A.	Geeson	Ph.D.	1983
R.K.M.	Jayanty	Ph.D.	1972	A.S.	Ali Al-Taie	Ph.D.	1984
R.S.	Brewerton	M.Sc.	1973	S.J.	Grubb	Ph.D.	1984
P.R.	Evans	M.Sc.	1973	D.M.	Sweeney	Ph.D.	1985
L.V.	Renny	Ph.D.	1973	F.A.A.	Al-Shamma	Ph.D.	1986
J.R.	King	M.Sc.	1974	S.	Campbell	Ph.D.	1986
S. J.	Atkinson	Ph.D.	1974	P.R.	Chalker	Ph.D.	1986
J.M.M.	Dugmore	Ph.D.	1975	D.E.	Gallagher	Ph.D.	1987
S.	Frost	Ph.D.	1975	R.A.	Ryder	Ph.D.	1988
K.M.	Byrne	Ph.D.	1976	L.M.R.	Cass	Ph.D.	1989

D.C.	Challinor	Ph.D.	1989
P.R.	Davies	Ph.D.	1989
S.	Yan	Ph.D.	1989
G.	Hawkins	Ph.D.	1991
R.J.	Holmes	Ph.D.	1991
B.	Afsin	Ph.D.	1992
D	Jones	Ph.D.	1992
G.D.	Savage	Ph.D.	1992
N.	Shukla	Ph.D.	1992
B.P.	Williams	Ph.D.	1992
T.S.	Amorelli	Ph.D.	1993
W.	Fancheng	Ph.D.	1993
A.J.	Strutt	Ph.D.	1993
A.W.	Owens	Ph.D.	1994
D.	Vincent	Ph.D.	1994
S.	Read	Ph.D.	1995
J.C.	Roberts	Ph.D.	1995
K.K.	Thomas	Ph.D.	1995
L.A.	Dollard	Ph.D.	1996
H.A.	Edwards	Ph.D.	1996
M.	Jahangir	Ph.D.	1996
A.M.	Shah	Ph.D.	1996
M.	Deakes	Ph.D.	1997
J.N.	O'Shea	Ph.D.	1998
H.	Griffiths	Ph.D.	1999
C.R.	Parkinson	Ph.D.	1999
R.V.	Jones	Ph.D.	2002

Index

RETURN TO: CHEMISTRY LIBRARY

100 Hildebrand Hall • 510-642-3753

LOAN PERIOD	1	2	3
4		1-MONTH U 6	

ALL BOOKS MAY BE RECALLED AFTER 7 DAYS.

Renewals may be requested by phone or, using GLADIS,
type **inv** followed by your patron ID number.

DUE AS STAMPED BELOW.

NON-CIRCULATING UNTIL:		
JUL 5 03	DEC 0 9 2005	
MAY 2 4 2003		
AUG 13 2004	MAY 1 8	
AUG 12 2005		FEB 25
	DEC 20	
NOV 20		
		APR 03

FORM NO. DD 10   UNIVERSITY OF CALIFORNIA, BERKELEY
2M 5-01   Berkeley, California 94720–6000